Springer Series in Synergetics

Editor: Hermann Haken

Synergetics, an interdisciplinary field of research, is concerned with the cooperation of individual parts of a system that produces macroscopic spatial, temporal or functional structures. It deals with deterministic as well as stochastic processes.

Dimensions and Entropies in Chaotic Systems

Quantification of Complex Behavior

Proceedings of an International Workshop
at the Pecos River Ranch, New Mexico,
September 11–16, 1985

Editor: G. Mayer-Kress

With 139 Figures

Springer-Verlag Berlin Heidelberg New York Tokyo

Dr. Gottfried Mayer-Kress

Center for Nonlinear Studies, Mail Stop: B258, Los Alamos National Laboratory,
Los Alamos, NM 87545, USA

Series Editor:
Professor Dr. Dr. h. c. Hermann Haken
Institut für Theoretische Physik der Universität Stuttgart, Pfaffenwaldring 57/IV,
D-7000 Stuttgart 80, Fed. Rep. of Germany

ISBN-13:978-3-642-71003-2   e-ISBN-13:978-3-642-71001-8
DOI: 10.1007/978-3-642-71001-8

2153/3150-543210

Preface

These proceedings contain the papers contributed to the International Workshop on "Dimensions and Entropies in Chaotic Systems" at the Pecos River Conference Center on the Pecos River Ranch in Spetember 1985. The workshop was held by the Center for Nonlinear Studies of the Los Alamos National Laboratory.

At the Center for Nonlinear Studies the investigation of chaotic dynamics and especially the quantification of complex behavior has a long tradition. In spite of some remarkable successes, there are fundamental, as well as numerical, problems involved in the practical realization of these algorithms. This has led to a series of publications in which modifications and improvements of the original methods have been proposed. At present there exists a growing number of competing dimension algorithms but no comprehensive review explaining how they are related. Further, in actual experimental applications, rather than a precise algorithm, one finds frequent use of "rules of thumb" together with error estimates which, in many cases, appear to be far too optimistic. Also it seems that questions like "What is the maximal dimension of an attractor that one can measure with a given number of data points and a given experimental resolution?" have still not been answered in a satisfactory manner for general cases. On the other hand, it has become more and more evident that there is a growing need in different disciplines for methods with which aperiodic and irregular data strings can be characterized. Thus it seemed to be natural to have a workshop where the status and the problems of these methods for the quantification of complex behavior could be discussed and expanded. We hoped that this workshop would help to systemize and unify the various aspects of these problems in order to provide reliable and economical methods for the wide class of users of dimension algorithms. We tried to bring together an interdisciplinary group of scientists with either theoretical or experimental expertise who worked for one week in a very relaxed and supportive atmosphere on the Pecos River Ranch.

The fact that a variety of computers was available on site made it possible to compare the algorithms directly and test conjectures and questions which came up in discussions, instantly and on site. Through telecommunication there was also the possibility of gaining access to mainframe computers on which larger algorithms could be run in order to update lectures and also to implement new results obtained from other speakers. The presence of computers also facilitated a very flexible schedule, such that the sequence of talks could be rearranged according to the intentions and agreements of the speakers, and even new or modified talks or discussions could be easily accommodated.

Thus the lectures of the workshop concentrate more on latest research results rather than on expository introductions. We hope that researchers and serious students will find in this volume a useful and comprehensive collection of papers for their own research into describing and quantifying chaotic systems.

This workshop was made possible by financial support of the Department of Energy and the NSWC Navy Dynamics Institute Program. I would like to thank A. Scott, D. Campbell and J.D. Farmer for their encouragement and support in the preparation of the workshop. I am also grateful to Frankie Gomez and the LANL-protocol office for their experienced support and Dr. H. Lotsch of Springer-Verlag for his efficient cooperation. My special thanks go to Marian Martinez for her enthusiasm and help with the new and unusual tasks and problems associated with the use of a new computerized system for the organization of this "workshop with a difference". I would also like to thank Akkana Peck for her support with the computer systems. Finally I wish to express my gratitude and respect for Patty Gunton, Celina Rael de Garcia, and their great staff of the Pecos River Ranch for creating that friendly and supportive environment during the workshop and also for their creativity in finding solutions to unforeseen problems.

Los Alamos, NM, December 1985 *G. Mayer-Kress*

Contents

Part I

Introduction

Introductory Remarks

G. Mayer-Kress

Center for Nonlinear Studies/Theoretical Division,
Los Alamos National Laboratory, Los Alamos, NM 87545, USA

The theory of nonlinear dynamics and chaotic attractors has been remarkably successful in providing a new paradigm for the understanding of complex and irregular structures in an enormous range of quite different systems [1,2,3,4,5]. The universality of its underlying principles, made it possible that the theory of chaotic dynamics can be applied not only to problems in basically all natural sciences but also in fields like medicine, economics and sociology.

For the quantitative description of erratic behavior observed in chaotic physical systems it has become clear that simple spectral analysis of the measured signals does not provide an adequate tool for characterizing the origin of the observed chaos. The measurement of other independent quantities like the Hausdorff-Besicovich - or "*fractal*" - dimension or the dynamical (Kolmogorov) entropy appears to be necessary for determining the "degree of chaos" which is present in the ex- perimental data.

The "*fractal*"- dimension, roughly speaking- provides some measure about how many relevant degrees of freedom are involved in the dynamics of the system under consideration. And naturally a system with many degrees of freedom is generally considered to be more complex than one with only a few degrees of freedom. In Figs.1,2 we can see that in a chaotic attractor a self similar structure is visible. It can also be seen however, that depending on where on the attractor one looks, the observed structure (for finite resolution) can be significantly different. Figs. 3,4 show the broken self similarity of a fractal object of dimension one with logarithmic corrections.

The dynamical- or Kolmogorov entropy provides a complementary measure based on information theoretic concepts which describes how many bits per second the dynamical system "*produces*". Thus it also provides insight about the predictability of the system.

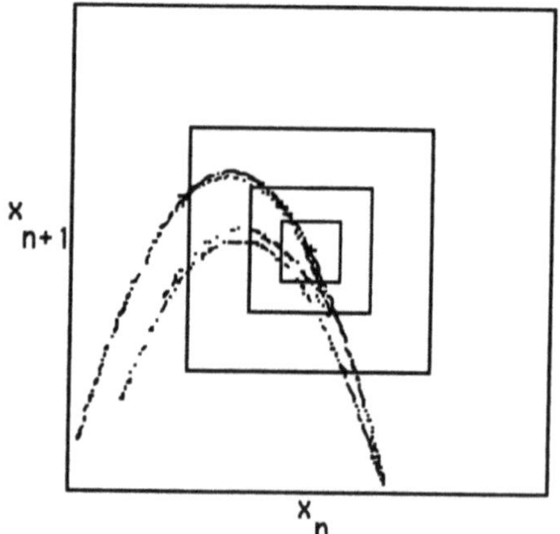

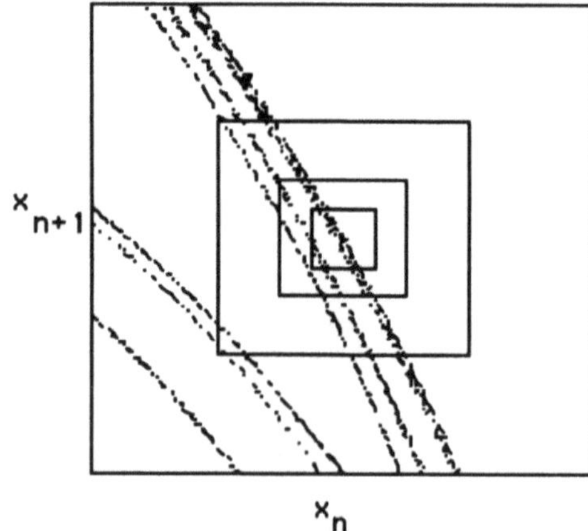

Fig. 1: Chaotic attractor of the Henon map. We have also shown a sequence of squares which are used for the calculation of the dimension-function.

Fig. 2: Blowup of the smallest box of the sequence in Fig.1

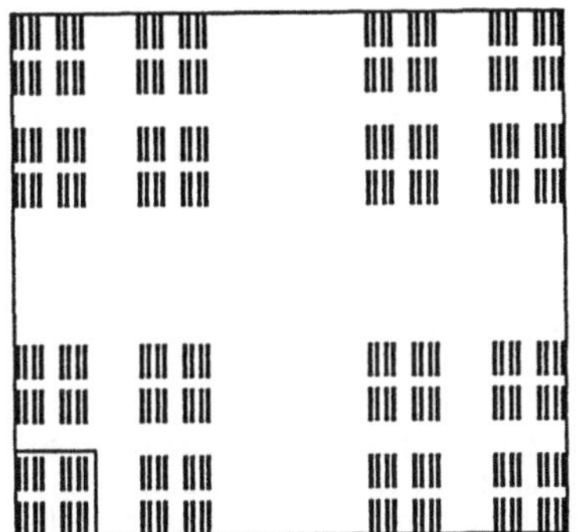 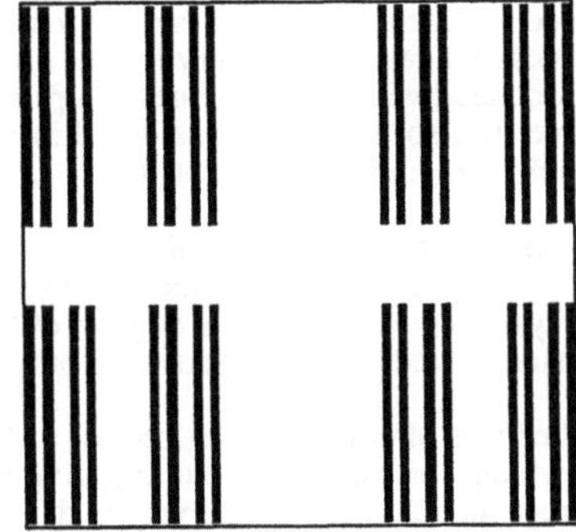

Fig. 3: Product of an approximation of the standard Cantor set (horizontal) with a fractal set of dimension 1 and zero (Lebesmeasure zero).

Fig. 4: Blowup of Fig. 3. Note the lack of self similarity of the vertical set which possesses logarithmic corrections of the dimension function.

The third concept which is discussed in these proceedings is that of *Lyapunov characteristic exponents*. They generalize the linear stability criteria for fixed points and limit cycles (*Floquet multipliers*) . Dynamical systems which have at least one positive Lyapunov exponent exhibit the property of *sensitive dependence on initial conditions* which means that arbitrary small perturbations in the state of the system at a given time will grow exponentially and therefore also restrict the long time predictability. It has been shown that there exist rigorous relationships between these quantities, that is not the topic of this volume.

In a previous volume of this series, H.HAKEN [6] pointed out that brains of a normal and healthy person appear to be more *"chaotic"* than those observed from people undergoing an epileptic seizure and that therefore *"higher degree of order does not necessarily imply a higher content of meaning"*. During the same conference on Schloss-Elmau the possibility was discussed, that the dynamics of the brain waves is actually generated by a *"strange attractor"* [7]. J.D.FARMER [8] expressed these ideas by associating chaos with creative thought . He remarked, somewhat poetically : *" Human beings have many of the properties of metastable chaotic solitary waves. (I say metastable because all of us eventually die and become fixed points.) Old age might be defined as the onset of limit cycle behavior. May your chaos be always of high dimension."*

I have shown recordings of brain - wave (EEG) data to several artists, and they were surprised why I called it chaotic. They intuitively perceived the ordered structure underlying these irregular data. If one looks at art objects, one recognizes that symmetries, periodic structures, and perfect "order" only occur on very limited occasions and often is considered to be boring. This judgement can also be heard with respect to "randomly generated art". And so most art objects have a well selected mixture of order and chaos: there exists repetition and recognition as well as surprise. I think that this quality is also very well observed in the graphical representations of strange attractors and Julia sets , that are generated with the help of nonlinear dynamical systems (see e.g. the pictures in [9]). In both cases some sort of self similarity is a key feature, a property which they have in common with all the fractal objects of B. MANDELBROT [10] and for which the dimension concepts discussed in this workshop are essential.

In an acoustical context I find it very interesting to listen to acoustical representations of data generated by low-dimensional chaotic systems like the logistic map. Here the *"higher order"* in the chaotic data becomes evident, and it is easy to distinguish this kind of *"chaotic music"* from randomly generated tones. Actually, when we talk about randomly generated we mean that these tones are generated by a random number generator of a computer, which in the language of nonlinear dynamics are chaotic of a very high dimension with a large dynamical (Kolmogorov) entropy; and the higher the dimension the better the approximation of *"pure random noise"*.

Thus it seems that our senses are able to distinguish between *"purely random"*, without any structure and form like in the original meaning of the "chaos" and the "tohu-wa-bohu" mentioned in the first chapter of the bible on the one hand and complex order or makroscopic *chaos* on the other hand.

The first type is related with emptyness and is often visualized as the completely randomized thermal motion of the molecules in a gas, where everything is homogeneous and no structures are visible [11]. The second type is associated with a rich and never repeating structure like turbulent water in a river or the coloring of marble. It also has been considered in ancient chinese cultures to represent the life energy *chi* as well as the principle *tao* itself [12].

In the modern sciences it was for a long time that only the first type s has been considered to be of scientific importance in the context of statistical mechanics. Today it is well known that simple nonlinear dynamical systems with only a few degrees of freedom can generate the kind of macroscopic chaos which is not characterized by simple statistical properties.

Thus it has become of considerable interest to find means and methods which enable us to measure quantities which correspond to our intuitive notions of uniform randomness and highly structured chaos. It appears that the classical way to detect non-periodic or random structures -namely the power spectrum- is not a good tool for separating these two kinds of chaos. However the concepts of fractional dimensions, entropy and Lyapunov exponents, which are some of the main topics of this workshop, appears to be good candidates for the quantification of the degree of chaos.

It became clear in all the talks of this session that the fractal dimension of a system is not suffient in order to characterize it. Procaccia discussed a new function $f(\alpha)$ which describes how the singularities of a given attractor with a certain order is distributed over a set with a given dimension. Other generalizations include logarithmic and additive corrections (Umberger), and lacunarity (Mandelbrot) of the dimension function. Mauldin presented some analytical results on graphs and random fractals. Aizawa disssssssced some interesting analogys between certain transitions in chaotic attractors and phase transitions of thermodynamics in which the dimensions are treated as a fluctuating quantity.

In the field of numerically and experimentally computing dimensions and entropies, it seems that a lot of progress has been made. Kostelich talked about a method to reduce the integer part of the dimension of an attractor such that a higher accuracy for the fractional part can be obtained. Fraser discussed the information theoretic aspects of chaotic time signals. He also explained the use of mutual information content as a very efficient tool for finding an optimal delay time for the reconstruction of an attractor from a time series. Fraser also demonstrated on site with the help of the available computer link, how for various datasets this algorithm can find the mutual information content for a number of time delays.

Wolf demonstrated in another computer demonstration his algorithm for computing Lyapunov exponents from a time series. Both of these algorithms are available for the scientific user. Vastano compared this method with some more recent approaches by which it is in principle possible to get all Lyapunov exponents simultaneously. This method appears however to be much more unstable with respect to changes of experimental parameters. During the Workshop, Wolf and Vastano obtained some new results on the reliability of their method. These results were stimulated by discussions and directly confirmed in computer simulations on the machines which were available at the conference center. Deissler reported a generalization of Lyapunov exponents to open flows.
In the session on reliability and accuracy various influences on the measured values of the dimension were discussed like filtering (Politi) embedding spaces and sampling rates (Caputo) and non-uniformity of the attractor (Holzfuss). Caputo also discussed a method of how the entropy of a system can be measured conveniently from a time series. Holzfuss introduced a systematic way to define error estimates for the calculated dimensions (see also the contribution by Caswell and Yorke). Somorjai talked about a series of statistical methods that could be applied to improve the accuracy at the calculation of dimensions but the overall impression was that with the present methods for realistic conditions a limit in reliable dimension calculations is reached at or below a value of ten.

In the session on spatio temporal chaos it became clear that it is extremely difficult to find satisfiing observables which can serve to quantify the spatio-temporal complexity of the system (Brandstaeter, Swinney). It turns out that there are significant problems in the attempt to carry over methods and notions from simple dynamical systems. Several experimental examples where spatial effects seem to be relevant in the generation of complex behavior have been described by Dougherty, Held and Martin.

In the session on experimental results and applications the extremely high precision and amount of control for some hydrodynamical experiments has been reported (Stavans, Ecke, Sreenivasan).
They found very good agreement between their experimental results and predictions from the theory of low-dimensional dynamical systems. Again there were no indications for reliable dimension estimates for high dimensional systems.

4

There were two talks (Sreenivasan, Cawley) in which the fractal dimension of the geometrical (as opposed to dynamical) properties of turbulent systems like clouds were described. Both speakers had the surprising result that for quite different measuring methods they found approximately the same dimension d = 1.3. This coincides with a value found by Mandelbrot and collaborators for the typical dimension of the boundary of clouds.

There were three talks in this session which came back to the old question of whether the brain exhibits some low-dimensional chaos (Albano, Babloyantz, Layne). It was evident that these investigations are only at the beginning and that one of the main problems with these kinds of data is the stationarity of the signal.

Nonlinear chaotic dynamics has become a very fascinating field and many speculations about its explanatory power and applicability have been found in the literature. There also have been many daring conjectures and speculations especially with respect to the description of chaotic data with the help of fractal- dimension calculations. We think there is a danger as in many exciting new fields, that wrong hopes and speculations are raised which cannot be fulfilled. Therefore I think that these concepts seriously have to be put into the proper context in order to prevent that the whole field loses its credibility. I think it is of considerable importance to investigate the possibilities as well as the limitations of dimensional analysis, which potentially has in it the power to become a tool for the empirical sciences of a value comparable to the spectral analysis of linear systems.

References

1   Evolution of Order and Chaos in Physics, Chemistry and Biology, ed. by H.Haken Springer Series in Synergetics, Vol.17, (Springer, Berlin, Heidelberg, New York 1982)
2   H. Haken, Advanced Synergetics, (Springer, Berlin, Heidelberg, New York 1983)
3   A.J. Lichtenberg, M.A. Lieberman, Regular and Stochastic Motion, (Springer, Berlin, Heidelberg, New York 1983)
4   P. Berge, Y. Pomeau, C. Vidal, *L'Ordre dans le Chaos,* (Hermann, Paris 1984, english translation: Wiley, 1986)
5   H.G. Schuster, Deterministic Chaos, (Physik Verlag, Weinheim, 1984)
6   H. Haken, in [1]
7   A. J. Mandell, private discussions
8   J.D. Farmer, in [1]
9   H.O.Peitgen, P.Richter, Schoenheit im Chaos, Mapart, Bremen, 1985
10 B.B. Mandelbrot, Fractal Geometry of Nature, (W.H. Freemanand Co., San Francisco,1982)
11 O.E. Roessler, private discussions
12 P. Rawson, L. Legeza, *Tao, Die Philosophie von Sein und Werden,* (Knaurs, Muenchen 1974)

Part II

General Theory, Mathematical Aspects of Dimensions, Basic Problems

The Characterization of Fractal Measures as Interwoven Sets of Singularities: Global Universality at the Transition to Chaos

I. Procaccia

Department of Chemical Physics, The Weizmann Institute of Science, Rehovot 76100, Israel

1. INTRODUCTION

The most dramatic event in the development of the modern theory of the onset of chaos in dynamical systems has been the discovery of universality [1]. Especially well known are the universal numbers α and δ, which in the context of period doubling pertain to the universal scaling properties of the 2^∞ cycle near its critical point, and the rate of accumulation of pitchfork bifurcations in parameter space respectively [1]. This type of universality is however local, being limited to behavior in the vicinity of an isolated point either in phase space or in parameter space. In this paper I wish to review some recent progress in elucidating the globally universal properties of dynamical systems at the onset of chaos. This progress has been achieved in collaboration with M.H. JENSEN, A. LIBCHABER, L.P. KADANOFF, T.C. HALSEY, B. SHRAIMAN and J. STAVANS [2-4]. In "global universality" we mean that an orbit in phase space has metric universality as a whole set or that a whole range of parameter space can be shown to have universal properties [5]. Examples that have been worked out recently include the 2^∞ cycle of period doubling, the orbit on a 2-torus with golden-mean winding number at the onset of chaos and the complementary set to the mode-locking tongues in the 2-frequency route to chaos. The approach used is however quite general, as will become apparent below.

The main theoretical idea that allows an investigation of globally universal properties is that the sets arising in dynamical systems should be considered as fractal measures, and that these measures should be characterized by their singularities [6] Considering some space $\vec{x}$ equipped with a measure $\mu(\vec{x})$, we box this space into boxes of sizes ℓ_i (i.e. not necessarily uniformly) and define

$$P_i = \int_{i\text{'th box}} d\mu(\vec{x}) \tag{1.1}$$

We are then interested in the scaling index α defined by

$$P_i \sim \ell_i^{\alpha} \tag{1.2}$$

as ℓ_i tends to zero. In typical cases [3] α takes on a range of values corresponding to different singularities in the measure. Moreover, one can introduce an index f by looking at the number of times α takes on a value between α' and $\alpha'+d\alpha'$, and writing it in the form

$$d\alpha' \rho(\alpha') \ell^{-f(\alpha')} .$$

The index $f(\alpha')$ can be then interpreted as the fractal dimension of the set of singularities of strength α'. We thus model fractal measures by interwoven sets of singularities of strength α, each characterized by its own dimension $f(\alpha)$. The connection to global universality is that the function $f(\alpha)$ (or equivalently $\alpha(f)$) are universal for the sets of interest in dynamical systems. It turns out that the scaling numbers α discovered previously [1,7] determine one point in these functions, and that point is actually associated with $f=0$, being an atypical singularity in the set.

In section II I review the theoretical background leading to this description of fractal measures. Section III reviews the theoretical results obtained to date concerning global universality via this formalism [3]. Section IV describes the results of the first experimental investigation along these lines [4]. In section V we summarize this paper and indicate some possible future avenues.

2. THEORETICAL BACKGROUND

A. GENERALIZED DIMENSIONS AND SINGULARITIES

The indices α and f are obtained by first considering a set of generalized dimensions (or mass exponents) D_q, that in ref. 7 were defined as

$$D_q = \lim_{\ell \to 0} \frac{1}{q-1} \frac{\log \Sigma \ p_i^q(\ell)}{\log \ell} \tag{2.1}$$

where here we have a uniform partition of space in mind. A more general definition of the set D_q is given in the next subsection. It has been shown [8-10] that D_0, D_1 and D_2 are the fractal, information and correlation dimensions [9] respectively. The parameter q takes on however all values $-\infty < q < \infty$. Since we assumed that $P_i \sim \ell^\alpha$ a number of time $d\alpha' \rho(\alpha')\ell^{-f(\alpha')}$, the sum ΣP_i^q can be estimated as

$$\sum_i P_i^q(\ell) = \int d\alpha' \rho(\alpha')\ell^{-f(\alpha')} \ell^{q\alpha'} \ . \tag{2.2}$$

In the limit $\ell \to 0$ the integral is dominated by the value of α' which minimizes $q\alpha' - f(\alpha')$, provided that $\rho(\alpha')$ is non zero. Thus we replace α' by $\alpha(q)$ which is defined by the conditions

$$\frac{d}{d\alpha'} (q\alpha' - f(\alpha')) \bigg|_{\alpha' = \alpha} = 0 \ . \tag{2.3}$$

We also have

$$\frac{d^2}{d\alpha'^2} (q\alpha' - f(\alpha')) \bigg|_{\alpha' = \alpha(q)} > 0 \ . \tag{2.4}$$

These two equations also mean that

$$f'(\alpha(q)) = q \tag{2.5}$$

$$f''(\alpha(q)) < 0 \ . \tag{2.6}$$

Using the dominant contribution in Eq. (2.2) we see that

$$D_q = \frac{1}{q-1} [q\alpha(q) - f(\alpha(q))] \ . \tag{2.7}$$

Consequently, if we know the set D_q we can find $\alpha(q)$ since

$$\alpha(q) = \frac{d}{dq} (q-1)D_q \ . \tag{2.8}$$

From D_q and $\alpha(q)$ we can evaluate $f(\alpha(q))$ via Eq. (2.7).

The intuitive meaning of these results is that q serves to pick a subset of singularities that determines D_q. Already from the definition (2.1) it is clear that as $q \to \infty$ the most concentrated parts of the measure are being stressed. When $q \to -\infty$ the most rarefied parts become dominant. In this sense q serves a role analogous to temperature in statistical mechanics, where at every temperature a different set of energy levels becomes important.

B. THE PARTITION FUNCTION

A more general definition of the set D_q for any set S embedded in d dimensional Euclidean space is obtained [3] by partitioning the set into N disjoint pieces, S_1, S_2...S_N, such that each piece has a measure p_i and it lies within a ball of radis ℓ_i, $\ell_i < \ell$. We consider then a partition function

$$\Gamma(q,\tau,\{S_i\},\ell) = \sum_{i=1}^{N} \frac{p_i^q}{\ell_i^\tau} \qquad (2.9)$$

Next one considers

$$\Gamma(q,\tau,\ell) = \begin{cases} \sup \ \Gamma(q,\tau,\{S_i\},\ell) & q>1 \\[2ex] \inf \ \Gamma(q,\tau,\{S_i\},\ell) & q<1 \end{cases} \qquad \text{and} \qquad (2.10)$$

$$\Gamma(q,\tau) = \lim_{\ell \to 0} \Gamma(q,\tau,\ell) \ . \qquad (2.11)$$

It has been argued [3] that the condition

$$\Gamma(q,\tau) = 1 \qquad (2.12)$$

fixes τ to be $\tau(q)$ where

$$\tau(q) = (q-1)D_q' \ , \qquad (2.13)$$

and

$$D_q' \lesssim D_q \ . \qquad (2.14)$$

The advantage of this definition is that now $D_{q=0}'$ is the proper Hausdorff dimension of the set, and that this definition allows us to consider non self-similar sets. The condition (2.12) can be also used as a scheme to generate the D_q' numerically. In the following we shall use D_q and D_q' interchangeably, since for the sets studied by us the equality sign in Eq. (2.14) seems to hold.

3. GLOBAL UNIVERSALITY AT THE TRANSITION TO CHAOS: THEORETICAL RESULTS

At this point three sets from dynamical systems theory were analyzed in detail along the lines of the new formalism described above. Another study on diffusion-limited aggregates has been done [2] but will not be reviewed here. The three sets considered are: (i) the 2^∞ cycle at the accumulation point of period doubling, (ii) the set of irrational winding numbers at the onset of chaos via quasi-periodicity, and (iii) the critical cycle elements at the golden mean winding number for the same problem. In all cases we calculate numerically the D_q and use Eqs. (2.8) and (2.7) to extract $\alpha(q)$, $f(q)$ and a plot of $f(\alpha)$. In all three cases we can find theoretically D_∞, $D_{-\infty}$ and thus $\alpha(q=\pm\infty)$.

3.A THE 2^∞ CYCLE OF PERIOD DOUBLING

Dynamical systems that period double on their way to chaos can be represented by one parameter families of maps $M_\lambda(\bar{x})$ where $M_\lambda : R^F \to R^F$ and F is the number of degrees of freedom. At values of $\lambda=\lambda_n$ the system gains a stable 2^n-periodic orbit. This period doubling cascade accumulates at λ_∞ where the system possesses a 2^∞ orbit. We generated numerically the set of elements of this orbit for the map $x'=\lambda(1-2x^2)$, with $\lambda_\infty \simeq .83700513$ [1]. The points making up the cycle are displayed in Fig. 1. The iterates of x=0 form a Cantor set, with half the iterates falling between $f(0)$ and $f^3(0)$ and the other half between $f^2(0)$ and $f^4(0)$. The most natural partition $\{S_i\}$ for this case simply follows the natural construction of the Cantor set, as shown in Fig. 1. At each level of the construction of this set, each ℓ_i is the distance between a point and the iterate which is closest to it. The measures p_i of these intervals are all equal.

With 2^{11} cycle elements we solved numerically $\Gamma=1$, thereby generating the D_q vs. q curve shown in Fig. 2. From these results we calculated $\alpha(q)$ from Eq. (2.8) and $f(\alpha)$ from (2.7). The curve $f(\alpha)$ is displayed in Fig. 3.

To understand the shape of the curve in Fig. 3 we first consider the end-points of the curve (for which f=0). These are associated with $q=\pm\infty$ and therefore we

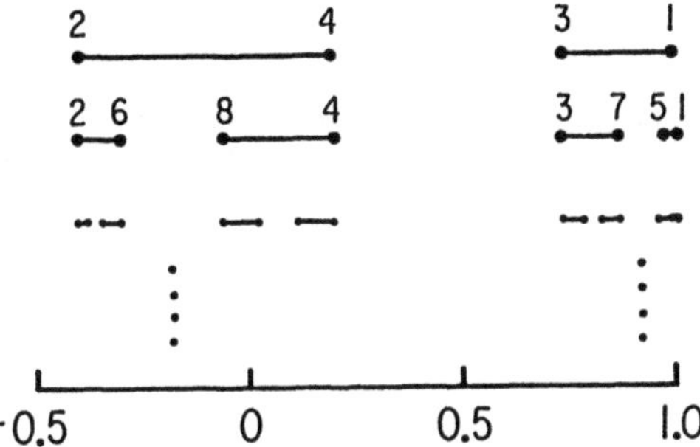

Figure 1. The construction of the period doubling attractor; the indices refer to the number of the iterate of x=0. The lines represent the scales ℓ_i.

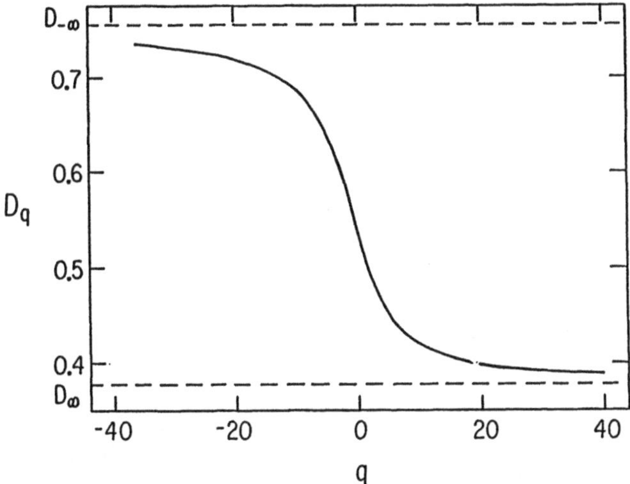

Figure 2. D_q vs. q calculated for the period doubling attractor of Fig. 1.

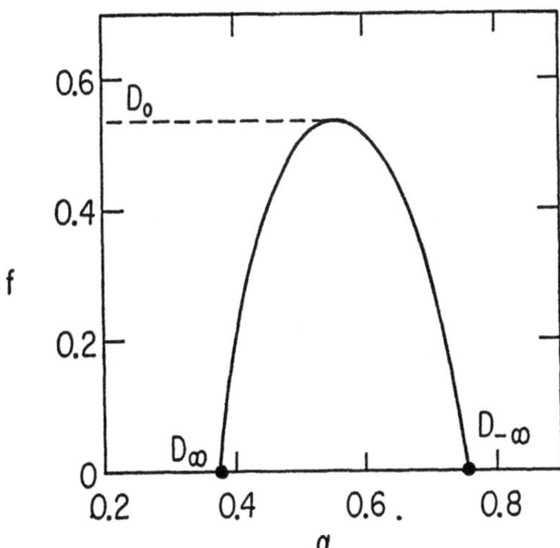

Figure 3. The function $f(\alpha)$ for the period doubling attractor of Fig. 1.

expect these two points to be determined by the most rarefied and the most concentrated intervals in the set. As has been shown by Feigenbaum these have scales $\ell_{-\infty} \sim \alpha_{pd}^{-n}$ and $\ell_{+\infty} \sim \alpha_{pd}^{-2^n}$ respectively, where $\alpha_{pd}=2.5029...$ is the universal scaling factor [1]. Since the measures there are simply $p_{-\infty} \sim 2^{-n}$, we expect these end points to be $\ell np_{-\infty}/\ell n\ell_{-\infty}$ and $\ell np_{\infty}/\ell n\ell_{\infty}$, respectively. These values are also $D_{-\infty}$ and D_{∞} so that we find

$$D_{-\infty} = \frac{\ell n2}{\ell n\alpha_{pd}} = .75551... \tag{3.1a}$$

$$D_{\infty} = \frac{\ell n2}{\ell n\alpha_{pd}^2} = .37775... \tag{3.1b}$$

These values are in extremely good agreement with the numerically determined endpoints of the graph. The curve $f(\alpha)$ is perfectly smooth. The maximum is at $D_0=.537..$ in agreement with previous calculations of the Hausdorff dimension for this set[1].Since the slope of the curve $f(\alpha)$ is q, $\alpha(q)$ will be very close to $D_{\pm\infty}$ even for $|q|\sim10$. However, Fig. 2 indicates that D_q is far from converged to $D_{\pm\infty}$ even for $q\sim\pm40$. Thus, the transformations (2.7-2.8) lead more easily to good estimations of $D_{\pm\infty}$ than do direct calculations of the D_q's.

It should be stressed that the universal number α_{pd} is determining a very atypical scaling behavior in the set, a behavior that is found on a set of dimension $f=0$. This is an edge point in the spectrum of singularities, and more typical behaviors are represented by the other α values along the curve $\alpha(f)$. Naturally, Feigenbaum was aware of the fact that α_{pd} does not exhaust the scaling structure of the 2^∞-cycle. Therefore he suggested [1] to examine the scaling functions which were generated by following the changes in the local scaling index everywhere. Although in principle this approach captures all the wanted information, it leads to a nowhere differentiable scaling function that is very difficult to work with. In the present approach we ask about the global <u>density</u> of scaling indices α of the same type. As a result we obtain smooth universal functions $f(\alpha)$.

3.B <u>MODE LOCKING STRUCTURE</u>

Dynamical systems possessing a natural frequency ω_1 display very rich behavior when driven by an external frequency ω_2. When the "bare" winding number $\Omega=\omega_1/\omega_2$ is close to a rational number the system tends to mode lock. The resulting "dressed" winding number, i.e. the ratio of the response frequency to the driving frequency, is constant and rational for a small range of the parameter Ω. At the onset of chaos the set of irrational dressed winding numbers is a set of measure zero, which is a strange set of the type discussed above. The structure of the mode locking is best understood in terms of the "devil's staircase", representing the dressed winding number as a function of the bare one. Such a staircase is shown in Fig. 4 as obtained for the map [12,13]

$$\theta_{n+1} = \theta_n + \Omega - \frac{K}{2\pi} \sin2\pi\theta_n \quad , \tag{3.2}$$

with $K=1$, which is the onset value above which chaotic orbits exist.

To calculate $D_{\pm\infty}$ analytically, we make use of previous findings that the most extremal behaviors of this staircase are found at the golden mean sequence of dressed winding numbers $F_n/F_{n+1} \to \omega^*=(\sqrt{5}-1)/2=.6108...$ where F_n are Fibonacci numbers, and at the harmonic sequence $1/Q \to 0$ [12]. The most rarefied region of the staircase is located around the golden mean. Shenker found that the length scales ℓ_j vary in that neighborhood as $\ell_{-\infty} \sim F_n^{-\delta} \sim \omega^{*n\delta}$, $\delta=2.1644...$ is a universal number [7]. The corresponding changes in dressed winding number are

$$p_{-\infty} \sim \frac{F_n}{F_{n+1}} - \frac{F_{n+1}}{F_{n+2}} \sim \omega^{*2n} \quad .$$

We thus conclude that

$$D_{-\infty} = \frac{\ell np_{-\infty}}{\ell n\ell_{-\infty}} = \frac{2}{\delta} = .9240... \quad . \tag{3.3a}$$

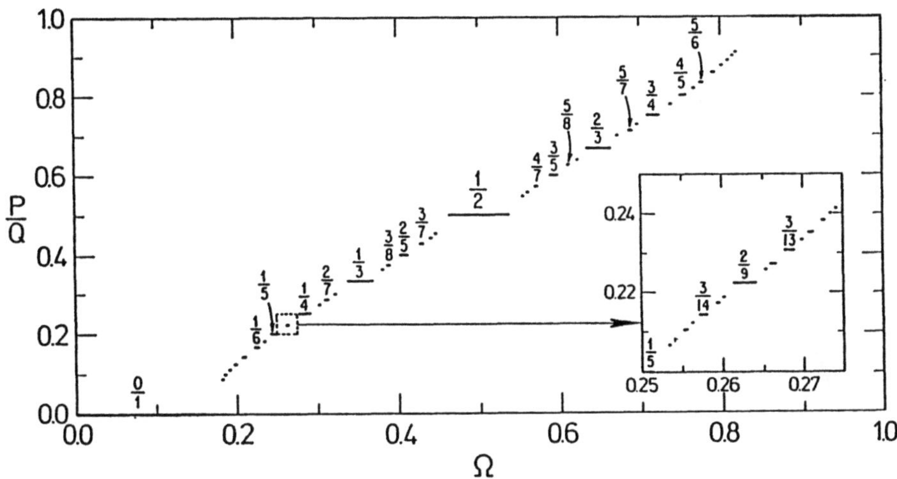

Figure 4. The "devil's staircase" for the critical map of Eq. (3.2). The dressed winding number is plotted versus the bare winding number.

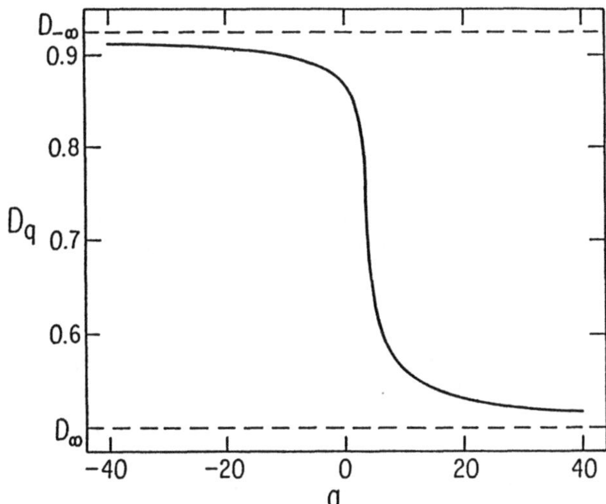

Figure 5. D_q vs. q for the staircase of Fig. 4.

For the 1/Q series it has been shown that changes in dressed winding number go as the square root of changes in bare winding number, i.e. that $p_i \sim \ell_i^{\frac{1}{2}}$ [12]. This series determines the most concentrated portion of the staircase (Fig. 4), which means that $p_\infty \sim \ell_\infty^{\frac{1}{2}}$ leading to

$$D_\infty = \frac{\ell n p_\infty}{\ell n \ell_\infty} = \frac{1}{2} \ .$$

(3.3b)

To construct the curve $f(\alpha)$ we generated 1024 mode-locked intervals following the Farey construction, which also defines the partition $\{S_i\}$ [13]. For each two neighboring intervals (see Fig. 4) we measured the change both in bare and in dressed winding numbers. The changes in bare winding numbers determined the scales ℓ_i of the partition $\{S_i\}$, whereas the changes in dressed winding numbers were defined to be the measures p_i. Solving then the equation $\Gamma=1$ we generated D_q as shown in Fig. 5 (for q>0 we accelerated the convergence as will be described shortly). Figure 6 shows $f(\alpha)$ for this case. Again the curve is smooth, in contrast to scaling functions found for the same problem by other authors [14].

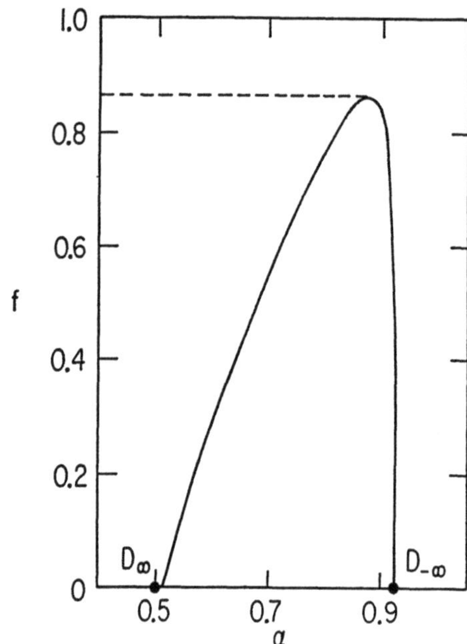

Figure 6. A plot of f vs. α for the mode locking structure of the circle map.

Note that the maximum on Fig. 6 gives the fractal dimension D_0 of the mode-locking structure as $D_0 \approx 0.87..$ in agreement with the predictions of Refs. 12 and 13. The right-most branch of the curve $f(\alpha)$ in Fig. 6 (i.e. for q<0) converges very rapidly within the Farey partition. This is, however, not the case for the left-most branch (i.e. for q>0). To improve the convergence of this portion of the curve substantially, we made use of the following trick. In general the partition function (2.1) will be of the form

$$\Gamma(\ell) = a e^{\gamma \ell n \ell} , \tag{3.4}$$

where a and γ are constants. The convergence is often slowed down by the prefactor a and by the logarithmic dependence on ℓ. However, by considering instead the ratio

$$\frac{\Gamma(\ell^2)}{\Gamma(\ell)} = e^{\gamma \ell n 2} \tag{3.5}$$

we find that a and ℓ do not appear in the equation. We thus determine $\tau(q)$ by requiring that $\Gamma(\ell^2)/\Gamma(\ell)(\tau,q)=1$. In general the numerator can be chosen to be of the form $\Gamma(\ell^b)$, where b is a constant. The right portion of the curve was generated with this method by calculating $\Gamma(\ell(1452))/\Gamma(\ell(886))=1$ (where $\ell(1452)$ and $\ell(886)$ are the maximal scales for partitions with 1452 and 886 intervals, respectively), and we observe that it passes through the point $(D_\infty,0)$. We found empirically that this method usually did not give reliable results for large values of $|q|$. Still, this method did successfully generate the entire curve in Fig. 6. We emphasize the ease of this measurement. The right-most branch of the f-α curve of Fig. 6 converges very rapidly, even when only 8-16 mode locked intervals are available.

3.C QUASIPERIODIC TRAJECTORIES FOR CIRCLE MAPS

Circle maps of the type (3.1) exhibit a transition to chaos via quasi-periodicity. A well-studied transition takes place at K=1 with dressed winding number equal to the golden mean, ω^* [15,16]. We have at this point studied the structure of the trajectory $\theta_1, \theta_2, \ldots, \theta_i, \ldots$. To perform the numerical calculation we chose $\theta_1 = f(0)$ and truncated the series θ_i at $i=2584=F_{17}$. The distances $\ell_i = \theta_{i+F_{16}} - \theta_i$ (calculated modulo 1) define natural scales for the partition with measures $p_i = 1/2584$ attributed to each scale. Figure 7 shows D_q versus q calculated for this set, and Fig. 8 shows the corresponding function $f(\alpha)$. Again the curve is

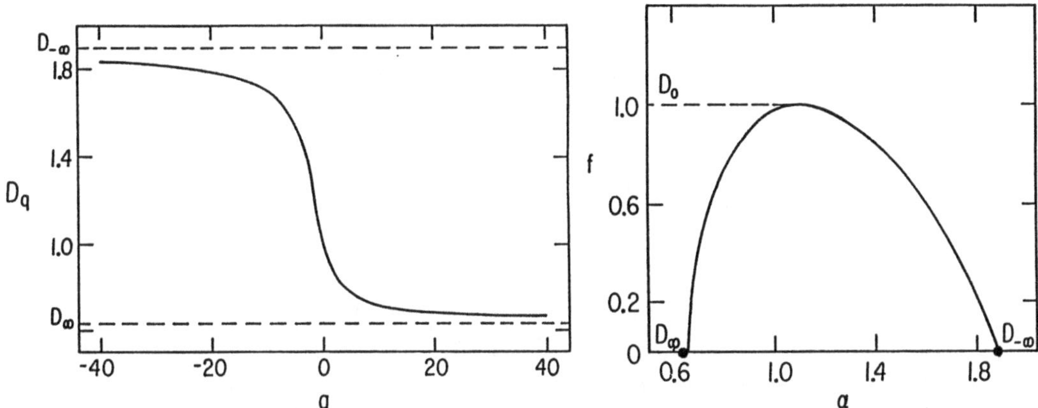

Figure 7. D_q vs. q for the critical trajectory of a circle map with golden mean winding number.

Figure 8. A plot of f vs. α for the golden mean trajectory of the circle map.

smooth. Shenker found for this problem that the distances around $\theta\sim0$ scale down by a universal scale factor $\alpha_{gm}=1.2885...$ when the trajectory θ_i is truncated at two consecutive Fibonacci numbers F_n,F_{n+1} [7]. This corresponds to the most rarefied region so that $\ell_{-\infty}\sim\alpha_{gm}^{-n}$. The corresponding measure scales as $p_{-\infty}\sim1/F_n\sim\omega^{*n}$ leading to

$$D_{-\infty} = \frac{\ell n\omega^*}{\ell n\alpha_{gm}^{-1}} = 1.8980.. \tag{3.6a}$$

The map (3.2) for K=1 has at $\theta=0$ a zero slope with a cubic inflection and is otherwise monotonic. The neighborhood around $\theta=0$, which is the most rarefied region of the set, will therefore be mapped onto the most concentrated region of the set. As the neighborhood around $\theta=0$ scales as α_{gm} when the Fibonacci index is varied, the most concentrated regime will scale as α_{gm}^3 due to the cubic inflection. This means that $\ell_\infty=\alpha_{gm}^{-3n}$ and $p_\infty=\omega^{*n}$, so that we obtain

$$D_\infty = \frac{\ell n\omega^*}{\ell n\alpha_{gm}^3} = .6326.... \tag{3.6b}$$

Figure 8 shows that the curve passes very close to the points $(D_\infty,0)$ and $(D_{-\infty},0)$. Again, however, we find that the dimensions D_q are far from $D_{\pm\infty}$ even for $q\sim\pm40$.

To check for universality it is important to investigate $f(\alpha)$ for a higher dimensional version of a circle map. We chose the dissipative standard map,

$$\theta_{n+1} = \theta_n + \Omega + br_n - \frac{K}{2\pi} sin2\pi\theta_n \tag{3.7}$$

$$r_{n+1} = br_n - \frac{K}{2\pi} sin2\pi\theta_n$$

and studied the critical cycle for b=0.5, again truncated at $i=2584=F_{17}$. We defined the scales by the Euclidean distances

$$\ell_i = \sqrt{(\theta_{i+F_{16}}-\theta_i)^2 + (r_{i+F_{16}}-r_i)^2} \tag{3.8}$$

We found that the convergence for the 2-d case was slightly slower than for the 1-d case. This is, however, to be expected,since it was found by Feigenbaum, Kadanoff and Shenker that the convergence of the scaling number α_{gm} is slower for the 2-d case than for the 1-d case [15]. To improve the convergence we again made use of the ratio trick as embodied in Eqs. (3.4)-(3.5). For this case we calculated the partition function for two consecutive Fibonacci numbers, $F_{16}=1597$ and $F_{17}=2584$ and found τ from the requirement $\Gamma(\ell(F_{17}))/\Gamma(\ell(F_{16}))(\tau,q)=1$ ($\ell(F_i)$ are the maximal scales for the partitions). This improves the convergence significantly and the f-α curve for this 2-d case coincide almost completely with the curve found for the 1-d case, and displayed in Fig. 8.

4. GLOBAL UNIVERSALITY AT THE TRANSITION TO CHAOS: EXPERIMENTAL RESULTS

We have used the ideas presented above to investigate the question whether an experimental orbit with golden-mean winding number at the onset of chaos in a forced Rayleigh-Bénard experiment is in the same universality class with the orbit discussed in section 3. The experiment has been described elsewhere [17] (and see also the paper by Stavans, Thomae and Libchaber in this volume). Basically one uses mercury as a fluid in a small Rayleigh-Bénard cell, the supports two convection rolls. The Rayleigh number is chosen in a range where the convection is oscillatory in time. A second frequency is obtained by passing an AC current sheet through the fluid. The nonlinearity of the coupling between the two oscillators is determined by the amplitude of the AC current.

The experimental orbit obtained at the point of breakdown of the 2-torus, which has a golden-mean winding number (with accuracy of 1 part in 10^4) is shown in Fig. 9. The basic question is whether this orbit, as a whole set, is in the same universality class as the orbit of the circle map or the map of the annulus considered in section 3. To answer this question we have to calculate its spectrum of scaling indices and their densities (the α-f curve) and compare with Fig. 8.

In order to calculate the α-f spectrum for the experimental orbit, we have to use an efficient algorithm directly on the time series, without relying on the "box counting" which is alluded to in the definition (2.1). To do this we make use of the fact that the critical orbit is topologically conjugate to a pure rotation with an irrational winding number, and is therefore ergodic. Thus having a time series

$$\{X_j\}_{j=1}^N$$

in phase space we can estimate $P_i(\ell)$ of Eq. (2.1) by relating it to the inverse of the time for recurrence of the orbit to within a distance ℓ from the point $\bar{x}_i$. Denoting this time as $m_i(\ell)$, we can estimate

$$p_i(\ell) \sim 1/m_i(\ell) \; . \tag{4.1}$$

Notice that here i can run over all the points in the time series. Recalling the partition function (2.9), we can estimate

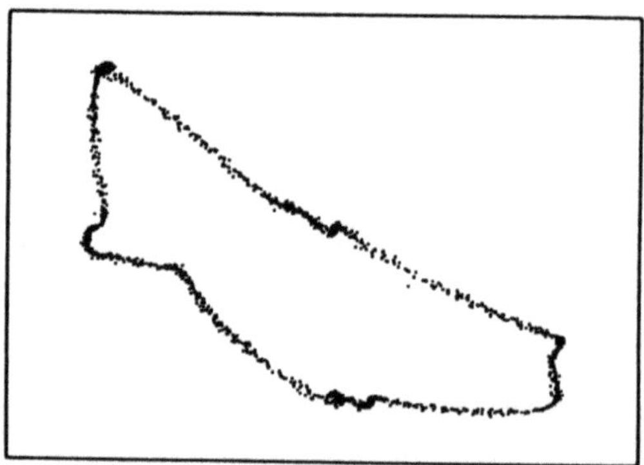

Figure 9. 2500 experimental points of a golden mean critical cycle at the onset of chaos in a forced Rayleigh-Bénard system.

$$<p_i(\ell)^{q-1}> \simeq <m_i(\ell)^{1-q}> \sim \ell^{\tau(q)} \quad , \tag{4.2}$$

where here the average is over all (or a subset of) the point of the time series.
The calculation of the set D_q can therefore easily be achieved by simply calculating
(once) $m_i(\ell)$ for all the points in the time series and then raising this $m_i(\ell)$ to
any desired value of q, and averaging over all the points. A calculation of the
$\alpha(f)$ curve for the circle map using D_q generated this way yielded results that
were indistinguishable from Fig. 8. We concluded that such an algorithm is
appropriate for quasi-periodic orbits.

On the basis of an experimental time series of 2500 points we calculated
$<m_i^q(\ell)>$ as required by Eq. (4.2). The $\alpha(f)$ curve was computed via Eqs. (2.7), (2.8)
with the result shown in Fig. 10. The errors bars indicate the uncertainty. We
concluded that up to these uncertainties the experimental orbit and the orbit of
the circle map seem to belong to the same universality class. We note in passing
that from the value of α_{max} one can read immediately α_{qm}, cf. Eq. (3.6a). To the
best of our knowledge this is the first direct measurement of this universal
(though atypical in the set) scaling number.

To conclude,we note that the raw experimental orbit in its reconstructed phase
space looks nothing like the orbit of the circle map. To the eye, it does not
appear to lie on a circle. It is twisted and contorted in a complicated way. Our
results demonstrate,however,that from the metric point of view these two sets <u>are
the same</u> within experimental accuracy. To date we are not aware of any other
approach that can lead to such a strong conclusion.

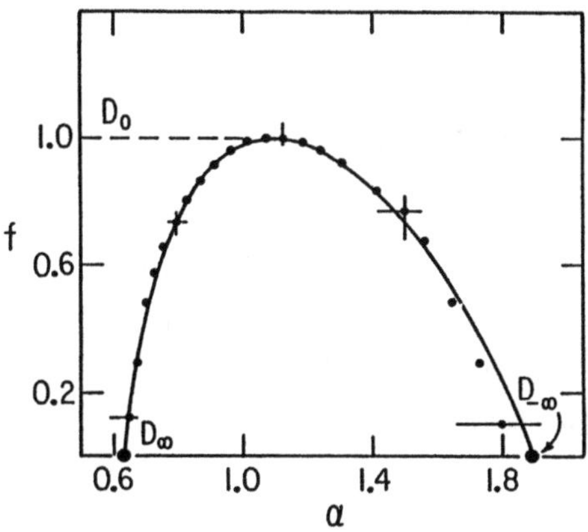

Figure 10.   A plot of f vs. α for the experimental orbit of Fig. 9, superimposed
on Fig. 8.   The bars indicate the uncertainty.

5. SUMMARY

The two main points of this paper are as follows: (i) To characterize strange sets
it is not sufficient to measure their dimension. A more complete characterization
is offered by the function D_q or the function $f(\alpha)$[6]. The latter displays the
information on all the scaling indices and their abundance in the sets. Sets that
differ only via a smooth change of coordinates have identical D_q and $f(\alpha)$ functions.
(ii) The formalism can be used to demonstrate global universality in sets encountered at the onset of chaos of dynamical systems. It has been demonstrated that
the ideas can be implemented to experimental data.

It should be clear at this point that the ideas presented above are not limited
to universal sets. As a method for characterization they are applicable to any
set. In particular, one can study the spectrum of singularities of generic strange

attractors. By doing so one might discover how these sets are organized, and what
determines their scaling structure. Preliminary studies show that for the general
case one should use algorithms that differ from the one discussed in section 4.
A survey of such algorithms will be published elsewhere.

Finally, the utility of these ideas for understanding strange sets outside
dynamical system theory should be stressed. An example concerning the harmonic
measure of diffusion-limited aggregates has been presented [2], but other examples
should and hopefully would be studied.

REFERENCES
1. M.J. Feigenbaum: J. Stat. Phys. 19, 25 (1978); 21, 669 (1979)
2. T.C. Halsey, P. Meakin and I. Procaccia: Phys. Rev. Lett., submitted
3. T.C. Halsey, M.H. Jensen, L.P. Kadanoff, I. Procaccia and B. Shraiman, Phys.
    Rev. A, submitted
4. M.H. Jensen, A. Libchaber, L.P. Kadanoff, I. Procaccia and J. Stavans, Phys.
    Rev. Lett., submitted
5. See also the paper by J.D. Farmer in this volume
6. The fact that fractal measures should be characterized by a spectrum of numbers
    rather than by one fractal dimension seems to have been stressed firstly by
    B.B. MANDELBROT. See for example J. Fluid Mech. 62, 331 (1979); Stat. Phys.
    13, Ann. Israel Phys. Soc. 225 (1977). In the context of turbulence this
    realization has led MANDELBROT to the often quoted inequalities between the
    intermittency exponent and the codimension of turbulence.
7. S.J. Shenker, Physica 5D, 405 (1982)
8. H.G.E. Hentschel and I. Procaccia, Physica 8D, 435 (1983)
9. P. Grassberger and I. Procaccia, Phys. Lett. 50, 346 (1983); Physics 9D, 189
    (1983)
10. I. Procaccia, Physics Scripta T59, 40 (1985)
11. P. Grassberger, J. Stat. Phys. 26, 173 (1981)
12. M.H. Jensen, P. Bak and T. Bohr, Phys. Rev. Lett. 50, 1637 (1983); Phys. Rev.
    A30, 1960 (1984); 30, 1970 (1984)
13. P. Cvitanovic, M.H. Jensen, L.P. Kadanoff and I. Procaccia, Phys. Rev. Lett.
    55, 343 (1985)
14. See for example P. Cvitanovic, B. Shraiman and B. Sodeberg, Nordita preprint
    (1985); M.J. Feigenbaum, Cornell preprint (1985)
15. M.J. Feigenbaum, L.P. Kadanoff and S.J. Shenker, Physica 5D, 370 (1982)
16. S. Ostlund, D. Rand, J.P. Sethna and E.D. Siggia, Phys. Rev. Lett. 49, 132
    (1982)
17. J. Stavans, F. Heslot and A. Libchaber, Phys. Rev. Lett. 55, 596 (1985)

Fractal Measures (Their Infinite Moment Sequences and Dimensions) and Multiplicative Chaos: Early Works and Open Problems

B.B. Mandelbrot

Physics Department, IBM Research Center, Mathematics Department,
Harvard University, Cambridge, MA 02138, USA

An infinite sequence of moments is needed to describe a fractal measure. This fact is widely known today, largely thanks to several speakers at this conference, who either refer to it, or push well beyond. Here, I propose to sketch the extensive early background in my work (before 1968) on the theory of turbulent intermittency. This old story matters, because my general procedure also brings forward a number of topics that have not been duplicated, and calls attention to interesting open issues.

1. TWO MAIN TRUNKS OF DEVELOPMENT AND BRANCHES: AN OUTLINE

Having discovered the need for an infinite sequence of moments shortly after the 1966 Kyoto Turbulence Conference, I reported it at the 1968 Brooklyn Symposium [1]. Recently, the telling term "multiplicative chaos" has been attached to the procedures that generate the fractal measures I studied, as well as variants, old or new. This explains the term "M-measure" to be used here.

Two "trunks" separated immediately. The first [2] involves discrete cascades, and fractals that are exactly renormalizable, because of an underlying hierarchical grid. The moments of orders 2/3, 2 and 4 were stressed in Orsay [3], and everything was summarized in Haifa [4]. The second trunk, involving continuous cascades, started at La Jolla [5].

A mathematical branch of the first trunk started in 1974 [6]. Some of my conjectures and theorems were proven or extended by J. PEYRIERE and J. P. KAHANE [7], which triggered other mathematics. Recently, KAHANE [8] proved corresponding conjectures in the second trunk.

The next major event was the rediscovery of results on M-measures by HENTSCHEL and PROCACCIA in 1982 [9], and the many rich developments that followed and are mostly beyond our scope here. Suffices to say that the growth of the main trunk has resumed [10,11]. PARISI and his coworkers [11] call the M-measures "multifractals", but _multi_ is redundant, since all fractals involve a multitude of dimensions, with the exception of the strictly self-similar sets.

2. ONE PARAMETER MODELS AND WOULD-BE CLASSES OF UNIVERSALITY

The models of intermittency available in 1968 seemed to manage with only one parameter, and to fall into two classes of universality: "all-or-nothing" and "lognormal".

The first models, independent of each other, were by KOLMOGOROV [12] and by BERGER and MANDELBROT [13]. My work concerned noise, but was soon modified to concern turbulence [14]. Then came NOVIKOV and STEWART [15]. The latter performed a recursive interpolation in a hierarchical cubic grid, hence involved self-similarities restricted to ratios the form b^k, with b an integer base b. The parameter b is <u>not</u> of immediate importance. Kolmogorov and I required no grid and allowed self-similarity of arbitrary r>0.

The parameter I featured was the fractal dimension D of the support of dissipation in fractally homogeneous turbulence. Novikov-Stewart featured the correlation exponent Q of the turbulent dissipation; their model being fractally homogeneous, this is the fractal co-dimension of the support of dissipation. Kolmogorov used one parameter μ, which specifies a log-normal distribution. In my "Kolmogorov-related" models, $\mu/2$ was to become the fractal co-dimension of the set on which dissipation concentrates. An excellent expository paper [16], which had the great merit of bringing my work to a wide public, stresses a parameter β, which again is <u>not</u> of immediate importance, but led to the term "β-model" often attached to fractally homogeneous turbulence.

Kolmogorov's model was enormously influential. Unfortunately, I found lognormality to be untenable as he stated it. (The words "Possible refinement..." in the title of [5] only reflect the difficulty then facing a negative comment on a parcel of Kolmogorov's work.) When a very great scholar stumbles in this way, something subtle is involved.

His basic idea is unchanged to this day: the idea of replacing <u>sums</u> of random processes by <u>products</u> that illustrate the notion of cascade. A physicist expects sums of random variables to be in the "domain of universality" of the Gaussian. So it seems safe to expect products of well-behaved strictly positive variables to converge to the lognormal, and this was proposed by GURVITCH and YAGLOM [17] to justify Kolmogorov's lognormality on very small scales. However, a step that seems harmless is incorrect in this instance: when a random variable x tends to a Gaussian, the moments of exp(x) <u>need</u> <u>not</u> tend to the moments of exp(G). This is a clear failure of universality, and its consequences are very interesting.

3. MULTIPLICATIVE CHAOS: MICROCANONICAL AND NONRANDOM

The M-measures are "singular" measures, i.e., continuous measures that fail to have a derivative. Examples of strict conservative M-measures abound in pure mathematics, and the new developments since 1968 resided in their use in science, and in their characterization by moments. I also introduced "mean conservative" M-measures; this concept raised altogether new issues.

A cascade process starts with a uniform measure. When the stages are discrete, the k-th stage multiplies the (k+1)st approximate measure by the k-th perturbation $P_k(\underline{x})$. Therefore, the k-th approximate measure of a domain Δ is $\mu_k(\Delta)=\int_\Delta \Pi_{h=1}^{k} P_h(\underline{s})d\underline{s}$, and one is interested in the limit $\mu(\Delta)=\lim_{k\to\infty}\mu_k(\Delta)$. The case $P_k(\underline{x})\geq 0$ is best understood (which is why – Section 6 – the most interesting new problems arise when $P_k(\underline{x})<0$ is allowed.) When the cascade proceeds in a grid of base b, the perturbations are called strictly conservative if $P_k(\underline{x})$ is constant over grid cells of side b^{-k}, and $\int_\Delta P_k(\underline{s})d\underline{s}=1$, with Δ any cell of side b^{-k}.

The B-measure of Besicovitch. This is my term for the special M-measure on a grid obtained when the perturbations are non-random, and $P_k(b^{k-1}\underline{x})=P_1(\underline{x})$, independently of k. $P_1(\underline{x})$ is the generator (="perturbator"?) of the measure. On the line, the generator is built from b "probabilities" p_β, satisfying $\sum p_\beta=1$, and $P_1(t)$ equals $b p_{t(1)+1}$ if $t=0.t(1)t(2)...t(k)$ in base b. Other perturbations at time t are $P_k(t)=b p_{t(k)+1}$. The integral $F_k(t)=\int_0^t \Pi_{h=1}^{k} P_h(s)ds$ is monotone non-decreasing, and is obtained by recursive interpolation. And $F(t)=\lim_{k\to\infty}F_k(t)$ is a self-affine non-random function of t. That is, the portion of F(t) over the interval $[(\beta-1)/b, \beta/b]$ is obtained from the portion F(t) over [0,1] by changing t in the ratio 1/b, and F in the ratio p_β, then translating. Reductions with unequal ratios are not similarities, but affinities [18], and F(t) is fully determined by the collection of affinities under which it is invariant. A generator for these affinities is a nondecreasing broken line with breaks located at multiples of 1/b. While "self-affine function" is a term used in my books, an explicit study is very recent [18] and it provides the proper framework here.

The Hentschel-Procaccia Measures. For many readers of this book, the first contact with the complexity of fractal measures came through [9], where HENTSCHEL and PROCACCIA introduce self-affine non-random fractal measures more general than the B-measures. In the 1-d case, the novelty is that the generator is a non-decreasing broken line with breaks located at arbitrary values of t, instead of multiples of 1/b.

$\underline{\text{The}}$ $\underline{\text{infinity}}$ $\underline{\text{of}}$ $\underline{\text{exponents}}$. The $\underline{\text{averages}}$ of the quantities $\mu^h(\Delta)$ over all subcells Δ of given size need not be derived in this section, because the argument is identical for the $\underline{\text{expectations}}$ of $\mu^h(\Delta)$ in the random measures in Section 4. In particular, the Hentschel-Procaccia measures involve nearly the same degree of generality as described in Section 4 for random weights in a hierarchical grid.

4. $\underline{\text{MULTIPLICATIVE}}$ $\underline{\text{CHAOS}}$; $\underline{\text{MICROCANONICAL}}$ $\underline{\text{IN}}$ $\underline{\text{A}}$ $\underline{\text{GRID}}$ $\underline{\text{AND}}$ $\underline{\text{RANDOM}}$.

The simplest random M-measure is obtained by randomizing, within each cell of side b^{-k}, the positions of the b^k values of $P_k(\underline{x})$.

$\underline{\text{"Microcanonical"}}$ $\underline{\text{M-measures}}$ [$\underline{2}$]. The perturbations are conservative, self-affine and stationary within cells. That is, the values of $P_k(\underline{x})$ within different cells of side b^{-k-1} are identically distributed random variables whose sum is 1. It is easiest to start with a random "weight" W satisfying W >0 and $<W>=1$, and to impose upon the weights W_β in different cells the condition that they must satisfy $\sum W_\beta b^{-d}=1$, i.e., $\sum W_\beta=b^d$. The resulting conditional weight will be denoted by $W_{(d)}$. The values of $P_k(\underline{x})$ in cells of side b^{-k-1}, taken jointly, are sample values of this $W_{(d)}$. Observe that $W_{(d)}<b^d$ and $<W_{(d)}>=1$.

The randomized B-measure is the microcanonical M-measure corresponding to W_d having b^d possible values of the form $b^d P_\beta$, with $\sum P_\beta=1$ and $\text{Prob}(W_d=b^d P_\beta)=b^{-d}$ for all β. (Strictly speaking, the assimilation requires that the relation $\sum i_\beta P_\beta=1$, with i_β integer ≥ 0 must be impossible unless $i_\beta=1$ for all β.)

$\underline{\text{The}}$ $\underline{\text{infinity}}$ $\underline{\text{of}}$ $\underline{\text{exponents}}$. Pick a cell of side b^{-k} at random. For all h>k, the measure $\mu_h(\Delta)$ satisfies $<\mu_h(\Delta)>=b^{-tk}= |\Delta|$, where $|\Delta|$ is the measure of Δ. Not unexpectedly, all the other moments $<\mu_k^h(\Delta)>$ are powers of $|\Delta|$. Their exponents, which I evaluated, are $m(h)=-\log_b<W^h> + dh$

Their being highly non-universal is well known today, but was a surprise in 1967. To evaluate the fractal dimension of the support of this measure, I introduced a procedure that was new at that time. I observed that a proportion of the measure between 1 and $1-\epsilon$ becomes, after sufficiently many stages $k(\epsilon)$, carried by a self-similar fractal set of codimension arbitrarily close to a quantity independent of ϵ, namely $c(1)=<W\log_b W>$.

This may be called the "ϵ-box dimension", the term "box dimension" itself denoting the classical form of fractal dimensions that part of our profession confusingly calls "capacity".

For the randomized B-measure,

$$\langle W\log_b W\rangle = \sum(1/b^d)b^d p_j \log_b b^d p_j = d + \sum p_j \log_b p_j = d - I_1.$$

Hence, the ϵ-box dimension of this measure is I_1, which is the entropy-information of the p_j. It was already well known, however [19], that I_1 is also the Hausdorff-Besicovitch dimension of the set of $t's$ for which the frequency of the digit β is $p_\beta+1$. This set is, loosely speaking, the support of most of Besicovitch measure. This made me conjecture that $\langle W\log_b W\rangle$ is a Hausdorff-Besicovitch codimension for every M-measure, and indeed it is [7].

5. <u>MULTIPLICATIVE CHAOS</u>: <u>CANONICAL</u>. <u>THE LITTLE KNOWN ROLE OF C(h) AS A CRITICAL CODIMENSION</u>. <u>CONTINUOUSLY PERTURBED MULTIPLICATIVE CHAOS</u>
The relations of conservation, $\sum W=b^d$, make a further detailed study of microcanonical cascades very cumbersome. Assuming that conservation only holds on the average makes everything simpler mathematically, and we shall see it yields a richer topic, worth of study on its merits. Anyhow, a low-dimensional cut through a microcanonical M-measure is characterized by partial, not strict, conservation. The reason is that overall conservation expresses that $\sum W_{(d)}=b^d$, the sum being carried over b^d variables, but a cut picks only $b^{d'}$ among these b^d variables. Call these new conditioned variables $W_{(d')}$. When $d'<d$, the $W_{(d')}$ are much less strongly correlated than the $W_{(d)}$. Thus, the model that picks uncorrelated weights and allows the W to be unconditioned and unbounded illustrates a cut through a microcanonical measure of extremely high dimension.

When $W>0$ and $\langle W\rangle=1$ is all that is assumed about W, the measures $\mu_k(\Delta)$ are no longer constructed by recursive interpolation. I showed that strange things may happen. For every domain Δ and $k<\infty$, the k-the approximate measure $\mu_k(\Delta)$ satisfies $\langle\mu_k(\Delta)\rangle=|\Delta|$. However, the seemingly obvious inference that $\langle\lim_{k\to\infty}\mu_k(\Delta)\rangle=|\Delta|$ <u>need</u> <u>not</u> hold. It <u>does</u> hold when $\langle W\log_b W\rangle<d$, but <u>does</u> <u>not</u> hold when $\langle W\log_b W\rangle>d$, and also [7] does not hold when $\langle W\log_b W\rangle=d$. In fact, $\langle W\log_b W\rangle\geq d$ is the necessary and sufficient condition for the cut to be empty almost surely. This result means that a question that seemed a contrived case of mathematical hairsplitting can sometimes become practical. After concrete application has retrained intuition, "hair-splitting" changes to "obvious". In the present case, it suffices to argue as if the measure reduced exactly to being supported by a fractal set of co-dimension $\langle W\log_b W\rangle$ in some high-dimensional space. There is a well-known rule about the effect of intersection upon dimension. Here, this rule shows that $d=\langle W\log_b W\rangle$ is a "critical" dimension: it separates the dimensions of spaces that almost surely miss our fractal, from the dimensions of spaces that hit it with positive probability.

What about the moments of $\mu_k(\Delta)$ when it is non degenerate? I discovered that they may $\to\infty$ as $k\to\infty$. For each space dimension, there is a "critical moment", and for each moment there is a critical space dimension,

$$C(h)=(h-1)^{-1}\log_b<W^h>,$$

such that moments are finite for $C(h)>d$ and infinite for $C(h)<d$.

<u>Generalization</u>. Once strict conservation has been abandoned in favor of mean conservation, the perturbation function $P_k(\underline{x})$ need no longer be constant over cells, hence need not be discontinuous. It can be any random function whose correlation range is b^{-k}. Moreover, the base b itself need no longer be an integer. For example, $P_k(\underline{x})$ may be the convolution of a white noise with a kernel having a typical radius of b^{-k}. The effect of this function upon the "texture" of a M-measure very much deserves to be investigated.

<u>The limit lognormal processes of La Jolla</u> [5]. Finally, mean conservation allows the perturbation index k to be made continuous. This was the point of the second trunk of early development mentioned in Section 1. I made $\log P_k(\underline{x})$ a lognormal process, as near as logic allows to Kolmogorov's original idea. There is a sketch in my 1982 book [p. 379]. I showed that $\mu/2$ is the ϵ-box codimension. Recently [8], it has been shown that the Hausdorff-Besicovitch codimension is also $\mu/2$.

<u>The term "Schutzenberger-Renyi Informations."</u> In the special cases of the Besicovitch measure and of related nonrandom fractal measures,

$$(h-1)^{-1}\log_b<W^h> \text{ becomes } d-I_h, \text{ where } I_h = (h-1)^{-1} \log_b\sum P_j^h.$$

Doyne Farmer noticed – <u>after</u> re-deriving I_h – that A. Renyi had called it a "generalized information". A precursor was M. P. Schutzenberger. There is a book that shows rigorously that I_h satisfies axioms that justify calling it "information". However, I happen to subscribe to Lebesgue's wariness of notions that serve no purpose besides being defined. Claude Shannon was not the first to write I_1, but the first to encounter I_1 in unexpected inequalities that inject entropy into the study of communication. In the study of fractal measures, I_1 was first encountered as a Hausdorff-Besicovitch dimension by Besicovitch and his students [19]. But there was no early counterpart for other I_h's.

On the scope of the term "fractal dimension". "Fractal dimension" should now be a generic notion, special cases of which are the box dimension ("capacity"), Frostman's capacity dimension, the ϵ-box dimension, the similarity dimension, the gap dimension, the Hausdorff-Besicovitch dimension, etc... However, some papers on M-measures follow a usage that restricts the generic term to the fractal dust that supports the M-measures. I feel the usage is misleading.

6. MULTIPLICATIVE CHAOS WITH WEIGHTS OF EITHER SIGN, AND A SURROGATE FOR BROWNIAN MOTION.

Open problems concerning multiplicative chaos are most numerous and obvious in the case when the weight W can take either sign. One new example [18] gives the flavor. On the line, one needs, in addition to the base b, a second base b">0 such that b-b">0 and is even; we shall write $H=\log_b b"$ so that $0<H<1$. The weight W will be two-valued: $W=\pm b/b"$. Strict conservation (of something like electric charge rather than mass!) is achieved by setting $W=+b/b"$ over $(b+b")/2$ cells of length b^{-1} and $W=-b/b"$ over the remaining ones. The sequence of + and - forms the generator. It may be fixed, yielding a non-random M-measure, or chosen each time at random under the above constraint, yielding a microcanonical M-measure. The functions $F_k(t)$ are no longer nondecreasing, and $F(t)=\lim_{k\to\infty}F_k(t)$ is shown in [18] to be a self-affine function, whose increment over an interval b^{-k} in the grid is $|\Delta F| = \pm|\Delta t|^H$, exactly. Similarly, fractional Brownian motion $B_H(t)$ (Wiener's Brownian motion if H=.5) satisfies $|\Delta B_H|\sim|\Delta t|^H$. However, the distribution of ΔF is not Gaussian but binomial. This makes F(t) a useful surrogate of $B_H(t)$. The exponent of the h-th <u>absolute</u> moment of ΔF is $m^+(h)=-\log_b<|W|^h> + h = hH$.

It is linear in h, which is the simplest possible behavior. (In the case of positive M-measures, m(h) linear in h corresponds to the M-measure that is homogeneous on a fractal dust). The critical exponent is the value of h for which $m^+(h)=hH=1$ is 1/H. To explore its significance, consider the h-variation of F, defined by $\int|\Delta F|^h = |\Delta t|^{hH-1}$, and let $\Delta t\to 0$.

When h>1/H, $\int|\Delta F|^h\to 0$, but when h<1/H, $\int|\Delta F|^h\to\infty$.

$\int|\Delta F|\to\infty$ expresses that F is <u>not</u> of bounded variation. With respect to $\int|\Delta F|^h$, F(t) behaves like $B_H(t)$. Observe that divergence occurs here <u>below</u> the critical h, and concerns the microcanonical case, while for the positive M-measure we know divergence occurs <u>above</u> the critical h, and is found only in the canonical case.

The corresponding canonical M-measure is obtained when W is binomial, with $Pr(W=b/b")=(b+b")/2b$ and $Pr(W=-b/b")=(b-b")/2b$. Now, ΔF is no longer binomial. Its h-th moment is finite when $h<1/H$, but infinite when $h>1/H$. (For example, moments of order $h>2$ are infinite when H takes the Brown value 0.5.) On both counts, the canonical version is very different from $B_H(t)$. But it is an exciting object for study, and I expect it to be useful; the little I know of its properties will be reported on elsewhere.

In the space of $d>1$ dimensions, we write $H=\log_b b"/d$, and we select $W=\pm b^d/b"=\pm b^{d(1-H)}$. Strict conservation now requires $W>0$ over $(b^d+b")/2$ cells and $W<0$ over the other cells. Again, microcanonical M-measure of a cell Δ, of side b^{-k} and of content $|\Delta|$, satisfies $|\mu_k(\Delta)| = b^{-Hkd} = |\Delta|^H$, and the critical value for the divergence of the h-variation is $h=1/H$.

REFERENCES

1. B. B. Mandelbrot, in Proceedings of the Symposium on Turbulence of Fluids and Plasmas (Brooklyn Poly, New York, 1968) p. 483 (Interscience, New York,1969).

2. B. B. Mandelbrot, J. Fluid Mech. 62:331 (1974).

3. B. B. Mandelbrot, in Turbulence and Navier Stokes Equation (Orsay, 1975). Lecture Notes in Mathematics. Vol. 565, p. 121 (Springer, New York, 1976).

4. B. B. Mandelbrot, in Statistical Physics Conference (Haifa, 1977) p. 225 (Bristol, Adam Hilger 1978).

5. B. B. Mandelbrot, in Statistical Models and Turbulence (La Jolla, 1972) Lecture Notes in Physics: Vol. 12, p. 333 (Springer, New York, 1972).

6. B. B. Mandelbrot, C. R. Acad. Sci. (Paris) 278A: 289 and 355 (1974).

7. J. Peyriere, C. R. Acad. Sci. (Paris) 278A:567 (1974). J. P. Kahane, C. R. Acad. Sci. (Paris) 278A: 621(1974). J.P. Kahane and J. Peyriere, Adv. Math. 22:131 (1976).

8. J. P. Kahane, C. R. Acad. Sc. (Paris) 301A (1985).

9. H. G. E. Hentschel and I. Procaccia, Physica 8D:435 (1983).

10. B. B. Mandelbrot, J. Stat. Phys. 34: 895 (1984).

11. R. Benzi, G. Paladin, G. Parisi and A. Vulpiani, J. Phys. 17A: 3521 (1984).

12. A. N. Kolmogorov, J. Fluid Mech. 13:82 (1962). Also A. M. Oboukhov, J. Fluid Mech. 13: 77 (1962).

13. J. M. Berger and B. B. Mandelbrot, IBM J. Res. Dev. 7: 224(1963).

14. B. B. Mandelbrot, IEEE Trans. Comm. Techn. 13: 71 (1965). Also Proc. Fifth Berkeley Symp. Math. Stat. and Probability 3:155 (1967). Also IEEE Trans. Inf. Theory 13: 289 (1967).

15. E. A. Novikov and R. W. Stewart, Isv. Akad. Nauk SSSR, Seria Geofiz. 3: 408 (1964).

16. U. Frisch, M. Nelkin and J. P. Sulem, J. Fluid Mech. 87:719 (1978).

17. A. S. Gurvitch and A. M. Yaglom, Physics of Fluids 10: 559(1967).

18. B. B. Mandelbrot, in Fractals in Physics (Trieste 1985) (Amsterdam, North-Holland, 1986).

19. P. Billingsley, _Ergodic Theory and Information_. (J. Wiley, New York, 1967).

On the Hausdorff Dimension of Graphs and Random Recursive Objects

R.D. Mauldin

Department of Mathematics, North Texas State University, P.O. Box 5116, Denton, TX 76203, USA

The purpose of this note is to present some recent results and techniques concerning the Hausdorff dimension of various objects. We will report on an estimate for the lower bound of the dimension of a wide class of graphs which includes the Weierstrass-Hardy-Mandelbrot functions,and also on the exact dimension of some objects constructed via random recursions.

It commonly occurs that one is able to obtain without too much difficulty an upper bound on the Hausdorff-Besicovitch dimension of some object,e.g., the fractal dimension of the object. However, the problem of showing that these bounds are fairly sharp, or, even better, actually is the Hausdorff dimension,is usually more intricate. In many cases one wants the exact Hausdorff dimension because there exist corresponding Hausdorff measures which can be used to make qualitative and quantitative statements about the problem at hand.

Let us recall the definition of Hausdorff dimension as presented by ROGERS [1].

Let X be a metric space with metric ρ. For each $\beta > 0$, $\epsilon > 0$ and subset E of X, set

$$\mu_{\beta,\epsilon}(E) := \inf \left\{ \sum_{G \in \mathcal{G}} (\text{diam}(G))^{\beta} \mid \mathcal{G} \text{ covers E and mesh } \mathcal{G} < \epsilon \right\}, \quad (1)$$

where mesh $\mathcal{G} < \epsilon$ means that if $G \in \mathcal{G}$, then diam $(G) < \epsilon$. For each β, if $0 < \epsilon_1 < \epsilon_2$,

$$\mu_{\beta,\epsilon_1}(E) \geq \mu_{\beta,\epsilon_2}(E).$$

So,

$$\lim_{\epsilon \downarrow 0} \mu_{\beta,\epsilon}(E) := \mu_\beta(E).$$

FUNDAMENTAL THEOREM. For each $\beta > 0$, μ_β is a measure defined on the Borel subsets of X. Moreover, for each $E \subset X$, there is a number $\alpha \geq 0$ such that

$$\mu_\beta(E) = \begin{cases} +\infty, & \text{if } \beta < \alpha \\ 0, & \text{if } \alpha < \beta \end{cases} \tag{2}$$

DEFINITION. The Hausdorff dimension of a subset E of X, denoted by $\dim_H(E)$, is the number α such that (2) holds.

Let us point out that one of the central problems in making numerical estimates of the Hausdorff dimension of an object lies in equation (1). One may obtain perhaps faulty estimates if one simply takes a regular grid of "boxes", finds how many are needed to cover E and uses the corresponding number to estimate $\mu_{\beta,\epsilon}(E)$. One must take into account that equation (1) states that one must first optimize the grid before estimating $\mu_{\beta,\epsilon}(E)$. This problem in itself means that the numerical methods employed must be rather sensitive.

1. <u>GRAPHS</u>. Consider the continuous function f which satisfies the functional equation:

$$f(x) = af(bx) + \cos x,$$

where $0<a<1$ and $1 \leq ab$. By iteration, we see that

$$f(x) = \sum_{n=0}^{\infty} a^n \cos(b^n x)$$

or, setting $\alpha = -\log a/\log b$,

$$f(x) = \sum_{n=0}^{\infty} b^{-\alpha n} \cos(b^n x).$$

These are the famous Weierstrass-Hardy functions which Hardy proved to be nowhere differentiable [2]. More generally, one can consider functions of the form

$$W_b(x) = \sum_{n=-\infty}^{\infty} b^{-\alpha n}\left[\cos(b^n x + \theta_n) - \cos \theta_n\right]$$

where $b > 1$ and the θ_n's are arbitrary. If each $\theta_n = 0$, then W_b satisfies the scaling law $f(x) = b^{-\alpha}f(bx)$. However, Mandelbrot suggested the introduction of the phases θ_n with the idea of using these functions to model various physical phenomena. Of course, the introduction of the phases also destroys any direct obvious scaling laws.

A problem of some interest is to calculate the Hausdorff dimension of the graphs of the functions W_b. FALCONER [3] and MANDELBROT [4] comment on this in their books. Also, comments and excellent computer studies of these functions appeared in the paper of BERRY and LEWIS [5]. These functions have appeared in several contexts at this conference. The general feeling seems to be the following:

CONJECTURE. $\dim_H(\text{Graph}(W_b)) = 2-\alpha$.

In [6], S. C. WILLIAMS and I prove the following:

THEOREM. Fix $0 < \alpha \leq 1$. There is a constant C such that if b is large enough, then, for any phases θ_n:

$$(2-\alpha)-C/\log b \leq \dim_H(W_b) \leq 2-\alpha.$$

Let us make some remarks concerning this theorem. First, the upper estimate is due to BESICOVITCH and URSELL [7]. Among other things, they showed that if a function f is Lipschitz of order α, then $\dim_H f \leq 2-\alpha$. It is straightforward to prove W_b is Lipschitz of order α, if $0 < \alpha < 1$. Second, the general idea behind the lower estimate is the following. If b is sufficiently large, one can recover enough scaling to define recursively a Cantor subset M of $\mathbb{R}$ such that the behavior of W_b over the intervals in the nth level of the construction of M is governed by the derivative of the nth approximating sum to W_b, and on each of these intervals, this derivative is numerically so large that the nth approximating sum is strictly increasing or decreasing very rapidly. In fact, this allows us to estimate how much time the function (at least the part of the function over the set M) can spend in a box with certain bounds on the edge length of the box. The final upshot is that there is a naturally defined, nontrivial, measure μ supported on the Cantor set M such that when one transfers μ to the graph of W_b, i.e.,

$$\tilde{\mu}(E) := \mu(\text{proj}_1(E \cap \text{Graph}(W_b))),$$

for all Borel sets $E \subset \mathbb{R}^2$, then $\tilde{\mu}$ witnesses the behavior of W_b over M. There is a constant C such that if b is sufficiently large and E is a Borel subset of $\mathbb{R}^2$ and $\dim_H(E) < (2-\alpha)-C/\log b$, then $\tilde{\mu}(E)=0$. Since μ is nontrivial, $\tilde{\mu}(\text{Graph}(W_b)) > 0$. Therefore, $\dim_H(\text{Graph}(W_b)) \geq (2-\alpha)-C/\log b$. This type of argument was used by Besicovitch and Ursell for a class of functions which exhibit much more lacunarity than the functions W_b.

We actually obtain the estimates given in the preceding theorem when the cosine function is replaced by more general functions. Also, one can "see" the nth level intervals which we use in the construction of the Cantor set by making graphs of the approximating nth partial sum to W_b and "blowing up" the intervals over which the sum is rapidly increasing or decreasing. Tony Warnock of Cray Research prepared the displays demonstrating this at the conference.

2. <u>Dimension of Objects Defined by Random Recursions</u>

S.C. Williams and I [8] have also developed this same technique of
constructing a measure which witnesses the dimensional complexity of
an object when the objects are constructed via a random recursion.
Before describing the general setting, consider the following.

EXAMPLE. We construct a Cantor subset of $[0,1]$ via a random algorithm
as follows. At level one we remove an open subinterval (x,y) of $[0,1]$
as follows. First, choose x at random (according to the uniform
distribution) on $[0,1]$. Second, choose y according to the uniform
distribution on $(x,1)$. We have two intervals $J(\langle 0 \rangle) = [0,x]$ and
$J(\langle 1 \rangle) = [y,1]$ remaining. This completes the recursion at level one.

If we have 2^n pairwise disjoint closed intervals $J(\langle e_1,\ldots,e_n \rangle)$
remaining at stage n, then we remove an open interval from each of
them according to the procedure given on level one scaled to that
interval. Thus, for each interval on the nth level, we will have two
closed subintervals remaining upon completion of the next level.
Finally, set

$$K = \bigcap_{n=1}^{\infty} \left[\bigcup_{\langle e_1,\ldots,e_n \rangle \in \{0,1\}^n} J(\langle e_1,\ldots,e_n \rangle) \right].$$

<u>THEOREM</u>. With probability one, the set K is a Cantor subset of $[0,1]$
and $\dim_H(K) = (\sqrt{5}-1)/2$.

Let us describe *RANDOM RECURSIVE CONSTRUCTIONS*.

Our general model is as follows: We fix a Euclidean space $\mathbb{R}^m$ and a
nonempty compact subet J of $\mathbb{R}^m$. We further require that J is the
closure of its interior in $\mathbb{R}^m$. We assume we have a probability space
$(\Omega.\Sigma,P)$ and are given a family of random subsets of $\mathbb{R}^m$,

$$J = \{ J_\sigma \mid \sigma \in \mathbb{N}^* := \bigcup_{n=0}^{\infty} \mathbb{N}^n \},$$

satisfying three properties.

(1) $J_\phi(\omega) = J$, for almost all $\omega \in \Omega$. For every $\sigma \in \mathbb{N}^*$ and for almost
all ω, if $J_\sigma(\omega)$ is nonempty, then $J_\sigma(\omega)$ is geometrically similar to J.

(2) For almost every ω and for every $\sigma \in \mathbb{N}^*$, $J_{\sigma*1}(\omega)$, $J_{\sigma*2}(\omega)$,
$J_{\sigma*3}(\omega),\ldots$ is a sequence of nonoverlapping subsets of $J_\sigma(\omega)$. (A and B
are nonoverlapping means int $A \cap$ int $B = \phi$.)

(3) The random vectors $T_\sigma = \langle T_{\sigma*1}, T_{\sigma*2},\ldots \rangle$, $\sigma \in \mathbb{N}^*$, are i.i.d.,
where $T_{\sigma*n}(\omega)$ equals the ratio of the diameter of $J_{\sigma*n}(\omega)$ to the
diameter of $J_\sigma(\omega)$ if $J_\sigma(\omega)$ is nonempty. (For convenience, let $T_\phi(\omega) = $
diameter of J.)

Such a system J we shall call a *random recursive construction*. Our constructions require only a "*stochastic ratio self-similarity*." We now define the random set K by

$$K(\omega) = \bigcap_{n=1}^{\infty} \left[\bigcup_{\sigma \in \mathbb{N}^n} J_\sigma(\omega) \right].$$

Our interest centers on the asymptotic properties of this random set K.

For convenience, let $0^0 = 0$. Then $\sum_{p=1}^{\infty} T_{\sigma*p}^0(\omega)$ counts the number of nonempty $J_{\sigma*p}(\omega)$, if $J_\sigma(\omega)$ is itself nonempty.

The main theorem of [8] is the following:

THEOREM. Suppose $E(\sum_{n=1}^{\infty} T_n^0) > 1$. Then with positive probability K is nonempty. Moreover, given that K is nonempty, then almost surely K has Hausdorff dimension α, where α is the least $\beta > 0$ such that

$$E(\sum_{n=1}^{\infty} T_n^\beta) \leq 1.$$

For instance, applying this theorem to the preceding example we have:

$$\phi(\beta) = E(\sum_{n=1}^{\infty} T_n^\beta) = E\,(T_1^\beta + T_2^\beta) = \int_0^1 \left[x^\beta + \frac{1}{1-x} \int_x^1 (1-y)^\beta dy \right] dx,$$

from which we find $\phi((\sqrt{5}-1)/2)=1$.

Let us make remarks about the proof of the theorem. First, set

$$\phi(\beta) = E(\sum_{n=1}^{\infty} T_n^\beta).$$

It can be shown that ϕ is nonincreasing, right continuous and $\phi(m) \leq 1$.

Now, the number α given by the theorem yields that a certain sequence of random variables $\{S_n\}_{n=1}^{\infty}$ forms a positive supermartingale. By the martingale convergence theorem, the S_n's converge to some real-valued random variable X. What this means is that for each n, $S_n(\omega)$ is an upper estimate of the Hausdorff measure of $K(\omega)$ with respect to Hausdorff's α-measure. So, almost surely $\alpha\text{-}m(K(\omega)) \leq X(\omega) < +\infty$. This means almost surely $\dim_H(K(\omega)) \leq \alpha$.

In order to obtain a lower bound, we show that there is enough self-similarity(albeit random) that one can construct a random measure $\mu(\omega)$ such that $(1)\mu(\omega)$ is supported on $K(\omega)$; (2) if $K(\omega) \neq \phi$, $\mu(\omega)(K(\omega)) > 0$ and (3) if E is a Borel subset of $\mathbb{R}^m$ and $\dim_H(E) < \alpha$, then

$\mu(\omega)(E)=0$. It follows from these conditions that if $K(\omega) \neq \phi$, then $\dim_H((K(\omega)) = \alpha$.

As a final note, we remark that in the deterministic case, i.e., if τ_ϕ is distributed as point mass at $(t_1, t_2, \ldots, t_n, 0, 0, 0, \ldots)$ where $t_1, t_2, \ldots, t_n > 0$, then our main theorem implies that K has Hausdorff dimension α where α satisfies the equation

$$t_1^\alpha + \ldots + t_n^\alpha = 1.$$

This deterministic result was obtained by P. A. P. MORAN [9]. Thus, our equation

$$E\left(\sum_{n=1}^\infty T_n^\alpha\right) = 1,$$

is a direct generalization of Moran's result.

References

1. Rogers, C. A., *Hausdorff Measures*. Cambridge University Press, 1970.
2. Hardy, G.H., "Weierstrass's non-differentiable functions," *Transactions American Mathematical Society* 17 (1916), 301-325.

3. Falconer, K.J., *The Geometry of Fractal Sets*. Cambridge Tracts in Mathematics vol.85, Cambridge University Press, 1985.

4. Mandelbrot, B. B., *Fractals: Form, Chance and Dimension*. San Francisco: Freeman, 1977.

5. Berry, M.V., and Z.V. Lewis, "On the Weierstrass-Mandelbrot fractal function." *Proceedings of the Royal Society of London* (1980) A370, 459-484.

6. Mauldin, R. D., and Williams, S. C., "On the Hausdorff dimension of the Weierstrass-Mandelbrot functions," preprint.

7. Besicovitch, A.S., and H.D. Ursell, "Sets of fractional dimensions, v: On dimensional numbers of some continuous curves." *Journal of the London Mathematical Society* (1937) (2), 32, 142-153.

8. Mauldin, R. D., and S. C. Williams, "Random Recursive Constructions:Asymptotic Geometric and Topological Properties," *Transactions American Mathematical Society*, (to appear).

9. Moran, P.A.P., "Additive functions of intervals and Hausdorff measure," *Proceedings of the Cambridge Phil. Society* 42 (1946). 15-23.

Chaos-Chaos Phase Transition and Dimension Fluctuation

Y. Aizawa

Department of Physics, University of Kyoto, Japan

1. Introduction

Recent studies on chaos have made clear that the concept of chaos is quite different from the probabilistic randomness. Much work has been especially done to understand the internal order in chaos such as topological and fractal ones. The discovery of some routes to chaos has also (contributed to the better understandings of order in germinal chaos, but hereafter the order immersed in the fully developed or grown-stage chaos should be elucidated.

Chaos, even in the small systems, has infinitely many internal fourier modes, and the order in it is expected to have some glassy-type structures such as is known in spin glass system. Then the spatially distributed weak attractive modes in glassy state would be reinterpreted as the temporal modes in chaos. In the many degrees of freedom systems, the analogy between chaos and glassy state seems to be very clear.

In the grown-stage chaos, the sharp bifurcation may not occur clearly, and it may be difficult to detect the change of order in chaos. However, when the special ordered mode in chaos works as a pacemaker which attracts the other modes into an entrained state, one can expect to observe the creation of new order and/or the destruction of old order. If so, this situation can be called chaos-chaos phase transition.

Generally, there seems to be a lot of causes which induce some structural changes in chaos. Following are typical familiar examples of the phase change in chaos. One is the fusion-type transition where several attractive basins merge, and another is the entrainment-type transition where the partial locking occurs on a special internal mode. Even in the small systems, both types of transition often appear, but in general the latter transition may be softened by the strong competition among several internal modes. The example studied in this paper is considered to be the entrainment type, and it is the typical diffusive or softened transition.

The purpose of this paper is to present several characteristics which describe the chaos-chaos transition. It is especially emphasized that the structural non-uniformity of the chaotic attractor is significant for the description of the chaos-chaos tansition.

2. Model system

The Lorenz model with periodic forcing is studied in this paper.

$$\dot{X} = \sigma(Y-X)$$

$$\dot{Y} = rX-Y-XZ$$

$$\dot{Z} = -bZ+XY+A\cos(Bt)$$

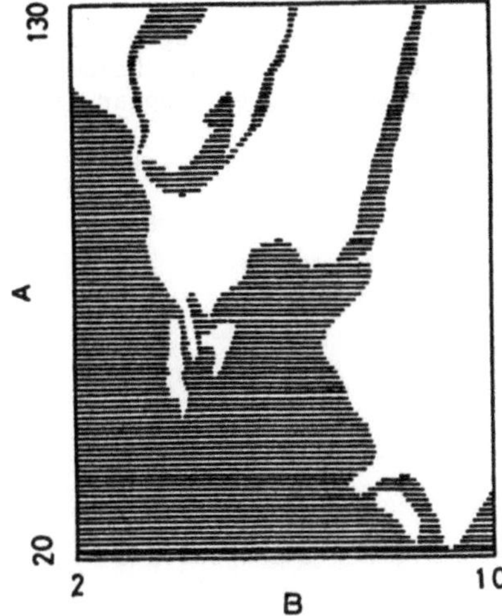

Fig.1 Chaotic and periodic response phase diagram

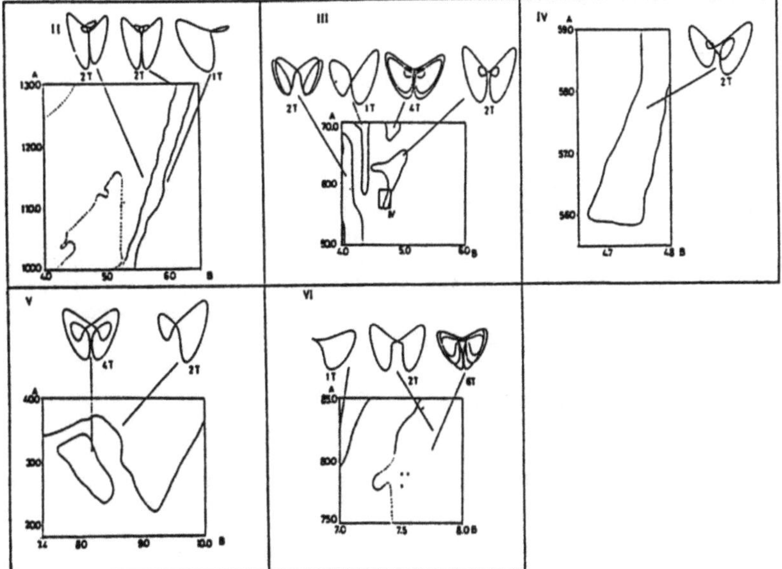

with σ=10, r=28, and b=8/3. The last term is the external perturbation and the bifurcation parameters are A and B. The Lorenz chaos is modulated by the perturbation. The variety of the response phases are shown in Fig.1, where many kinds of bifurcation phenomena are observed and some of them were systematically studied before. Some ergodic and topological natures of the response phases were made clear in the previous paper.[2] The main reason why the same model is studied here again is to elucidate the mechanism of the chaos-chaos phase change.

In our model system, there exist two types of chaos-chaos transitions, as were mentioned in §1; one is the fusion type and the other is the internal structural change. The former case is easily understood as the collapsing or crisis of two attractive basins. The type studied in the present article is the latter one, which seems to be more interesting than the former, since there has not yet been theoretical approach till now.

35

In this paper the special attention is payed in the parameter regime around A≈60 and B≈6, where the different types of chaos are observed above and below the borderline of A≈60. As is shown in Fig.2, small spiral mode comes to appear above the borderline.

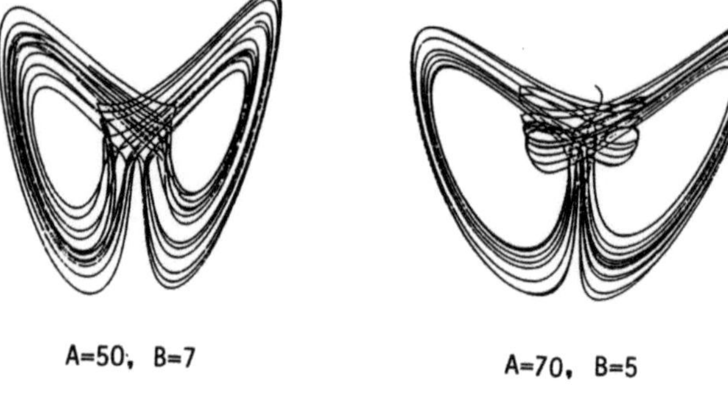

Fig.2 x-z projected orbits

3. Some evidences of chaos-chaos transition

In this section, several chaos parameters are applied in order to understand the transition mechanism.[1,3] In what follows, the bifurcation parameter B is fixed as B=6.

(A) Auto-regressive analysis

From the theory of the time series analysis, one can derive some useful parameters that characterize chaos. Let $\alpha(t)$ be a real time series for long but finite duration ($0<t<T$), and denote the correlation function by $A(\tau)=<\alpha(t+\tau)\alpha(t)>$, where $< >$ stands for the time average. One of the most important theoretical problems is to construct the most reliable statistical predictor $\beta(t)$ which simulates the original process $\alpha(t)$. By means of the linear auto-regressive analysis, the predictor $\beta(t)$ is realized by,

$$\beta(t)= \int_{\tau_m}^{t} B(t-\tau)\beta(\tau)d\tau+\varepsilon(t) \quad ,$$

where the kernel $B(t)$, the markovian time τ_m and the white gaussian noise $\varepsilon(t)$ are uniquely determined by the minumum criterion of the Final Prediction Error.[4]

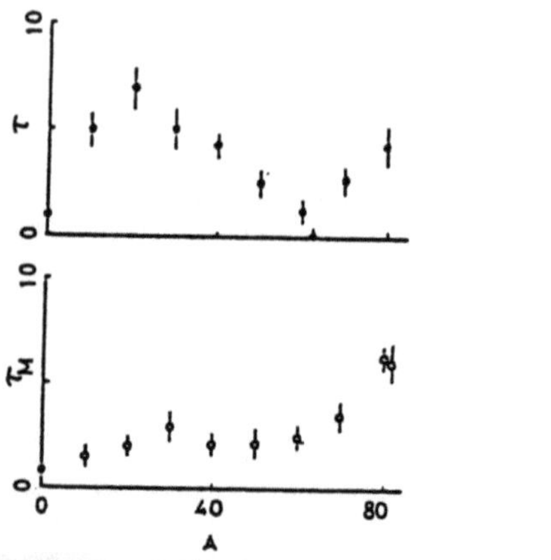
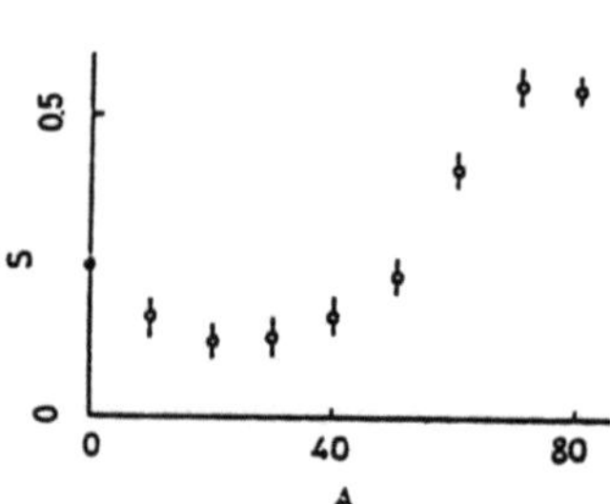

Fig.3 Correlation time, markovian time and stiffness

The markovian time τ_m and the relative intensity of noise or irrelevant term $S=\langle\varepsilon^2(t)\rangle/\langle\alpha^2(t)\rangle$ are two characteristics. τ_m is not always equal to the relaxation time of the correlation function, but is the parameter which describes the induction period for the loss of the initial information. The second parameter S denotes the stiffness of chaos. The predictor $\beta(t)$ decays out almost monotonically to zero if the noise part is negligibly small, but the monotonical regressive nature is disturbed and the essential chaotic behavior is generated if the noise level is large enough. Therefore, it is reasonable to say that the parameter S describes the attractor's strength or stiffness.

Figure 3 shows some statistical quantities of Y-variable of our model system. τ and τ_m are the correlation time and the markovian time, and S is the stiffness. Above the critical point A ≃ 61, the correlation function reveals oscillatory damping and the correlation time is estimated from its envelope. The appearence of the oscillatory decay is in accordance with the onset of the small spiral mode that is shown in Fig.2.

(B) Partial locking in recurrence phase

Level crossing analysis is applied to detect the phase transition. Let us consider the poincare map on the special section surface at Z=27. Then denoting the successive recurrence times by t_1, t_2, t_3,...., the recurrence phase on the section is defined by $\psi_i=Bt_i$ (mod. 2π). The information entropy H_I of the phase distribution $P(\psi)$ is estimated by,

$$H_I = \frac{-1}{\ln(2\pi)} \int_0^{2\pi} P(\psi)\, \ln P(\psi)\, d\psi$$

which measures the complexity and the mixing rate of the phase information. When the phase recurrence occurs uniformly, the information entropy takes maximum $H_I=1$.

Figure 4 shows the singularity induced by the chaos-chaos transition. As is shown in Fig.4-(a), some recurrence phases are forbidden above the critical point.

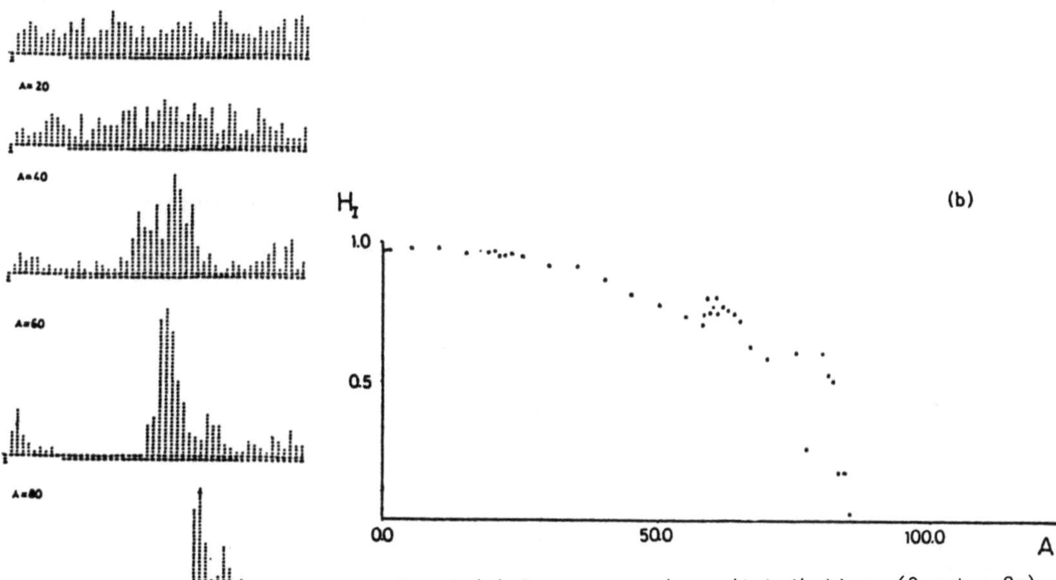

Fig.4 (a) Recurrence phase distribution $(0 \leq \psi \leq 2\pi)$

(b) Information entropy

This figure clearly shows that the chaos-chaos transition studied here is originated from the partial phase locking on a certain internal mode.

(C) Lyapunov exponents and dimension
Let us consider the 3-dimensional stroboscopic map F at the integral time $t=1,2,3,\ldots$,

$$F: \quad (X_i, Y_i, Z_i) = (X_{i+1}, Y_{i+1}, Z_{i+1})$$

and determine the Lyapunov spectra (K_1, K_2, K_3) in the ordinary manner, $\Sigma K_i < 0$ and $K_1 > K_2 > K_3$. The entropy H is estimated by the sum of positive K_i, and the dimension D_A by,

$$D_A = \begin{cases} 3 - (K_1+K_2+K_3)/K_3 & (K_1+K_2 > 0) \\ 2 - (K_1+K_2)/K_2 & (K_1+K_2 < 0) \end{cases}$$

From the definition[5], it is clear that D_A is a continuous function of K_i, although the differentiability might be lost. Even when D_A is not differentiable at $K_1 + K_2 = 0$, one can not expect the occurrence of the chaos-chaos phase transition, since such abnormality merely comes from the definition. This kind of abnormality is discussed in the next section. To avoid the confusion, we denote the dimension parameter by $D_A = 3 - (K_1+K_2+K_3)/K_3$ in this section.

Figure 5 shows the Lyapunov exponents and the dimension. The singularity at the critical point $A \simeq 61$ is mainly created by the third Lyapunov exponent K_3, of which derivative seems to jump remarkably. The same singularity appears in D_A. It may be accidental that $K_1 + K_2$ is almost zero at the critical point.

(D) Entanglement of chaotic orbits
The nature of chaotic orbits must be characterized not only by the instability parameter such as the Lyapunov exponents, but also by the topological parameter. The linking coefficient is the typical topological one.[3] Let us consider two orbits C_1 and C_2 whose coordinates are denoted by $r_1(t)$ and $r_2(t)$, $r = (X, Y, Z)$. The total linking number during $0 < t < T$ is defined by,

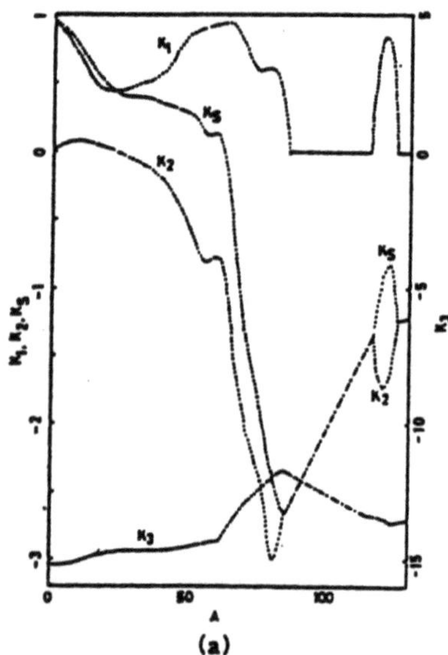

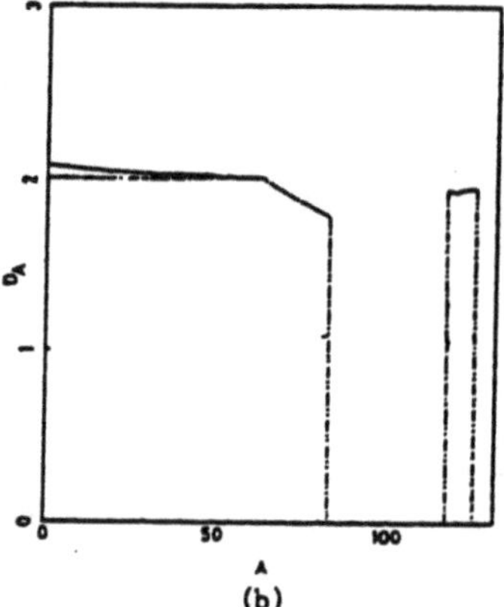

Fig.5 Lyapunov exponents and dimension

$$L(C_1, C_2; T) = \frac{-1}{4\pi} \int_0^T \int_0^T \frac{(r_2-r_1, \ dr_1 \times dr_2)}{|r_2 - r_1|^3}$$

When both orbits are closed and their periods are T_1 and T_2, the long-time behavior of the linking is estimated by,

$$L(C_1, C_2; T) \simeq n_L \cdot T^\nu \qquad \text{with } \nu=2.$$

$$(T >> T_1, T_2)$$

n_L is called the linking coefficient, and ν is the linking index.[3] If both orbits C_1 and C_2 are chaotic, the above estimation of $\nu = 2$ is surmised to be true. In fact, many numerical calculations support this conjecture, though there is no mathematical general proof so far.

In the non-autonomous case, the estimation is extended, and the linking coefficient per unit cycle T_p of the external force is defined by,

$$L(C_1, C_2, NT_p) \simeq n_0 \ N^2$$

where $T_p = 2\pi/B$ for our model.

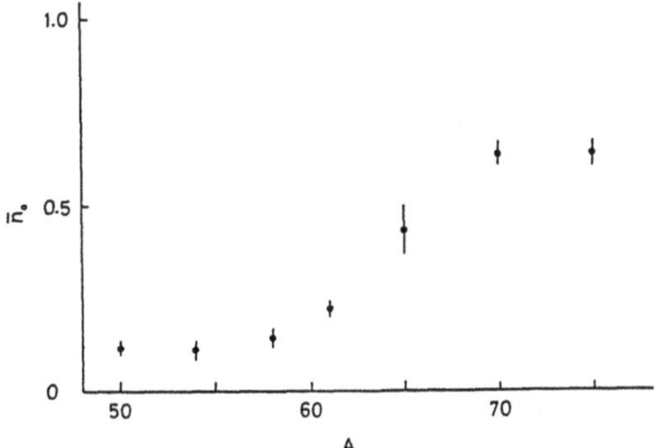

Fig.6 Linking coefficient per cycle

Figure 6 shows the remarkable change of the linking. The coefficient n_0 can be unity when C_1 and C_2 are fully entrained on limit cycles with a simple loop, but in the present case n_0 is always less than unity, since our example is still partial locking even after the critical point.

4. Dimension and its fluctuation

The dimension is a parameter which describes the cantorian structure of chaotic attractor, but the totality of an attractor can not be characterized completely by a single parameter. For instance, let us consider the 3-dimensional generalization of baker's transformation $X_{n+1}=F(X_n)$ of unit cube as is illustrated in Fig.7,

$$X_{n+1} = 2X_n$$

$$Y_{n+1} = Y_n/\lambda_y \qquad\qquad \text{for } 0 \leq X_n \leq 1/2$$

$$Z_{n+1} = Z_n/\lambda_z$$

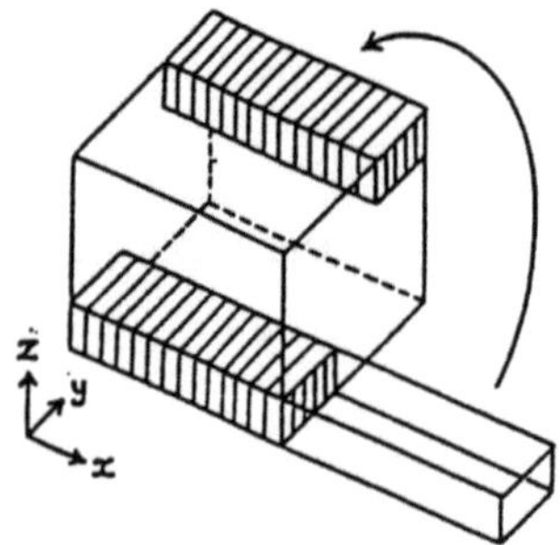

Fig.7 Generalized baker's
transformation

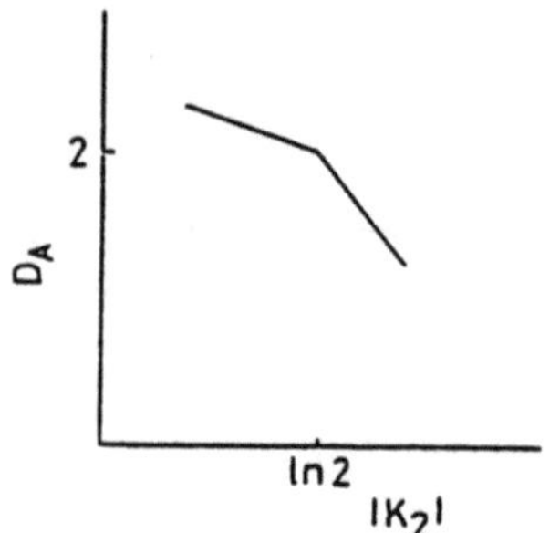

Fig.8 Deceptive anomaly
of dimension

$$X_{n+1} = 2X_{n-1}$$

$$Y_{n+1} = (Y_n + \lambda y - 1)/\lambda_y \qquad \text{for} \quad 1/2 < X_n \leq 1$$

$$Z_{n+1} = (Z_n + \lambda z - 1)\lambda_z$$

Here we assume $\lambda_z > \lambda_y > 1$ and $\lambda_y \lambda_z > 2$. The Lyapunov exponents $(K_1, K_2, K_3) = (\ln 2, -\ln\lambda_y, -\ln\lambda_z)$. For an example where $\lambda_z = 3$ fixed, the dimension is parametrized by $\ln\lambda_y$, and a remarkable singularity occurs at $\lambda_y = 2$ as shown in Fig.8. However, this anomaly does not imply the occurrence of the structural change in attractor, but only comes from the incompleteness of the estimation. In fact, in the case of $\lambda_y > 2$, the above transformation must be characterized by two dimension parameters; $D_1 = 1 + \ln2/\ln3$, and $D_2 = 1 + \ln2/\ln\lambda_y$. Each one describes the cantor structure of the projected attractor into X–Z or X–Y plane. This simple consideration suggests that the dimension should be spectra of which components are defined in projected sub-spaces.[1]

Now, we have to discuss the essential difference between the deceitful singularity and the true chaos–chaos phase transition. Here we use an analogy with the critical phenomena in statistical mechanical systems, especially with the second order phase transition in spin system. The critical point is characterized not only by the anomalous change of the magnetization order parameter, but also by the remarkable enhancement of the magnetization fluctuation. Generally, the fluctuation is correlated to the susceptibility which describes the linear response properties under the appropriate perturbation. At the present time, the best definition of the fluctuation has not yet been obtained for the chaos–chaos transition, but in what follows I will try to present one definition of it in a practical manner.[1]

First, let us define the local Lyapunov exponents as follows,

$$K_i(T_0, X_0) = \frac{1}{T_0} \ln \frac{|(dT)^{T_0} e_i|}{|e_i|}$$

where $(dT)^{T_0}$ is the T_0 time iterations of the mapping in the tangent space, and e_i is the Lyapunov eigen-vector for i-direction at the phase point X_0. When T_0 goes to infinity, the function K_i yield the ergodicity of the multiplicative case.[6] Essential point is that K_i is locally defined at every phase point, and that one can expect the existence of the mean value,

$$K_i(T_0) = \langle K_i(T_0, X_0) \rangle$$

where $\langle \ \rangle$ stands for phase average. The fluctuation of the local exponent is defined by

40

$$k_i{}^2(T_0) \equiv \langle (K_i(T_0, X_0) - K_i(T_0))^2 \rangle.$$

The fluctuation is the function of the observation time T_0, and when T_0 goes to infinity the fluctuation becomes zero. However, the dynamical or transient fluctuation can be detected during the finite time observation. The same idea is used for the local dimension parameter $D_A(T_0, X_0)$ defined by,

$$D_A(T_0, X0) = 3 - \sum K_i(T_0, X_0)/K_3(T_0, X_0)$$

and its fluctuation,

$$d_A{}^2(T_0) = \langle (D_A(T_0) - D_A(T_0, X_0))^2 \rangle$$

where $D_A(T_0) = \langle D_A(T_0, X_0) \rangle$.

Though the fluctuation defined above is the parameter that describes the reliability in measuring each statistical quantity, the important point is that these fluctuations characterize some structural non-uniformity of an attractor. Indeed, in many cases the structure of the chaotic attractor is not uniform.

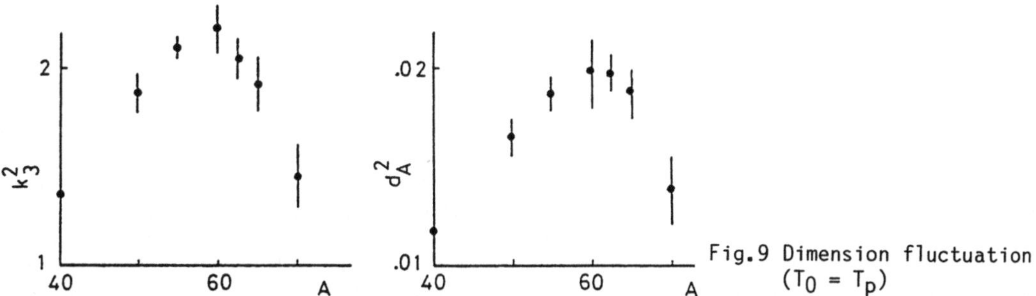

Fig.9 Dimension fluctuation $(T_0 = T_p)$

When the chaos–chaos transition really occurs, one can expect that the enhancement of fluctuations is observed because of the strong non–uniformity of the attractor. In the case of the generalized baker's transformation, the structure of the attractor is always uniform, so that the fluctuation level is zero. This implies that the observed singularity is quite deceptive and no chaos–chaos transition. On the other hand, the transition in our forced Lorenz system is not deceptive at all. Figure 9 shows the fluctuation enhancement near the transition point. The remarkable enhancement is observed in the dimension fluctuation, and its behavior is similar with the third exponent's fluctuation. As is clear from the definition, our fluctuation is not divergent even at the critical point, but the result obtained here strongly supports the similarity between the chaos–chaos phase transition and the critical phenomena.

The concept of local dimension is already discussed by Beardon [7], but it has not yet been applied sucessfully to the dynamical system. Main difficulty seems to arise from the fact that the attractor is a connected set. Future problem is to pursue the new local concept which characterizes the non–uniform structure of attractor.

References
[1] Y. Aizawa: Prog. Theor. Phys. 68 (1982), 64
[2] Y. Aizawa and T. Uezu: Prog. Theor. Phys. 69 (1982), 1862
[3] Y. Aizawa and T. Uezu: Prog. Theor. Phys. 67 (1982), 982
     Y. Aizawa: Prog. Theor. Phys. 70 (1983), 1249
[4] H. Akaike: Ann. Inst. Statist. Math. 22 (1970), 203, 219; 23 (1971), 163
[5] J. L. Kaplan and J. A. Yorke: Lecture Note in Math. 730 (1979), 228
[6] V. I. Oseledec: Trans. Moscow Math. Soc. 19 (1968), 197
[7] A. F. Beardon: Proc. Camb. Phil. Soc. 61 (1965), 679

Hausdorff Dimensions for Sets with Broken Scaling Symmetry

D.K. Umberger[1], G. Mayer-Kress, and E. Jen[2]

Center for Nonlinear Studies, MS B258, Los Alamos National Laboratory, Los Alamos, NM 87545, USA
[1] Also at Dept. of Physics, Univ. of Arkansas, Fayetteville, AR 72701, USA
[2] Permanent address: Dept. of Mathematics, Univ. of Southern California, Los Angeles

Based on Hausdorff's original approach to fractional dimensions, we study systems which are not sufficiently characterized by their "fractal" or scaling dimension. We construct informative examples of such sets and relate them to sets observed in the context of dynamical systems.

1. Introduction

Fractal sets which are not adequately characterized by their Hausdorff-Besicovitch dimensions occur in various physical contexts. One such class of fractals consists of sets which have only either zero or infinite d-dimensional Hausdorff measures [1]. Many random fractals such as those generated by brownian motion in the plane fall into this class [1,2]. In the brownian motion example, it can be shown that the set generated by the motion has a Hausdorff-Besicovitch dimension of 2 and yet is not area filling [2]. Although this set is fundamentally different from an ordinary 2 dimensional set such as a square, the dimension alone gives no indication of this. There is however a more general formalism due to HAUSDORFF [3] which is sensitive to such differences. In this formalism, the Hausdorff measure of a set is defined in terms of functions, called *gauge functions,* that tend to zero as a length scale ϵ tends to zero.

The functional form of the particular gauge function which yields a nonzero and finite (referred to throughout simply as finite) Hausdorff measure is used to characterize the set. In the case where a set has the Hausdorff-Besicovitch dimension d_0 and a finite d_0-dimensional Hausdorff measure, the gauge function of the set is ϵ^{d_0}. If the set has a dimension d_0 and its corresponding d_0-dimensional Hausdorff measure is either zero or infinite, this gauge function is not a strict power law but is modified by a multiplicative correction which can in many cases be expanded in a product of powers of iterated logarithms in $1/\epsilon$. Thus, sets can be described in terms of many exponents which can be viewed as higher order *corrections* to the usual Hausdorff-Besicovitch dimension. These corrections can be used to distinguish sets with integer dimensions, but nonfinite Hausdorff measures from *ordinary,* nonfractal sets.

A second class of sets which are not adequately described by their Hausdorff-Besicovitch dimensions are *fat* fractals [4,5,6,7]. These are fractal sets which have integer dimensions and finite volumes (Lebesgue measures). The gauge functions of these sets are pure power laws with integer exponents which are equal to their dimensions. The gauge functions of ordinary sets such as intervals, squares, and 3-tori also have this property. Thus it appears that the Hausdorff scheme is inadequate for describing this class of fractals. This apparent failure has led to other formulations for characterizing these sets [4,5,6]. In these treatments, some coarse-grained Lebesgue measure which approximates the true measure of a set is introduced. This coarse-grained measure depends upon some length scale ϵ. The manner in which it approaches the actual Lebesgue measure of the set as $\epsilon \to 0$ is used to characterize the set. Since the Lebesgue measure of a fat fractal is just a nonzero multiple of its Hausdorff measure, it is natural to wonder whether or not the Hausdorff formalism can be extended in such a way that fat fractals can be distinguished from ordinary sets.

The purpose of this paper is to discuss the manner in which the two classes of sets just described can be characterized. In particular we use Hausdorff's gauge function approach together with the notion of *capacity* dimension [8] to arrive at a simple scheme for describing these sets. We first present the general formulation, where a set's dimension is determined by the scaling properties of its intrinsic gauge function. We then present some simple examples of thin fractals (fractals that are not fat) which illustrate how logarithmic corrections to pure power laws can arise. Next, we examine the characterization of fat fractals in terms of additive corrections to their intrinsic gauge functions. We then define the notion of a *metadimension* which arises from this type of corrections. We will see that this notion is applicable to thin fractals as well.

2. Gauge functions as dimensions

We begin by introducing the Hausdorff-Besicovitch dimension of a set S embedded in a D-dimensional space. Choose an $\epsilon > 0$ and consider every countable covering of S whose elements are D-dimensional cubes of side ϵ or smaller [8,9]. For each of these coverings, form the quantity $\sum_m \epsilon_m^d$, where the sum is over all of the elements of a particular cover and d is some positive real number. Find the infimum of this quantity over all such covers to get the coarse-grained d-dimensional Hausdorff measure of S, $\nu^{(d)}(\epsilon) = \inf \sum_m \epsilon_m^d$. Then the d-dimensional Hausdorff measure of the set is defined by

$$\nu^{(d)} = \lim_{\epsilon \to 0} \nu^{(d)}(\epsilon). \tag{1}$$

Now, there exists a critical value of d, say d_0, such that $\nu^{(d)}$ is infinite for all $d < d_0$ and zero for all $d > d_0$ [3]. This critical exponent is called the Hausdorff-Besicovitch dimension of the set.

The concepts just introduced were motivated by the desire to have a generalization of the notion of size which is applicable to sets having nonfinite Lebesgue measures. When d_0 is an integer and $\nu^{(d_0)}$ is finite, the d_0-dimensional Lebesgue measure is just equal to $\nu^{(d_0)}$ [10]. When d in Eq.(1) is not an integer, it may be considered to be a generalization of the concept of dimension, and $\nu^{(d)}$ may be viewed as a generalized *volume* corresponding to that dimension. The definition of d_0 given above implies nothing about the finiteness of its associated d_0-dimensional Hausdorff measure. Thus, there can exist sets for which this generalization of size is inadequate. HAUSDORFF [3] recognized this and suggested for linear sets a more general measure based upon arbitrary *gauge functions* λ of a non-negative argument ϵ satisfying

$$\lambda(\epsilon) > 0, \tag{2a}$$

$$\lambda(\epsilon_1) < \lambda(\epsilon_2) \quad \text{for } \epsilon_1 < \epsilon_2, \tag{2b}$$

$$\lambda(\epsilon) \to 0 \quad as \quad \epsilon \to 0, \tag{2c}$$

$$\lambda(\epsilon) \to \infty \quad as \quad \epsilon \to \infty, \tag{2d}$$

and

$$\begin{vmatrix} \lambda(\epsilon_1) & \epsilon_1 & 1 \\ \lambda(\epsilon_2) & \epsilon_2 & 1 \\ \lambda(\epsilon_3) & \epsilon_3 & 1 \end{vmatrix} > 0 \quad \text{for } \epsilon_1 < \epsilon_2 < \epsilon_3. \tag{2e}$$

As HAUSDORFF [3] already noted, these conditions are only to be satisfied close to the origin, but it is convenient to consider a class of functions for which this holds in general. In our examples and generalizations later on we shall also consider cases, for which this distinction becomes relevant, i.e. we want to restrict the functions to small values of the argument. Let $\lambda(\epsilon)$ be any function satisfying Eqs.(2). Then, define the coarse-grained Hausdorff measure of a set S with respect to λ by $\nu_\lambda(\epsilon) = inf \sum_m \lambda(\epsilon_m)$. This coarse-grained measure is defined in the same way as the d-dimensional measure except that ϵ_m^d has been replaced by its generalization $\lambda(\epsilon_m)$. Then the Hausdorff measure with respect to λ is just

$$\nu_\lambda = \lim_{\epsilon \to 0} \nu_\lambda(\epsilon). \tag{3}$$

Note that the d-dimensional Hausdorff measure is recovered from this definition by choosing ϵ^d for the gauge function. The particular gauge function λ that gives a finite Hausdorff measure is called the intrinsic gauge function of the set (or simply the gauge function of the set).

The rate at which the intrinsic gauge function of a set vanishes as $\epsilon \to 0$ is associated with the *dimension* of the set. This is easy to see when the gauge function is a pure power law, since the function scales to zero faster for larger values of d_0. Thus we say that the more quickly a set's intrinsic gauge function vanishes with ϵ, the larger is the dimension the set. In fact, HAUSDORFF [3] refers to a set's intrinsic gauge function itself as the dimension of the set.

An example of a gauge function which corresponds to a dimension that is *between* two power laws is

$$\lambda_1(\epsilon) = \epsilon^{d_0} \left[\log(1/\epsilon)\right]^{-d_1}. \tag{4}$$

It is easy to see that when $d_1 > 0$, this function goes to zero faster than ϵ^{d_0} and slower than $\epsilon^{(d_0+\delta)}$ for any $\delta > 0$. If d_1 is chosen to be negative, this function vanishes slower than ϵ^{d_0} and faster than $\epsilon^{(d_0-\delta)}$. To carry this a bit further, note that for any $d_0, d_1, \delta > 0$, we can find a function that goes to zero faster than the function given in Eq.(4) and slower than the function $\epsilon^{d_0} \left[\log(1/\epsilon)\right]^{-(d_1+\delta)}$. An example of such a function is

$$\lambda_2(\epsilon) = \lambda_1(\epsilon) \left[\log^2(1/\epsilon)\right]^{-d_2}$$

where $\log^2(x) = \log\log(x)$ and we have chosen $d_2 > 0$. In his 1919 paper, Hausdorff presents a general expansion of gauge functions that allows for a large number of vanishing rates. It has the form

$$\lambda(\epsilon) = \epsilon^{d_0} \prod_{k=1}^{n} \left[\log^k (1/\epsilon)\right]^{-d_k} \tag{5}$$

where $\log^1(x) = \log(x)$, $\log^k(x) = \log\log^{k-1}(x)$, and n is finite. Furthermore, he supplies an algorithm for constructing a linear Cantor set having a gauge function of this form for any finite n and any set of d_k's provided that the gauge function vanishes slower than ϵ. It should be kept in mind that the form given in Eq.(5) is just one possible expansion. The gauge function of a set is strictly defined as that function $\lambda(\epsilon)$ which satisfies Eqs.(2) and yields a finite Hausdorff measure. However, the expansion is motivated by the study of general scaling properties of functions. It is a special case of what HARDY [11] calls the logarithmico-exponential scale. In cases where the expansion of the form given by Eq.(5) does not exist for a finite n, the expansion can still be used to obtain an approximate gauge function for the set since it is guaranteed that a set's gauge function scales to zero slower at a rate which is *between some two power laws*.

Some comments are in order. First note that the gauge function of a set that is embedded in a D-dimensional space cannot vanish faster than ϵ^D [3]. Otherwise the set would have a dimension that is greater than the space it is embedded in. Secondly, the intrinsic gauge function of a set is not unique: only an equivalence class of such gauge functions can be defined. We see this as follows: Let $\lambda(\epsilon)$ be a known gauge function of the set S. Choose any number $A > 0$ and any gauge function $\xi(\epsilon)$. Then $\lambda^*(\epsilon) = A\lambda(\epsilon)(1 + \xi(\epsilon))$ is also an intrinsic gauge function of S. A serves only to change the normalization of the measure. The factor $\xi(\epsilon)$ has no effect on the Hausdorff measure, and, in the sense defined by HARDY [11], it does not affect the scaling rate of the gauge function. This, together with the fact that intrinsic gauge functions are determined (in the Hausdorff formalism) only in the $\epsilon \to 0$ limit makes the definition of a unique intrinsic gauge function impossible

3. Problems with the Application of the Formalism

The general Hausdorff formalism is difficult to apply in both real and numerical experiments. This is easy to imagine when one considers the definition of the coarse-grained Hausdorff measure. First of all one needs to consider countable covers rather than just finite ones. Secondly, the cover which minimizes the measure given in Eq.(2) must be found before the measure can be estimated. Thus, finding intrinsic gauge functions, or even the Hausdorff-Besicovitch dimension, in any real situation, is intractable. When dealing with bounded sets, it is much easier to use the *capacity* notion of dimension [8]. Suppose we modify the definition of the d-dimensional Hausdorff measure of a set in the following manner: Choose an $\epsilon > 0$ as before, but only consider uniform coverings, i.e., coverings whose elements are all D-dimensional cubes of the same length ϵ. Then form the sum $\sum_m \epsilon^d$ over all elements of a given ϵ-cover. Since the set is assumed to be bounded, the sum involves a finite number of equal terms. Then for all such ϵ covers take the infimum of these sums. This infimum is just that particular sum with the smallest number of terms $N(\epsilon)$. Then the capacity version of the d-dimensional measure is

$$\mu^{(d)} = \lim_{\epsilon \to 0} N(\epsilon)\epsilon^d . \tag{6}$$

The value of d, say d_0, for which this measure is infinite for all $d < d_0$ and zero for all $d > d_0$ is usually called the capacity dimension of the set. The obvious extension of the definition of the Hausdorff measure with respect to a gauge function λ is

$$\mu_\lambda = \lim_{\epsilon \to 0} N(\epsilon)\lambda(\epsilon). \tag{7}$$

We call the function λ which yields a finite μ_λ the intrinsic gauge function of the set. To avoid confusion, we call the gauge functions of the Hausdorff formalism *Hausdorff gauge functions*, while referring to those of the capacity version simply as gauge functions. n general, $N(\epsilon)$ is a step function which diverges as epsilon goes to zero. Note, however, that we can always find smooth functions, which approximate $N(\epsilon)$ and have the same scaling behavior.

Since $N(\epsilon)$ diverges as $\epsilon \to 0$, it is necessarily true that the intrinsic gauge function of any bounded set must vanish with ϵ. Thus, from general scaling considerations, we expect this function to be expandable, in most cases, in the form given by Eq.(5) [11]. Furthermore, such a λ will scale to zero no faster than ϵ^D, with D being the dimension of the embedding space.

The foregoing formulation has several practical advantages which are well known [8]. First, the covers to be dealt with are all finite. Second, the members of each cover considered are identical to each other. These properties make it easier to evaluate such quantities as the capacity dimension of a set and, as we shall see, its gauge function. In the rest of this paper, we will be working exclusively with bounded sets and the capacity framework for describing these sets.

<u>4. Examples</u>

We now present some simple examples of Cantor sets which have gauge functions of the form described by Eq.(5). Before doing so, we introduce some notation that is common to all of the examples we consider, and we outline the method which will be used to determine the gauge functions of these sets. We are interested in simple linear Cantor sets that are constructed in the following manner. Start with a unit interval and in the first stage of construction delete a fraction $h_1 < 1$ from its middle. The resulting stage 1 set is composed of $N_1 = 2^1$ intervals of length $\epsilon_1 = [1-h_1]/2$. In the second stage of construction delete a fraction h_2 from the middle of each of these intervals. The stage 2 set is composed of $N_2 = 2^2$ intervals of length $\epsilon_2 = [1-h_2]\epsilon_1/2$. Next delete a fraction h_3 from the middle of each remaining interval and repeat this process *ad infinitum*. The limit set is a Cantor set and its nth approximant is a set for which

$$N_n = 2^n, \tag{8a}$$

$$\epsilon_n = \frac{1}{2^n} \prod_{k=1}^{n} [1-h_k], \tag{8b}$$

and

$$L_n = N_n \, \epsilon_n \tag{8c}$$

where N_n is the number of segments in the set, ϵ_n is the length of each segment, and L_n is the set's Lebesgue measure. We refer to h_k as the *hole function* of a set.

We evaluate the gauge function of these sets as follows: We use the intermediate sets that are generated by the foregoing construction as a sequence of coverings for the limit set. That is, we examine the behavior of $N(\epsilon)$ for the sequence of length scales $\left\{ \epsilon_n \right\}$. We then use

$$\lambda(\epsilon) \approx \frac{\mu_\lambda}{N(\epsilon)}. \tag{9}$$

This approximation is based on Eq.(7) and the fact that μ_λ is finite. Knowledge of the value of μ_λ is unimportant, since we are only interested in the functional form of $\lambda(\epsilon)$. At any rate, we can always normalize the measure to 1 so it is clearly of no consequence. We can equate the expression for the gauge function in Eq.(9) with the expression of Eq.(5) in a convenient manner. Letting $u = \log(1/\epsilon)$, $\theta(u) = -\log \lambda(\epsilon)$, and $\log^0 x = x$, Eq.(5) becomes

$$\theta(u) = \sum_{k=0}^{n} d_k \, \log^k u. \tag{10}$$

Taking the negative log of both sides of Eq.(9), using Eq(8a), and restricting ourselves to the coverings that correspond to the ϵ_n we get

$$\theta(u_n) = \log(1/\mu_\lambda) + n \, \log 2. \tag{11}$$

where $u_n = \log(1/\epsilon_n)$ and n is to be regarded as a function of the u_n. Now taking the negative log of Eq.(8b) results in

$$u_n = n \log 2 - \sum_{k=1}^{n} \log[1 - h(k)]. \tag{12}$$

So we can find the gauge functions of our examples by solving Eq.(12) for n as a function of u_n, substituting into Eq.(11), and comparing the results with Eq.(10).

46

<u>Example 1:</u>

In order to show how this calculation can be done explicitly, we now present an example of a two parameter family of Cantor sets which have the first iterated logarithmic factor in their gauge functions. In the general Cantor set construction outlined in the previous paragraph, choose as the hole function the two parameter family of functions $h_k(b,c) = b + c/k$ where $0 < b, |c|, |b+c| < 1$. To solve for n as a function of the u_n, use the fact that for

$$n \gg 1, \quad u_n \approx A + n\,\log[2/1-b] + \frac{c}{(1-b)}\sum_{k=1}^{n} 1/k$$

where A is some constant that depends on b and c. Replace the sum over k by $C + \log n$ (C is the Euler-Mascheroni constant), and solve for n recursively keeping only the terms in u_n which diverge as $n \to \infty$. The result is

$$n \approx \frac{1}{\log[2/(1-b)]}[u_n - \frac{c}{1-b}\log u_n].$$

Plugging this into Eq.(11) we find that λ has the form of Eq.(5) with

$$d_0 = \frac{\log 2}{\log(2/1-b)} \tag{13a}$$

and

$$d_1 = \frac{-c\,\log 2}{(1-b)[\log(2/1-b)]}. \tag{13b}$$

and $d_k = 0$ for all $k > 1$.

<u>Example 1.1:</u> The classic Cantor set is obtained by choosing $c = 0$ and $b = 1/3$. As is well known, the gauge function for this set is a pure power law with exponent $d_0 = \log 2/\log 3$.

<u>Example 1.2:</u> When we set $b = 1/3$ and let c be nonzero, we get a perturbed classic Cantor set whose gauge function is modified from the unperturbed case by a logarithmic correction. It is important to note that in this example we still have $d_0 = \log 2/\log 3$. The reason for this is that the perturbation from the hole function of the classic Cantor vanishes asymptotically. However,this perturbation does not vanish fast enough to make its presence completely unfelt. Thus a logarithmic correction to the power law in the set's gauge function is produced. If we repeat the above analysis on a set generated by $h_k(c) = 1/3 + c^k$ for some $0 < c < 1$, we would not have observed a logarithmic correction since this perturbation vanishes rapidly with the stage of construction (we will return to the effects of this kind of perturbation in the next section). Returning to the original example 1, we note that, when we set $b = 0$ in the expression for $h_k(b,c)$ and choose $0 < c < 1$, we obtain a dimension 1 object that has zero linear extent or 1-dimensional Lebesgue measure. Figs.(1) and (3) are pictures of such sets. Fig.(1b) shows a portion of Fig.(1a) that has been magnified by a factor of 1000. Note that the relative size of the holes appears to be shrinking making the set appear to have positive measure. This should be compared to the corresponding figures for the classic Cantor set and a fat Cantor-set (Figs.(2) and (5) respectively).

From these simple examples we see that a logarithmic gauge function is introduced when a set does not exhibit exact asymptotic self-similarity i.e., when the scale invariance of a set is not exact. We should expect that these corrections occur in nonlinear dynamical systems which have multiple scalings associated with them. For example, in determining the fractal dimension of basin boundaries in asymmetric tent maps, TAKASUE [12] observed a small oscillation in his $\log N(\epsilon)$ vs. $\log \epsilon$ plot which he attributed to the presence of two scalings associated with the map's asymmetry. Such an oscillation would be seen if the gauge function for the set possessed the first two logarithmic corrections. To see this, we plot $\log 1/\lambda(\epsilon)$ vs. $\log \epsilon$ for such a

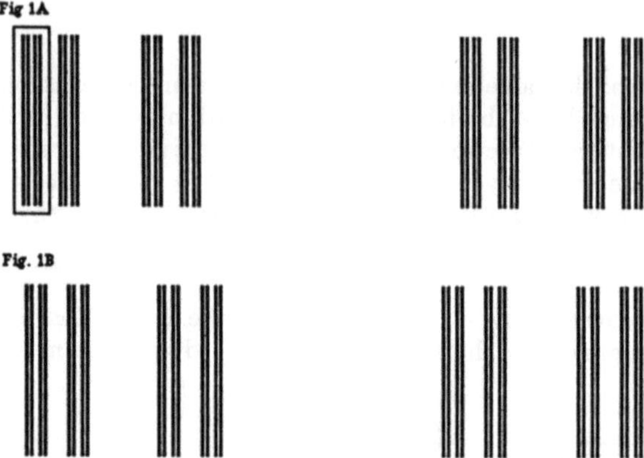

Fig.1. A perturbed classic Cantor set generated by the hole function of example 1.2 with $b = 1/3$ and $c = .1$. Fig. 1a is the a representation of the entire set and Fig. 1b is a blowup of the region outlined in the box in Fig. 1a. The slowly vanishing perturbation gives rise to a logarithmic correction to the pure power law in the set's intrinsic gauge function.

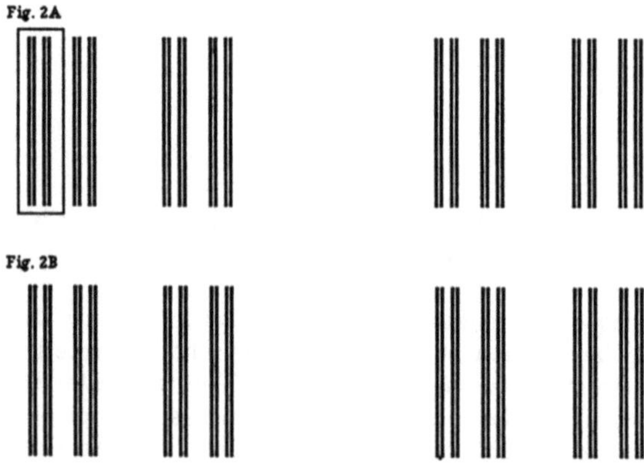

Fig. 2. Same as Fig. 1 but for the classic Cantor set obtained from example 1.1 with $b = 1/3$ and $c = 0$. The gauge function for this set is a pure power law.

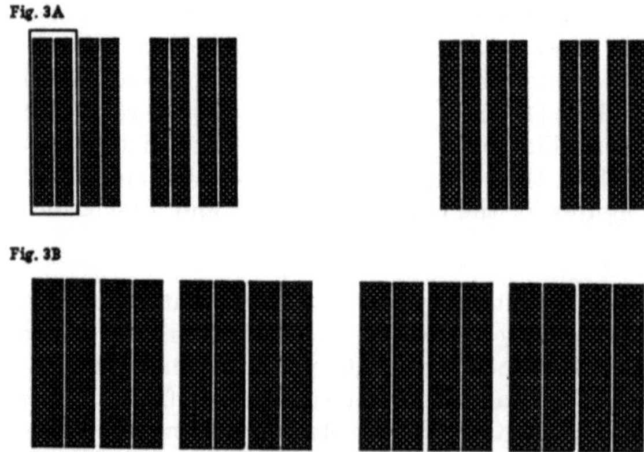

Fig.3. A dimension 1 thin fractal obtained by choosing $b = .0$ and $c = 1/3$. Note from the blowup that the set looks like it may in fact be fat. There is a logarithmic correction, however, that ensures that the Lebesgue measure of this set is zero.

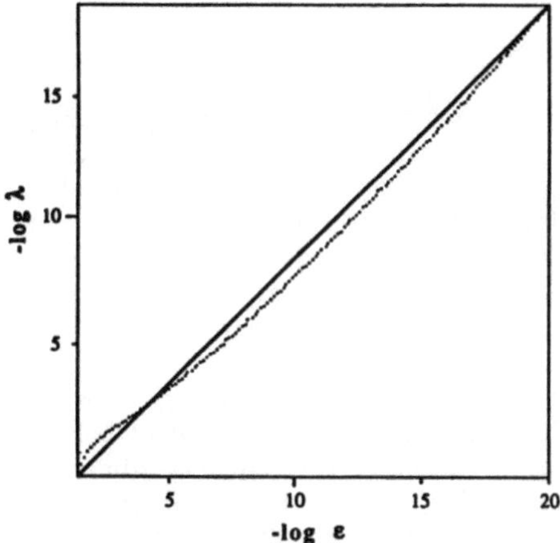

Fig. 4. A plot of $-\log\lambda(\epsilon)$ vs. $\log\epsilon$ where λ is given by Eq.(11) with $d_0 = 1$, $d_1 = -3$, $d_2 = 1.7$ and $d_k = 0$ for all $k > 2$. The straigth line represents the case where $d_0 = 1$ and no logarithmic corrections occur.

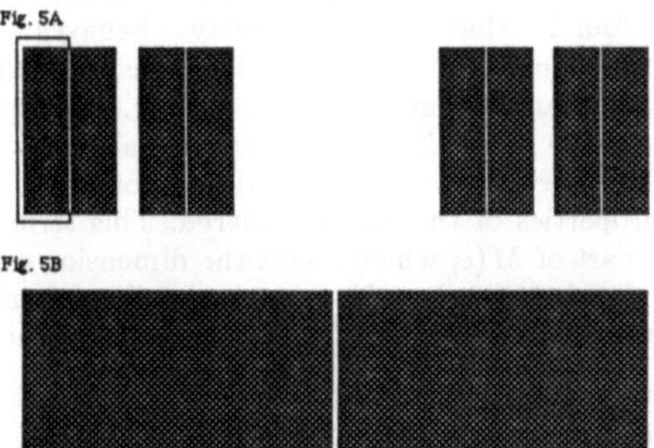

Fig.5. A fat Cantor set generated by the hole function of example 2 with $c = 1/3$. Hardly any levels of this set are resolvable since the hole sizes scale very rapidly to zero. This set has an additive power law correction to its gauge function with an exponent $\delta = \log3/\log2$.

gauge function in Fig.(4). The straight line is what we would see when there is no logarithmic correction. Note that a single oscillation about the line is induced by the logarithmic corrections. This should be compared to the figures in [12].

5. Characterization of Fat Fractals : Metadimensions

We now turn our attention to the characterization of fat fractals. As stated earlier the Hausdorff gauge functions of fat fractals are necessarily pure power laws with integer exponents. Since this is also true of ordinary sets which have finite Lebesgue measure, other methods for characterizing this class of fractals must be employed. In [4] and [5] the discussion is limited to sets which derive their fractal properties from simple *holes* embedded in them. The coarse-grained Lebesgue measure which is chosen is the measure of the complement of all holes having a diameter larger than a size ϵ. Letting $\mu_h(\epsilon)$ be this measure, the exponent

$$\beta = \lim_{\epsilon \to 0} \frac{\log[\,\mu_h(\epsilon) - \mu_h(0)]}{\log\epsilon}$$

is used to characterize the set. In [6], a fat fractal is *fattened* by centering balls, each having a radius ϵ, on each point of the set. The coarse-grained measure of interest, which we denote by $\mu_f(\epsilon)$, is the measure of the fattened set, i.e, the measure of the union of all the balls. Then an exponent $\alpha = \lim\limits_{\epsilon \to 0} \dfrac{\log[\mu_f(\epsilon) - \mu_f(0)]}{\log \epsilon}$ is used to characterize the set. This second scheme makes use of a method originated by Cantor for estimating the volume of sets.[1] Since the Hausdorff formalism is a generalization of this, it is natural to wonder whether or not fat fractals can be described by generalizing the Hausdorff formalism. It would be nice, for example, if a unique intrinsic Hausdorff gauge function could be defined which can distinguish between fat fractals and ordinary sets. As stated earlier, this appears to be impossible. This is not true of the capacity formalism, however, and this fact together with the numerical results of [4] suggest such a generalization within this framework.

In [4], it is conjectured that bounded chaotic orbits of certain area-preserving maps are fat fractals. The numerical procedure used for estimating β for these orbits is that of box counting. This procedure has been used in many studies to estimate the capacity dimension of strange attractors in dissipative systems [15]. In these studies, a fixed grid of squares ϵ on side is placed on a portion of the phase space accessible to a given orbit. The number of squares needed to cover a subset of the orbit, say $M(\epsilon)$, is computed and it is assumed that $M(\epsilon) \approx N(\epsilon)$. Then the capacity dimension is extracted by fitting a straight line to a plot of $\log M(\epsilon)$ vs. $\log \epsilon$, this being motivated by the definition given in Eq.(6). In [4], the area of the closure of a given chaotic orbit, say $\mu_g(\epsilon)$, is estimated by computing $M(\epsilon)$ for an entire orbit and constructing $\mu_g(\epsilon) = M(\epsilon)\epsilon^2$. It is found that this quantity behaves like $\mu_g(\epsilon) \approx \mu_g(0) + \kappa \, \epsilon^\gamma$ for ϵ small compared to the diameter of the orbit. The constants $\mu(0), \kappa$ and, γ are all positive, and it is conjectured that γ is an estimate of β when $\gamma < 1$. The scaling of the measure $\mu_g(\epsilon)$ implies that $M(\epsilon) \approx \epsilon^{-2}[\mu(0) + \kappa \, \epsilon^\gamma]$. The first term in brackets is just what we would expect to get for the behavior of $M(\epsilon)$ of an ordinary 2-dimensional set. The second term, however, arises from the fractal properties of the sets considered. This term can be viewed as higher order correction to that part of $M(\epsilon)$ which yields the dimension of the set. This suggests the following description of fat fractals: Let $N(\epsilon)$ again be the minimum number of D dimensional squares of side ϵ needed to cover a d_0 dimensional fat fractal, S. Then define a quantity δ by

$$\delta = \lim_{\epsilon \to 0} \frac{\log[\, N(\epsilon)\epsilon^{d_0} - \mu_0]}{\log \epsilon}. \tag{14}$$

where $\mu_0 = \lim\limits_{\epsilon \to 0} N(\epsilon)\,\epsilon^{d_0}$. We call δ a *metadimension* of the set S [16]. It describes the rate at which the measure estimated from an optimal uniform ϵ cover converges to the true measure of the set as ϵ tends to zero. By definition, $0 < \delta \leq \infty$. When $\delta \gg 1$, very small changes in ϵ decrease the error in the estimate of a set's true measure dramatically. When $\delta \ll 1$, very large decreases in ϵ are required to make this error small.

Some caution must be exercised in interpreting the exponent γ of [4] as being equivalent to δ. As pointed out in [4], any nonfractal set which is bounded by a smooth curve will have the power law scaling of the measure described above with $\gamma = 1$. This is an artifact of using a fixed grid for a cover. However when $\gamma < 1$, we expect that γ does reflect the fractal nature of the set and equals δ. These issues are discussed in detail elsewhere [17,18].

An equivalent way of viewing the higher order scaling discussed in the foregoing paragraph is that fat fractals can be distinguished from ordinary sets by additive corrections to their gauge functions. This is accomplished by first noting that a unique intrinsic gauge function for a set can be defined through Eq.(7). We simply let the intrinsic gauge function of a set be defined as

$$\lambda(\epsilon) = \frac{\mu_\lambda}{N(\epsilon)}. \tag{15}$$

where μ_λ can be normalized to unity. Thus if we find, for some set of interest, $N(\epsilon) \approx \epsilon^{-d_0}[\mu_0 + \kappa \epsilon^\delta]$ (with $d_0, \mu_0, \kappa, \delta > 0$, we get

$$\lambda(\epsilon) \approx \epsilon^{d_0} [1 - \frac{\kappa}{\mu_0} \epsilon^\delta].$$

where we have kept only the lowest order terms in ϵ. The first term in the brackets is associated with the dimension. The second term in brackets is a correction that vanishes as $\epsilon \to 0$. This example has only a power law correction but there is no reason that it cannot scale to zero in some other fashion. This is a point which we return to later.

<u>Example 2:</u>

We now give an example of a fat Cantor set for which δ is easily computed. Consider the Cantor set construction described in Sec.(2) for a one-parameter family of hole functions $h_k(c) = c^k$ for $0 < c < 1$. For a fixed c, it is easy to verify that the one-dimensional Lebesgue measure of the generated set is given by $0 < \mu_0 = L_\infty = \lim_{n \to \infty} L_n < 1$. To calculate δ for this set, we use $\delta = \lim_{n \to \infty} \dfrac{\log [L_n - L_\infty]}{\log \epsilon_n}$ where the ϵ_n correspond to the subsequence of covers defined in Sec.(2). This limit is the same as $\lim_{n \to \infty} \dfrac{\log[1 - \prod_{k=n+1}^{\infty} (1 - c^k)]}{n \ \log 1/2}$ which is easily evaluated to give

$$\delta = \log (1/c)/\log 2. \tag{16}$$

Thus for $n \gg 1$, we have $L_n \approx L_\infty (1 + \kappa \ \epsilon^\delta)$ which yields (from Eq.(8c))

$$N(\epsilon) \approx L_\infty \ \epsilon^{-1} [1 + \kappa \ \epsilon^\delta]. \tag{17}$$

Therefore, for this fat Cantor set, the gauge function has an additive correction which is a power law in ϵ. Had we done the same calculation for an interval or a collection of intervals, no such additive correction would have been observed. Thus we conjecture that a fat fractal of dimension n is distinguished from a regular set of the same dimension by an additive correction to ϵ^n in its gauge function. We call this additive correction the *metadimension* to distinguish it from the dimension given in Eq.(5).

Given the above considerations, it is natural to wonder whether or not thin fractals can also exhibit this type of higher order scaling behavior. It turns out that they can. To see this, consider the following Cantor set construction. Choose a two parameter family of hole functions given by $h_k(b,c) = b + c^k$ where $0 < b, c, b + c < 1$. As stated earlier, this set has an intrinsic gauge function which has no multiplicative corrections to a power law. This can be verified by performing the calculation outlined in Sec.(2). The capacity dimension of the set is the same as that for the sets generated by choosing the one parameter family of gauge functions $h_k(b,0)$, i.e., d_0 is given by Eq.(13a). Now if we modify our definition of δ by replacing μ_0 in Eq.(14) by the set's d_0 dimensional measure (with d_0 the capacity dimension set), we find that δ is the same as that of the fat Cantor set discussed above, i.e., it is given by Eq(16). Thus the notion of a metadimension is applicable to thin fractal sets as well.

The above considerations suggest the following definition: Let S be some bounded set with an intrinsic gauge function $\lambda(\epsilon)$ and a measure μ_λ. Furthermore, suppose that $\lambda(\epsilon)$ can be expanded in the form of Eq.(5) for some finite n. Construct the function $\lambda_m(\epsilon) = N(\epsilon) \lambda(\epsilon) - \mu_0$. Then $\lambda_m(\epsilon)$ is a function which vanishes as $\epsilon \to 0$ and describes the rate at which the measure μ_λ is approached with decreasing length scale. Then define the exponent δ by

$$\delta = \lim_{\epsilon \to 0} \frac{\log \lambda_m}{\log \epsilon}. \tag{18}$$

There are no geometrical restrictions which require that this function vanish slower than some power in ϵ as there was for $\lambda(\epsilon)$. Thus $\lambda_m(\epsilon)$ can be anything asymptotically. For example, it can have the following logarithmic-exponential[11] form:

$$\lambda_m(\epsilon) = \prod_{j=1}^{l} \left[\exp^j (1/\epsilon) \right]^{-q_j} \epsilon^{d_0} \prod_{k=1}^{n} \left[\log^k (1/\epsilon) \right]^{-d_k} \tag{19}$$

where $\exp^j(x)$, is the jth iterated exponential, and $q_j > 0$. Many other possibilities for the form of $\lambda_m(\epsilon)$ can be found in HARDY [11].

6. Conclusions

We have discussed the characterization of fractal sets that do not exhibit exact scale invariance. This lack of scale invariance leads to corrections to a pure power law of a set's intrinsic gauge function. We expect these corrections to occur in systems which have many scalings associated with them. Two types of these corrections occur. The first type is multiplicative in nature and can be approximated by an expansion in terms of powers of iterated logarithms. These represent relatively strong perturbations away from scale invariance. These types of corrections occur in many random fractals. A specific example of this is the two dimensional random walk discussed in TAYLOR [2]. There, although the set generated by the walk is statistically self-similar, it is not exactly so. Thus a logarithmic correction to a pure power law occurs in the set's gauge function which results in a 2-dimensional set that has zero Lebesgue measure and is thus not area filling. Other cases, in which these corrections seem to appear naturally are associated with fractal basin boundaries as discussed in [12,13] .

A second kind of correction to a pure power law that can be seen in fractals is an additive one. These occur in sets whose deviation from exact scale invariance are relatively weak. Fat fractals are a class of such sets in which these types of corrections occur. It is the deviation from exact scale invariance of these sets which distinguish them from ordinary sets. Examples of such sets occur in many dynamical sytems.

We addressed the question "Why were we born to suffer and die ?" which is the major theme in the work of Kilgore Trout. We showed that this question can be addressed in the mathematical framework that was presented by Felix Hausdorff in his 1919 paper. We then presented numerical evidence which indicates that only God can answer this question.

Acknowledgements

We are deeply indebted to W. Beyer, who insisted that not everything is well understood about dimensions, and who persuaded us to read Hausdorff's original paper. We also would like to thank J.D. Farmer and especially R. Eykholt for useful discussions.

References

1   B. Mandelbrot, *The Fractal Geometry of Nature* (W. H. Freeman and Co., San Francisco, 1982).

2   S.J. Taylor, Pr. of the Cambridge Philosophical Society **60** , 253 (1964)

3   Felix Hausdorff, Math. Ann. **79** , 157 (1919)

4   D.K. Umberger and J.D. Farmer, Phys. Rev. Lett. **55** , 661 (1985)

5   J. D. Farmer, Phys. Rev. Lett. **55** , 351 (1985).

6   C. Grebogi, S. W. McDonald, E. Ott, and J. A. Yorke, Phys. Lett. **110A** , 1 (1985).

7   R. Ecke, J. D. Farmer, and D. K. Umberger, unpublished (1985).

8   J. D. Farmer, E. Ott, and J. A. Yorke, Physica **7D** , 153 (1983).

9   The original definition of the Hausdorff measure utilizes arbitrary covers. We have chosen the formulation in terms of D dimensional squares for simplicity.

10   The definition of the Hausdorff measure which we choose guarantees this, however in the general case (see previous reference) the Lebesgue measure and the Hausdorff measure differ by a nonzero factor.

11   G. H. Hardy, *Orders of Infinity,* Cambridge Tracts in Mathematics and Mathematical Physics no. 12, Cambridge University Press (1954).

12   S. Takasue, Ph.D. dissertation, Tokyo University, 1983

13   S. W. McDonald, C. Grebogi, E. Ott, and J. A. Yorke, to appear in Physica D

14   B. Mandelbrot : this volume

15   D. A. Russell, J. D. Hanson, and E. Ott, Phys. Rev. Lett. **45** , 1175 (1980).

16   D.K. Umberger and R. Eykholt, unpublished (1985).

17   J.D. Farmer and D.K. Umberger, to appear Phys. Rev. Lett. (1985).

18   R. Eykholt and D.K. Umberger, unpublished (1985).

Scaling in Fat Fractals

J.D. Farmer

Center for Nonlinear Studies, MS B258, Los Alamos National Laboratory,
Los Alamos, NM 87545, USA

Fat fractals are fractals with positive measure and integer fractal dimension. Their dimension is indistinguishable from that of nonfractals, and is inadequate to describe their fractal properties. An alternative approach can be couched in terms of the scaling of the coarse grained measure. For the more familiar "thin" fractals, the resulting scaling exponent reduces to the fractal codimension, but for fat fractals it is independent of the fractal dimension. Numerical experiments on several examples, including the chaotic parameter values of quadratic mappings, the ergodic parameter values of circle maps, and the chaotic orbits of area-preserving maps, show a power law scaling, suggesting that this is a generic form. This paper reviews several possible methods for defining coarse-grained measure and associated fat fractal scaling exponents, reviews previous work on the subject, and discusses problems that deserve further study.

Introduction

The fact that the dimension can take on noninteger or fractional values is perhaps the property most strongly associated with the word fractal. Yet, as described in MANDELBROT's book [1], there is a wide class of fractals whose dimension takes on integer values. While dimension remains a very important property of these sets, it gives very little information about their fractal nature. This paper describes an alternative to dimension, couched in terms of the scaling properties of the coarse-grained measure, which provides a good number to summarize the most basic properties of a fat fractal. More important, this number provides a quantitative language to describe several physically relevant properties of fractals that manifest themselves in different types of mathematical models. Since these are likely to occur in experimental contexts as well, computing the scaling exponent described here provides a useful alternative to computing dimension.

As an illustration of the problem that I am referring to, consider the basic Cantor set with slightly modified geometry. In particular, starting with the interval [0,1], delete the middle third, leaving two pieces. Departing from the usual construction, delete the middle *ninth* of each of these, leaving four pieces. Continue by deleting the middle 1/27 of each of these, and so on (see Fig./ (5) of reference [9]). In contrast to the usual example, obtained by deleting the middle third at each step, the set resulting from the limit of the procedure described above has positive Lebesgue measure. This immediately implies that its fractal dimension is one. From the point of view of dimension, this set is indistinguishable from the simple line interval. Nevertheless, it is a Cantor set, topologically indistinguishable from that obtained by the usual construction, and certainly deserves the name fractal. I will refer to fractals with positive volume (Lebesgue measure) as *fat fractals*, and those with zero volume as *thin fractals* [3,4].

As discussed by MANDELBROT [1], a fractal can be defined as "a set for which the Hausdorff Besicovitch dimension strictly exceeds the topological dimension". Since the topological dimension of the fat Cantor set constructed above is zero, this makes it clear that it is a fractal. It would be nice, though, to be able to describe its "fractalness" in a more quantitative manner. Furthermore, there is no known numerical method to compute topological dimension, making this definition difficult to apply in the experimental arena.

Besides their dimensional properties, another salient feature of fractals is the variation of their observed size or volume with the scale of resolution. An alternative way to define a fractal might be as a set whose observed volume depends on the scale of resolution over a range of scales. This loose concept can be formulated precisely in a variety of different ways, according to the definitions used for "volume" and "resolution". In this paper I discuss several possible definitions. Once a precise definition of measure is established, the real interest is in its scaling properties. It appears that one particularly common asymptotic behavior is that of a power law, in which case the resulting scaling exponent provides a good way to characterize fractal properties. For thin fractals this scaling exponent is simply the fractal codimension, but for fat fractals it is independent of the dimension. The basic concepts discussed here, including applications to several examples, were introduced in references [2-4].

In all of the definitions that follow, $\mu(\epsilon)$ refers to the measure of a set resolved at finite resolution ϵ, providing a precise way to speak of the "observed volume" discussed above. It can be defined in any one of several ways, as described below. For the discussion that follows it is convenient to separate μ into two parts,

$$\mu(\epsilon) = \mu(0) + f(\epsilon), \tag{1}$$

where $\mu(0)$ is the asymptotic measure in the limit as $\epsilon \to 0$. In general $f(\epsilon)$ is not smooth and satisfies no special constraints except that it is monotone nondecreasing with $f(0) = 0$. As $\epsilon \to 0$, however, $f(\epsilon)$ often approaches a smooth function. One form that seems to be singled out is

$$\mu(\epsilon) \approx \mu(0) + A\,\epsilon^B \tag{2}$$

where A and B are constants. Of these, the scaling exponent B is independent of the units used for ϵ, and is a more important quantity.

One method to define $\mu(\epsilon)$ is as follows: For a fixed grid of D-dimensional cubes that are ϵ on a side, let $\mu(\epsilon) = N(\epsilon)\epsilon^D$, where $N(\epsilon)$ is the number of cubes needed to cover the set, i.e., the number of them containing part of the set. $\mu(\epsilon)$ is just the usual notion of coarse grained measure used in statistical mechanics. A scaling exponent can be defined as:

$$\gamma = \lim_{\epsilon \to 0} \frac{(log\ \mu(\epsilon) - \mu(0))}{\log \epsilon} \tag{3}$$

This definition, presented in [4], is somewhat similar to the definition of capacity [5]. The principle difference is that there is an extra factor of ϵ^D in the denominator, and also the asymptotic measure $\mu(0)$ is subtracted in the numerator. When $\mu(0) = 0$ (a thin fractal), $\gamma = D - d_f$, where d_f is the capacity, or for simplicity, the "fractal dimension" [5]. Thus, for thin fractals γ is simply the "fractal codimension". When $\mu(0) > 0$, however, γ is generally unrelated to the fractal dimension. γ is always in the range $0 \leq \gamma \leq 1$.

An alternate approach to defining an exponent follows. As stated here, this definition only applies to fractals that might be called "holistic", i.e., those that can be constructed by removing holes from a solid object (e.g. Cantor sets), but it can be extended to general fractals [6]. $\mu(\epsilon)$ is defined in the following way: Let $h(\epsilon)$ be the total measure of all holes that are greater than ϵ in diameter. Let

$$\mu(\epsilon) = 1 - h(\epsilon). \tag{4}$$

(The value one in this definition is arbitrary; any fixed value will do.) Applying this definition of μ to Equation (3) yields another exponent, whose value will be called β, and can be anywhere in the range $0 \leq \beta \leq \infty$.

Yet another alternative has been suggested by GREBOGI et al. [7]. Given some set S, they define $\mu(\epsilon)$ as the Lebsegue measure (D-dimensional volume) of the "fattened" set consisting of all the values that are within ϵ of S. Again, by plugging this μ into Equation (3) another exponent emerges, whose value they call α, in the range $0 \leq \alpha \leq 1$.

Yet another approach to these same ideas can be phrased in terms of a reexamination of the original ideas of HAUSDORFF [8]. This approach, in which these scaling exponents are given the name *metadimensions*, is taken by UMBERGER et al. [9] in this proceedings.

The first question that immediately comes to mind is: Are these definitions equivalent? The first difference that is apparent between these different definitions is their range of allowed values: For nonfractal sets α and γ take on the value one, whereas β is infinite. For fat fractals, providing $\beta \leq 1$ in one dimension it can be proved [6] that $\alpha = \beta = \gamma$, but when $\beta > 1$, $\alpha = \gamma = 1 \neq \beta$. In this sense β is more discriminating, in that it gives information about a larger class of fractals, and is more useful for demonstrating that a given object is a fractal. In higher dimensions the situation is less clear, but my conjecture is that, at least for "typical" examples the relations given above continue to hold. It would be nice to formulate a precise notion of "typical" and prove something in this regard rigorously, or alternatively, to pick an example that seems general enough to be called typical and compute the values for each definition. There is a strong analogy here to the situation for the equality of different definitions of fractal dimensions [5,10]. I am sure that there are many other ways to define scaling exponents closely related to those below, which should emerge as this general concept is applied in other contexts.

Another question that arises is whether any of these definitions is preferable to the others. My feeling is that this is largely a matter of convenience: The definition of choice is the one that is easiest to apply to the problem at hand. There are some technical problems with the definitions, for example, the fact that μ depends on the nature of the grid in the definition of γ. Some other advantages and disadvantages will emerge in the discussion of applications given below.

Applications

The definitions of α and γ given above are quite general and are applicable to any set. They are most useful, however, when applied to fat fractals, since in this case they give information that is not already contained in the fractal dimension. In this sense, a list of applications amounts to a list of fat fractals. The scaling exponents discussed here can be used to test whether a given object is indeed a fat fractal, and if so, to give some quantitative information about its characteristics.

MANDELBROT [1] has given several examples of fat fractals. These include the vascular system, the branching structure of bronchia in the lung, rivers, and botanical trees. Some of the ideas given here have been applied to the vascular system in a quantitative way by GREBOGI et al. [7]. They also suggest that the structure of porus materials and clusters formed by ballistic aggregation may provide other examples of fat fractals. In addition, we have demonstrated that the following are fat fractals:

Chaotic parameter values beyond the period-doubling transition to chaos [3]

Just beyond the period-doubling transition to chaos in dissipative systems, chaotic and periodic orbits are interwoven in a very complex way. For mappings of the interval, such as $x_{t+1} = \lambda x_t (1 - x_t)$, it is widely believed that there is a stable periodic interval arbitrarily close to any value of λ. At the same time, both numerical evidence [3,11] and an analytic proof [12] demonstrate that the chaotic parameter values are of positive measure in λ. The immediate conclusion is that the chaotic parameter values (defined here as those values with positive Lyapunov exponents) form a fat Cantor set, with the stable periodic intervals playing the role of the holes. Numerical computations of both β [3] and α [7] indicate that this fat fractal follows the scaling law of Equation (2). Furthermore, as described in [3], there is some evidence that the scaling exponent is universal, although the evidence is at this point not strong enough to be convincing. The computed values in both references [3] and [7] are within a few percent of 0.44. If this behavior is indeed universal, this universality is *global*, and much stronger than that of the period-doubling transition itself.

If the scaling exponent is indeed universal, the value depends on the order of the maximum, just as it does for the period-doubling transition. Certain limits are clear: As the order of the map goes to infinity, the set of chaotic values is a thin Cantor set. (For example, consider a flat top map such as a trapezoid.) In this case the scaling exponent must take on a value equal to one minus the fractal dimension. For a zeroth order maximum, such as the tent map, the stability interval of all periodic orbits has zero width, which implies that β must be infinite. The behavior in between is not at all clear. For example, in the limit of an infinite order critical point, does β go smoothly into the fractal codimension of the chaotic set, or is there a "first order phase transition", i.e., does β take a jump? In the zeroth order limit, is there typically a region where $\beta > 1$ before it goes to infinity, so that α and γ take on the value 1 over an entire interval? These questions deserve further study.

It seems likely that these same scaling properties also apply to higher dimensional examples. (Presumably in higher dimensions the only relevant case is that of a quadratic critical point.) Nothing is known rigorously, but numerical studies suggest something like the following: In higher dimensions the Cantor set is instead the Cartesian product of a Cantor set and R^n, where $n = D - 1$, D being the dimension of the parameter space. Following a generic path through the parameter space, the set of chaotic parameter values moving along this path form a fat Cantor set. Presumably this Cantor set satisfies the same scaling properties as for one-dimensional maps, at least approximately. Although from numerical studies it seems clear that this picture is basically correct, it remains to be seen if it typically applies in the limit as $\epsilon \to 0$, where some of the periodic structure known to exist in one dimension may be truncated.

Whether approximate or exact, as described in reference [3], the scaling exponent provides a means of quantifying the sensitive dependence on parameters that occurs when two qualitatively different types of behavior such as this are interwoven in such as close fashion. It would be nice to determine whether this exponent is indeed universal, as it would describe the entire transition zone beyond the onset of chaos.

Arnold Tongues [13]

A somewhat similar situation occurs in circle maps, or more generally, for flows on the torus. As observed in the famous "Huygen's clocks" experiment, when two oscillators are coupled together it is often the case that one entrains or "mode locks" the other. As parameters are varied, there is typically an alternation between locked (periodic) and unlocked (ergodic) behavior. There are two essential parameters, one which controls the winding number (and can be thought of as the ratio of the frequencies of the two oscillators), and another which controls the nonlinearity. For zero nonlinearity, almost everything is ergodic, but as the nonlinearity parameter is increased, numerical evidence indicates that there is a critical surface where almost everything is mode locked. The mode locked regions are called Arnold tongues. It can be proven that each rational ratio of the winding number generates a stable periodic interval, which occurs arbitrarily close to any parameter value.

In the terminology of this paper, the set of ergodic parameter values below the critical surface forms a fat fractal. At the critical surface there is a transition from a fat to a thin fractal. The ergodic parameter values here play the same role as the chaotic parameter values do beyond the period-doubling transition.

Numerical experiments performed by ECKE et al. [13] indicate that the Arnold tongues obey the same power-law scaling conjectured in Equation (2). Near criticality, the exponent takes on a value of 0.13, which within experimental error is equal to the critical fractal codimension computed by JENSEN et al. [14]. As the nonlinearity decreases the exponent smoothly increases, though remaining below one, in contradiction of the predictions of GREBOGI et al. [7]. This indicates that the transition from a fat to a thin fractal at the critical surface is a second order phase transition. One advantage of computations involving the fat fractal scaling exponent is that there are no cross-over phenomena near the critical region as there are in computations of the dimension [14].

Chaotic Orbits of Hamiltonian Systems [4]

An interesting question, originally stimulated by the ergodic hypothesis, is the nature of chaotic orbits in Hamiltonian systems. Numerical experiments with low-dimensional dynamical systems typically show chaotic orbits that appear to cover part but not all of the energy surface (in smooth systems). In particular, around elliptic fixed points there are islands of stability bounded by KAM surfaces, which exclude the chaotic orbits that surround them. The result is that the closure of typical chaotic orbits are punctuated by an infinite number of holes, with an assortment of different sizes. This suggests that these orbits are fractals. Evidence from numerical experiments also indicates that these orbits have positive measure, indicating that these are in fact fat fractals.

To test this conjecture, in reference [4] we partitioned the energy surface of several different area-preserving mappings into a square grid. Choosing an initial condition on the largest chaotic orbit, we counted the number of squares visited by the orbit, using grids with different values of ϵ. In all cases, we observed the power-law scaling conjectured in Equation (2). This illustrates a very interesting property of Hamiltonian chaotic orbits, suggesting the existence of a global renormalization scheme. In addition, the fact that we had a clear cut scaling law allowed us to extrapolate our measurements, giving us an accurate value for the area, and strong evidence that our calculations were not meaningless because of finite resolution.

General Remarks

It is not really very surprising that fat fractals exhibit scaling properties in the general form given in Equation (1). What is surprising is that the power-law scaling of Equation (2), which was originally inspired by the power-law scaling observed for thin fractals, is so widespread. While it is possible to construct examples with different asymptotic scaling properties, the differences are either minor (e.g. logarithmic corrections to a power law) or the constructions are so heavily contrived that they feel very unnatural (e.g. with $f(\epsilon) = e^{1/\epsilon}$) [6]. (Note: Many different examples *asymptotically* scale as a power law, even though there are higher order corrections. For example, setting $mu(\epsilon) = e^{\epsilon}$ in Equation (1) gives an asymptotic power-law scaling with an exponent $B = 1$.) The fact that power-law scaling is so ubiquitous seems to indicate that it is a generic form. It would be nice if this could be deduced from general principles.

As already mentioned, the existence of scaling of the coarse-grained measure with the resolution ϵ might provide an alternate way to define the word "fractal". The existence of $\alpha < 1$ or $\gamma < 1$ is not, however, an adequate definition, since it is easy to construct examples of Cantor sets that do not satisfy either of these relations. $\beta < \infty$ is better, but it is still possible to construct Cantor sets with $\beta = \infty$, although they begin to seem much more diabolical than those that merely have $\beta > 1$. The existence of $\beta < \infty$ or α or $\gamma < 1$ can be viewed as *sufficient* but not necessary conditions that something be a fractal. From an experimental view, it is quite clear that the class of fractals with $\beta < \infty$ is much larger than the class of fractals with noninteger fractal dimension, so that in this respect β or any of the other similar exponents discussed here are more useful than the dimension. Measuring these exponents provides a useful tool for quantifying the properties of a more general class of fractals. This may be especially useful in extending renormalization theory to apply to global rather than merely local properties.

Whether to use α, β, or γ depends very much on the application at hand. In the experiments that we did on Hamiltonian mappings, γ was the logical choice, because we needed to follow a single orbit and the computation was easy to perform using a grid. On the other hand, this approach, which is similar to the box counting approach to computing dimensions, requires a large amount of computer memory, and does not work very well in higher dimensional spaces. Furthermore, using γ makes it necessary to argue that there is no dependence on the grid used for the coarse graining. For the computations with quadratic maps described in reference [3], β was a natural choice because the fat fractal under consideration was a Cantor set, and I had a numerical algorithm that allowed me to compute accurate values for the boundaries of the periodic intervals. By using α rather than β, however, GREBOGI et al. [7] were later able to get results of comparable accuracy using a much simpler numerical method. The definition in terms of α is the closest in spirit to the pointwise definition for fractal dimension [5,10], and may prove to be the most generally useful for experimental computation of these exponents. Development of a good, general purpose experimental algorithm is an interesting problem that would extend the applicability of these results and that I hope will be addressed in the future.

In conclusion, the use of scaling exponents for a coarse-grained measure in whatever form provides a potentially useful tool to characterize a large class of fractals that up until very recently have received little attention. I expect that the examples of fat fractals discussed here are just a sample of what is a very large class of objects. I hope that the work discussed here will stimulate mathematicians to develop a rigorous theory to test some of the conjectures presented here. In addition, I hope it will stimulate numerical experimentalists to explore other applications and to develop good numerical algorithms for computing the values of the scaling exponents. Finally, I hope that these scaling relations can be observed in an experimental context.

Acknowledgements

I would like to thank Robert Eykhold and David Umberger for valuable discussions.

References

1. B. Mandelbrot, *The Fractal Geometry of Nature*, (W.H. Freeman, San Francisco, 1982).
2. J.D. Farmer, in *Fluctuations and Sensitivity in Nonequilibrium Systems*, edited by W. Horstemke and D. Kondepudi (Springer, New York, 1984) p. 172.
3. J.D. Farmer, Phys. Rev. Lett. **55**, #4, (1985) 351.
4. D.K. Umberger and J.D. Farmer, Phys. Rev. Lett.**55**, #7 (1985) 661.
5. J.D. Farmer, E. Ott, and J.A. Yorke, Physica **7D** (1983) 153.
6. R. Eykholt and D.K. Umberger, unpublished.

7. C. Grebogi, S.W. McDonald, E. Ott, and J.A. Yorke, Physics Letters A, **110**, #1 (1985) 1.

8. F. Hausdorff, Math. Ann. **79** (1919) 157.

9. D.K. Umberger, G. Mayer-Kress, and E. Jen, "Hasudorff dimensions for sets with broken scaling symmetry", this issue.

10. L-S. Young, Ergodic Theory and Dynamical Systems, **2** (1982) 109.

11. E.N. Lorenz, Tellus **6**, 1 (1964) and Ann. N.Y. Acad. Sci. **357**, 282 (1980).

12. M. Yakobson, Commun. Math. Phys. **81**, 39 (1981).

13. R. Ecke, J.D. Farmer, and D.K. Umberger, unpublished.

14. M.H. Jensen, P. Bak, and T. Bohr, Phys. Rev. Lett. **21**, 1637 (1983), and Phys. Rev. **A30**, 1960 (1984).

Part III

Numerical and Experimental Problems in the Calculation of Dimensions and Entropies

Lorenz Cross-Sections and Dimension
of the Double Rotor Attractor

E.J. Kostelich and J.A. Yorke[1]*

Department of Mathematics and [1]Institute for Physical Science
and Technology, University of Maryland, College Park, MD 20742, USA

Abstract

A Lorenz cross-section of an attractor in $\mathbf{R}^n$ with $k > 0$ positive Lyapunov exponents is the transverse intersection of the attractor with an $n-k$ dimensional plane. We outline a numerical procedure to compute Lorenz cross-sections of chaotic attractors with $k > 1$ positive Lyapunov exponents and apply the technique to the attractor produced by the double rotor map, two of whose numerically computed Lyapunov exponents are positive and whose Lyapunov dimension is 3.64. The pointwise dimension of the Lorenz cross-sections is computed approximately as 1.64. This numerical evidence supports a conjecture that the pointwise and Lyapunov dimensions of typical attractors are equal.

The visualization of attractors has been an important tool in the study of dynamical systems. A Poincaré return map is often the first step in reducing the phase space dimension of an attractor. However, this reduction may not be sufficient to visualize high dimensional systems. Recently Lorenz[1] suggested a technique to obtain meaningful two-dimensional cross-sectional representations of higher dimensional chaotic attractors with one positive Lyapunov exponent. Given a map (which perhaps arises by taking a return map for the flow produced by a system of ordinary differential equations), a Lorenz cross-section is a further slice of the attractor. We assume that there is an ergodic natural measure[2] defined on the attractor with respect to which almost every point on the attractor generates the same Lyapunov exponents.[3] We make the following definition.

Definition. Let $F: \mathbf{R}^n \to \mathbf{R}^n$ be a map whose attractor has $k > 0$ positive Lyapunov exponents. A *Lorenz cross-section* of the attractor for F is the intersection of a plane (or surface) L of dimension $n-k$ with the attractor. The plane L must be transverse to the unstable manifold at typical points of the intersection.

We have developed a numerical procedure to compute Lorenz cross-sections of attractors with $k > 1$ positive Lyapunov exponents and have applied it to an attractor arising from a four-variable map which models the motion of a kicked double rotor. Since it is a map, there is no Poincaré return map available. Numerical computations show that the double-rotor attractor has two positive Lyapunov exponents, and its Lyapunov dimension[4] is approximately 3.64.

Details of the derivation of the double-rotor map will be given elsewhere.[5] Two massless rods are connected as shown in Fig. 1. The first rod pivots about P_1 (a fixed point), and the other pivots about P_2 (the moving end of the first rod). Friction is present at both pivots. Gravity and air resistance are ignored. A mass is attached to the first rod at P_2, and equal masses are attached to each end of the second. At

---

* Current address: Department of Physics and Center for Nonlinear Dynamics, University of Texas, Austin, Texas 78712

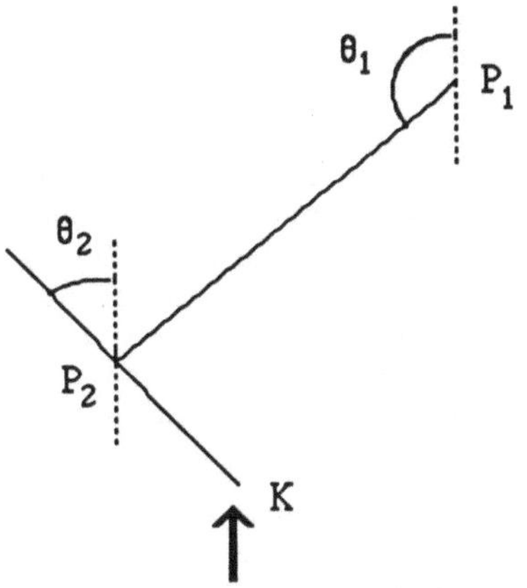

Figure 1. The double rotor.

integer times the second rod is kicked at the end marked K. The kick is always from the same direction and changes the angular velocity of the ith rod in proportion to $\kappa \sin \theta_i$ for some constant κ. The double-rotor map gives the position $\theta_i^{(n)}$ and angular velocity $\dot\theta_i^{(n)}$ of each rod instantaneously after the nth kick and is given by

$$\begin{bmatrix} \theta_1^{(n+1)} \\ \theta_2^{(n+1)} \end{bmatrix} = M_\theta \begin{bmatrix} \dot\theta_1^{(n)} \\ \dot\theta_2^{(n)} \end{bmatrix} + \begin{bmatrix} \theta_1^{(n)} \\ \theta_2^{(n)} \end{bmatrix}$$

$$\begin{bmatrix} \dot\theta_1^{(n+1)} \\ \dot\theta_2^{(n+1)} \end{bmatrix} = M_{\dot\theta} \begin{bmatrix} \dot\theta_1^{(n)} \\ \dot\theta_2^{(n)} \end{bmatrix} + \begin{bmatrix} c_1 \sin \theta_1^{(n+1)} \\ c_2 \sin \theta_2^{(n+1)} \end{bmatrix}$$

where the 2×2 matrices M_θ, $M_{\dot\theta}$ and the constants c_1, c_2 depend on the lengths of the rods, the attached masses, friction at the pivots, and the size of the kick. The angles θ_1, θ_2 (mod 2π) measure the position of each rod, and the angular velocities are given by $\dot\theta_1$, $\dot\theta_2$.

Fig. 2 is a schematic illustration of the idea behind the numerical procedure for a map in the plane with one positive Lyapunov exponent. Each curve represents an example of what the unstable manifold might look like. L is a fixed line that intersects the attractor. We iterate the map until a point (x′) falls close to L. We back

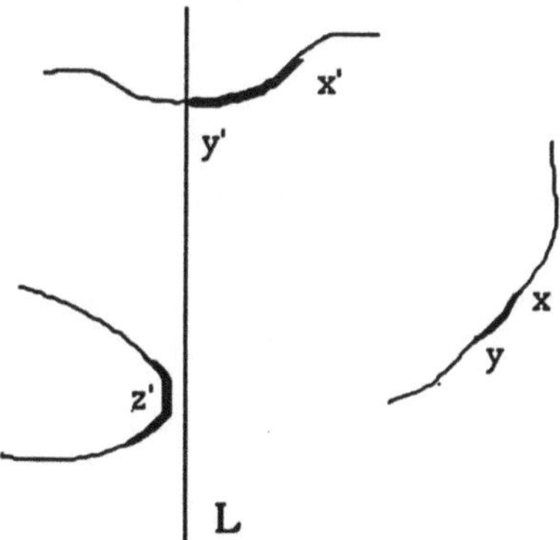

Figure 2. Schematic diagram of the shadowing procedure for the case of a planar map with one positive Lyapunov exponent.

up m iterates to a point $\mathbf{x}$ and try to determine whether there is a point $\mathbf{y}$ near $\mathbf{x}$ whose mth iterate, $\mathbf{y}'$, lies on L. In the diagram, the unstable manifold at $\mathbf{x}'$ crosses L, so in principle we can locate $\mathbf{y}$. Because points typically separate rapidly along the unstable manifold, $\mathbf{y}$ is much closer to $\mathbf{x}$ than $\mathbf{y}'$ is to $\mathbf{x}'$. We try to minimize $||\mathbf{y}-\mathbf{x}||$ by backing up as many iterates as possible. (For example, our numerical procedure to locate $\mathbf{y}$ would be attempted perhaps $m = 13$ iterates back from L rather than 2 iterates back.) Of course, if $\mathbf{x}$ maps to the point $\mathbf{z}'$ instead, we expect our procedure to fail for every m, because the unstable manifold curves away from L at $\mathbf{z}$. We use the past history of the trajectory to compute the direction of the unstable manifold at $\mathbf{x}$, from which we construct an approximation of the unstable manifold in a neighborhood of $\mathbf{x}'$. We use this to determine whether the attractor intersects L near $\mathbf{x}'$, and if so, to calculate $\mathbf{y}$.

The same idea is used in the higher dimensional case of the double rotor, except that L is a two-dimensional hyperplane and we approximate the unstable manifold at $\mathbf{x}'$ by a two-dimensional surface. We locate $\mathbf{y}$ using a Newton method to solve a nonlinear system of equations. (Note that the transverse intersection of each two dimensional sheet of the unstable manifold with L is a single point.) The procedure is attempted whenever a point $\mathbf{x}'$ falls within either 0.02 or 0.04 of L, depending on the location of L within the attractor. As before, we try to find $\mathbf{y}$ as many iterates back from L as possible. For 50 to 90 percent of the points $\mathbf{x}'$ which fall near L, depending on the location of L within the attractor, we can compute a point $\mathbf{y}$ such that $\mathbf{y}'$ lies within 10^{-9} of L. On the average, $m = 7.5$, i.e., we successfully perturb points 7.5 iterates back from L. We have developed a procedure to estimate how far the computed point $\mathbf{y}'$ is from the "true" unstable manifold at L. Full details will be published elsewhere.[6]

Fig. 3 shows a Lorenz cross-section of the double-rotor attractor for $c_1 = 0.3536$, $c_2 = 0.5000$, and

$$M_{\dot\theta} = \begin{bmatrix} 0.7496 & 0.1203 \\ 0.1203 & 0.8699 \end{bmatrix}, M_\theta = \begin{bmatrix} -5.800 & -6.602 \\ -6.602 & -12.40 \end{bmatrix}.$$

The cross-section is the intersection of the attractor with a 2-plane chosen by fixing two coordinates. In Fig. 3, θ_1, θ_2 are fixed at 0, so that the picture is a slice of the attractor along the (θ_1, θ_2) plane.

Suppose the Lyapunov exponents of the attractor are ordered so that $\lambda_1 \geq \lambda_2 \geq \cdots \geq \lambda_n$, and let k be the index such that $\sum_{i=1}^{k} \lambda_i \geq 0$ but $\sum_{i=1}^{k+1} \lambda_i < 0$. The *Lyapunov dimension*[4] of the attractor is defined by

$$d_L = k + \frac{\lambda_1 + \cdots + \lambda_k}{|\lambda_{k+1}|}.$$

The Lyapunov exponents for the double-rotor attractor whose cross-section is shown in Fig. 3 are 1.1720, 0.11377, -0.48053, -1.2655, so $d_L \approx 3.64$.

The *pointwise dimension* is defined in the following way. Let $\mathbf{x}$ be a point on the attractor and let $B_\varepsilon(\mathbf{x})$ be the ball of radius ε about $\mathbf{x}$ with natural measure $\mu(B_\varepsilon(\mathbf{x}))$. The pointwise dimension at $\mathbf{x}$ is defined by

$$d_P(\mathbf{x}) = \lim_{\varepsilon \to 0} \frac{\log \mu(B_\varepsilon(\mathbf{x}))}{\log \varepsilon}. \tag{1}$$

If this limit is the same for almost every $\mathbf{x}$, then we say that the attractor has pointwise dimension d_P. In other words, $\mu(B_\varepsilon(\mathbf{x})) \sim \varepsilon^{d_P}$ for small ε. The Lyapunov dimension is a dynamical notion; pointwise dimension is a geometrical one. The

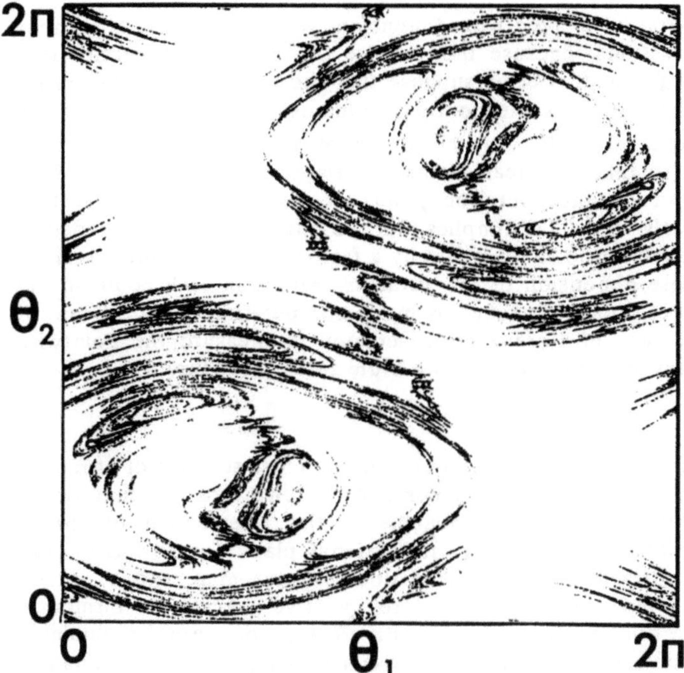

Figure 3. Lorenz cross-section of the double-rotor attractor. The plane of the paper is the 2-plane defined by $0 \le \theta_1 < 2\pi$, $0 \le \theta_2 < 2\pi$ with $\theta_1 = \theta_2 = 0$. The picture is a plot of θ_1 (horizontal axis) against θ_2.

Lyapunov exponents of an attractor are easy to compute when the generating equations are known. However, it is impractical to compute pointwise dimension except for low-dimensional attractors because the required number of iterations of the map grows exponentially with d_P.

It is conjectured[4] that $d_P = d_L$ for typical chaotic attractors. In order to test the conjecture numerically, it is necessary to compute the pointwise dimension as accurately as possible. We have computed the pointwise dimension $\overline{d_P}$ of the Lorenz cross-section instead of the pointwise dimension of the entire attractor, since the latter computation is impractical, as we explain below. Heuristically, one thinks of the unstable manifold as a two-dimensional sheet through typical points on the attractor when the attractor has two nonnegative Lyapunov exponents. Hence, we have tested whether $\overline{d_P} + 2 = d_L$, where $\overline{d_P}$ is the pointwise dimension of the cross-section as described below and d_L is the Lyapunov dimension of the entire attractor.

The pointwise dimension $\overline{d_P}$ of the Lorenz cross-section was measured by choosing 2,048 points $\{x_n\}$ on the cross-section at random from the total collection of 260,000 computed points. The limit (1) was estimated for each x by computing $\mu(B_\varepsilon(x))$ for a decreasing sequence of ε values, and the results were averaged over the collection $\{x_n\}$ to give $\overline{d_P}$. The process was repeated for four different samples of 2,048 points, giving an average value of $\overline{d_P} = 1.64 \pm 0.007$ for the Lorenz cross-section shown in Fig. 3. Similar results are obtained for other Lorenz cross-sections of this attractor[6]. This numerical result supports the conjecture that $d_P = d_L$, because the cross-section inherits the fractal structure of the attractor, which is apparent only in directions transverse to the unstable manifold.

The numerical computation of pointwise dimension would be infeasible without the use of Lorenz cross-sections, because they reduce the number of iterates that must be generated by a large factor. Let us suppose that the pointwise dimension d_P of the attractor is $\overline{d_P}+2$, where $\overline{d_P}$ is the pointwise dimension of the Lorenz cross sections as discussed above. By (1), on the order of $\varepsilon^{-(\overline{d_P}+2)}$ points must be generated on the attractor so that the mean distance between nearest neighbors is ε. The mean nearest neighbor distance between points on the Lorenz cross-section shown above is approximately 10^{-3}. This implies that the number of points needed to measure $\overline{d_P}$ to the same resolution ε as d_P is less by a factor of 10^6, and this saving is independent of the pointwise dimension of the attractor. Although each point on the Lorenz cross-section requires about 10^3 times as much computation as one iterate of the double rotor map, the use of Lorenz cross-sections to measure the pointwise dimension is about 10^3 times less expensive than a brute force procedure.

Conclusion

A numerical procedure is outlined to compute Lorenz cross-sections of attractors with more than one positive Lyapunov exponent. A Lorenz cross-section of an attractor with $k > 0$ positive Lyapunov exponents is the intersection of the k dimensional unstable manifold with an $n-k$ dimensional plane (or surface). The method is applied to the attractor of the kicked double-rotor map, two of whose numerically determined Lyapunov exponents are positive. Accurate numerical estimates of the pointwise dimension of the Lorenz cross-section can be obtained. These results support a conjecture that the pointwise and Lyapunov dimensions of typical attractors are equal. The use of Lorenz cross-sections to estimate pointwise dimension reduces the number of map iterations by a factor of 10^3 over a brute force approach.

Acknowledgment

This research was sponsored by the Department of Energy Office of Basic Energy Sciences, the Air Force Office of Scientific Research, and the National Science Foundation. E. Kostelich was supported in part by a National Science Foundation Graduate Fellowship.

REFERENCES

1. E. N. Lorenz, *Physica* **13D**, 90 (1984).

2. J. D. Farmer, E. Ott and J. A. Yorke, *Physica* **7D**, 153 (1983).

3. J. P. Eckmann and D. Ruelle, *Rev. Mod. Phys.* **54**, 153 (1983).

4. J. L. Kaplan and J. A. Yorke, in *Functional Differential Equations and Approximation of Fixed Points*, H. O. Peitgen and H O. Walther (eds.), pp. 228-237. Springer Lecture Notes in Mathematics, Vol. 730. (New York: Springer-Verlag, 1979).

5. E. Kostelich, C. Grebogi, E. Ott and J. A. Yorke, in preparation.

6. E. Kostelich and J. A. Yorke, in preparation.

On the Fractal Dimension of Filtered Chaotic Signals

R. Badii[1] and A. Politi[2]

[1] Institut für Theoretische Physik, Schönberggasse 9,
CH-8001 Zürich, Switzerland
[2] Istituto Nazionale di Ottica, Largo E. Fermi 6, I-50125 Firenze, Italy

1 Introduction

Much progress has been done in the reconstruction of the geometry of strange attractors, from experimental single time-series, exploiting embedding techniques [1], which make possible, for instance, the estimation of fractal dimensions and metric entropies. A particularly relevant aspect of these procedures, which has not yet been pointed out, concerns the role of filtering. In fact, not only any measurement of experimental signals is to some extent filtered, due to the finite instrumental bandwidth, but often an explicit intervention of the observer is present as well, motivated by the need of "cleaning" the system's output from the presence of noise.

As far as linear filters are concerned, this topic is easily recognized to belong to the class of linear-response problems with the main difference of dealing with chaotic, rather than random, sources. As a consequence, the interest is moved from power spectra and correlation functions analysis to fractal dimensions and Lyapunov exponents of the filter's output versus the input signal. A useful instrument of investigation is, in our case, the Kaplan-Yorke (KY) relation [2] expressing the dimension in terms of the Lyapunov exponents. As a result, an increase in the fractal dimension is to be expected, when the filter's bandwidth is decreased. Moreover, chaos-chaos transitions [3] can be identified, in correspondence of which the structure changes from self-similar to self-affine [4] or the Lyapunov dimension D_L crosses an integer value. The latter case is particularly appealing, since the attractor does not appear continuous along all directions, instead presenting a very lacunar [4] texture. This richness of geometries offers the opportunity of non--trivial tests of the KY conjecture whose validity has been verified even for a self-affine set. Finally, the effect of filtering on the nonuniformity is numerically studied for the Duffing attractor, indicating a large increase of the nonuniformity factor [5].

The paper is organized as follows. In the second Section, a general discussion is first done to derive the relation linking the Lyapunov dimension of the filtered signal to the poles of the filter. A numerical test is then performed for a simple flow evaluating the dimension function [6] $D(\gamma)$ for $\gamma = 0$, which corresponds to the information dimension, and for $\gamma = 1$

In Section 3, the discrete example of the baker map is extensively discussed, and a geometrical analysis of the critical regions is provided in a fully analytic way.

In Section 4, the main results are summarized and some open questions pointed out. Finally, in the Appendix, an indipendent explanation of the reason for the dimension increase is obtained by formally solving the filter equation.

2 General Results

Let us consider an N-dimensional dynamical system

$$\dot{\underline{X}} = \underline{F}\,(\underline{X}) \quad , \tag{1}$$

and indicate with x(t) the measured component of the vector $\underline{X}$ on which the embedding procedure is performed. As a first approach to the problem, we suppose the signal x(t) to be processed by a simple low-pass filter, whose action is modelled by the additional equation

$$\dot{z} = -\eta z + x \quad , \tag{2}$$

where z(t) is the output signal on which dimensional analysis will be performed and η is the filter's bandwidth. Being z(t) coupled to system (1) through x(t), the embedding on this variable will faithfully reconstruct the attractor of the complete system (1)+(2). Indicating with $\lambda_1 \geqslant \lambda_2 \geqslant \ldots \geqslant \lambda_N$ the Lyapunov exponents of (1), the KY conjecture provides an estimation for the information dimension D_1 as

$$D_L = j + \sum_{k=1}^{j} \lambda_k / |\lambda_{j+1}| \quad , \tag{3}$$

where j is the largest integer for which the sum in (3) is non-negative. In order to see how this is modified by the introduction of the filter, notice that the addition of eq. (2), while yielding a new exponent $-\eta$, does not affect the previous ones. Therefore, the Lyapunov dimension D_L remains unaffected as long as η is larger than $|\lambda_{j+1}|$. For a detailed analysis of the dependence of D_L on the cut-off η, we specialize on a three-dimensional dynamical system, with $\lambda_1 > 0 > \lambda_3$.

Depending on the value of η, three regions can be individuated

a) $\quad D_L = 2 + \lambda_1 / |\lambda_3| \quad , \quad \lambda_1 < |\lambda_3| < \eta$

b) $\quad D_L = 2 + \lambda_1 / \eta \quad , \quad \lambda_1 < \eta < |\lambda_3| \tag{4}$

c) $\quad D_L = 3 + (\lambda_1 - \eta)/|\lambda_3| \quad , \quad \eta < \lambda_1 < |\lambda_3| \;.$

More in general, if dealing with an order-k filter, it is necessary to compare the real parts of the k-poles with the sequence of Lyapunov exponents. A piecewise differentiable function, analogous to eq. (4), derives, exhibiting a larger number of different behaviours.

In Fig. 1, the function $D_L(\eta)$ is plotted (solid line) for the case of the Duffing equation

$$\ddot{x} + \gamma \dot{x} - x + 4x^3 = A \cos(\omega t) \quad , \tag{5}$$

with $\gamma = 0.154$, A = 0.18 and $\omega = 1.17$, yelding a chaotic attractor with Lyapunov exponents $\lambda_1 = 0.161$, $\lambda_3 = -0.315$ (i.e. $D_L = 2.51$).

Clearly, large values of η (region a) correspond to a fast response of z(t) to the input x(t), thus leaving the dimension unaffected. In region b, we observe a hyperbolic growth with decreasing η and, finally, D_L increases linearly in region c. Notice that the largest deviation of D_L from its primitive value is 1, obtained in the limit $\eta \to 0$. However, the case $\eta = 0$ is non-physical, as the response becomes generally unbounded. In particular, two critical regions T_1 and T_2 can be identified, in which second order transitions between different chaotic states [3] occur. The meaning and implications of these transitions will be discussed afterwards, both in the case of a differential equation and a map. To perform a direct measurement of the information dimension, we employed the Dimension Function method [6]. This is based on the observation that mutual distances among neighbouring points

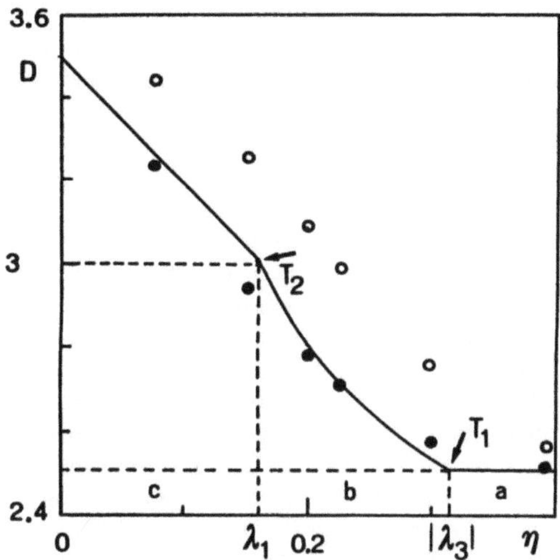

Fig. 1 Lyapunov dimension, $D_L(\eta)$ vs. η (solid line), for the Duffing equation (5). Regions a, b, c correspond to the three different behaviours of eq. (4). T_1 and T_2 represent the two transition points. Full dots are the numerical estimates of the information dimension D(0), and open dots of D(1).

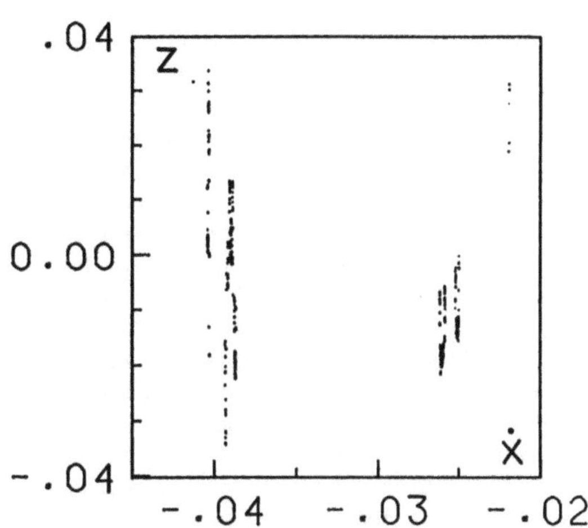

Fig. 2 Section of the Poincaré section for the filtered Duffing equation with η = 0.15. The variables $\dot{x}$ and z have been computed at x = 0 and t = $2\pi k/\omega$

on the attractor decrease when the number n of the latter ones increases. More specifically, we indicate with $\delta(n)$ the distance between a point $\underline{X}$ and its nearest-neighbour (chosen among n points) and compute the moment of order γ of the nearest-neighbour distribution P (δ,n).

The Dimension Function $D(\gamma)$ is, then, defined through

$$\langle \delta^{\gamma}(n) \rangle \equiv \int_0^{\infty} \delta^{\gamma} P(\delta,n)d\delta \sim K\, n^{-\gamma/D(\gamma)} , \qquad (6)$$

where K is a prefactor, usually irrelevant for dimension calculations. It has been proved that D(0) coincides with the information dimension D_1, at least when self-similarity is asymptotically established [6]. In Fig. 1, we report the values of D(0) (full dots) and D(1) (open dots) superimposed on the curve (4), computed at some values of η. The difference $\tilde{\chi}$ = D(1) - D(0) gives a rough estimate of the "non-uniformity factor" of the system [5]. It is evident that $\tilde{\chi}$ starts increasing already before the first critical point T_1 and saturates rapidly around a nearly constant and large value, when $\eta \rightarrow 0$.

A good agreement with the theoretical prediction is generally found except in the two transition regions. This discrepancy could be attributed, at this stage, to a sort of critical slowing down: that is, the nearer we are to the transition point, the smaller are the length scales at which the asymptotic behaviour sets in. Furthermore, due to the inequality $D_L \geqslant D_1$ [7], we must expect complete agreement at the point T_1, for an infinitely large number of points n, while a finite deviation in region T_2 cannot be a priori excluded. In fact, the equality $D_1 = D_L$ has been only proved for 2-d maps [8], while we have a 3-d Poincarè section. In addition, as shown by V. SHTERN [9] , a breakdown of the KY relation occurs whenever the set is Cantor in more than one direction (hence, self-affine). To clarify these aspects of the problem, we study a 2-d section of our 3-dimensional Poincaré section (taken at $t = 2\pi k/\omega$), defined by x = 0. This was done by first determining all points in the Poincaré section, whose x-coordinate lies between $-\varepsilon$ and ε ($\varepsilon = 10^{-3}$), and then suitably refining the $\dot{x}$ and z estimates. This is achieved by exploiting the continuity of the attractor along the expanding direction, which can be easily determined by integrating the linearized flow. A further linear extrapolation of the $\dot{x}$, z coordinates on the chosen section x = 0 yield very accurate measurements (the error is now of order 10^{-6}). The resulting picture, for η = 0.15, is reported in Fig. 2. Due to the extreme nonuniformity, a large fraction of the points has not been plotted in the most dense regions, in order to obtain a clean picture, which still faithfully reproduces the attractor's transversal section. By a first inspection, the set appears to be Cantor-like in the x-direction, while no definite conclusion can be drawn with regard to the z-direction: we expect it to be continuous, notwithstanding the visible gaps which are reasonably imputable to the nonuniformity. This fact contributes to enhance computational difficulties. To circumvent these problems and make firmer statements, we turn to a discrete model.

3 Filtered Baker Transformation

In this Section we analyze a discrete model which, apart from allowing more computational speed and accuracy, can be analytically solved. The dynamical system we consider is the following version of the baker equation

$$x_{n+1} = x_n/3 + 2/3[2y_n]$$

$$y_{n+1} = 2y_n \bmod 1 \qquad , \tag{7}$$

where $[y]$ is the largest integer smaller than or equal to y. Obviously, y cannot be taken as a "good" embedding variable as it is decoupled from x; equally, x does not allow a faithful reconstruction of the dynamics, because its future evolution cannot be inferred from its past history alone. It is therefore necessary to choose some linear combination w of the two variables: the optimal choice is obtained for

$$w = (3x + 2y)/5 \tag{8}$$

which yields a continuous and invertible mapping from (x_n, y_n) to (w_n, w_{n+1}). Moreover, w turns out to be bounded between 0 and 1. Therefore we are led to the recursive relation

$$w_{n+2} = w_{n+1}/3 + 2\left[(3w_{n+1} - w_n)\bmod 1\right]/3$$

for the variable w, on which filtering is performed. In analogy with eq.(2), we write

$$z_{n+1} = \alpha z_n + w_{n+1} \quad , \tag{9}$$

where z is the "output" signal to be analyzed, and $0 < \alpha < 1$ is the filter's multiplier: that is, a Lyapunov exponent $\ln \alpha = -\eta$ is added. Similarly to the case of the Duffing equation, we take a section of the attractor to get rid of the continuous direction, everywhere parallel to the eigenvector (in x,y,z coordinates)

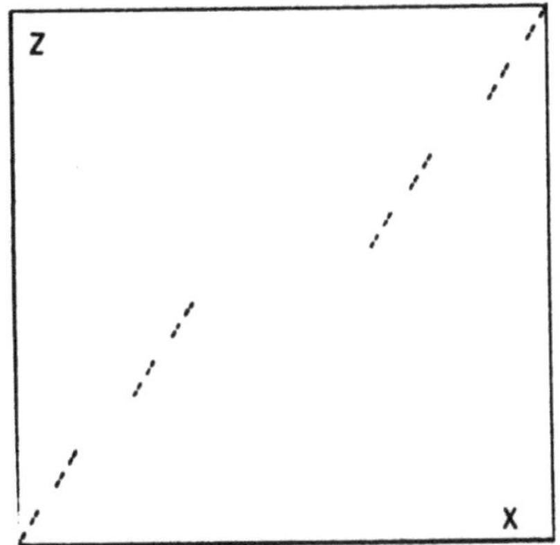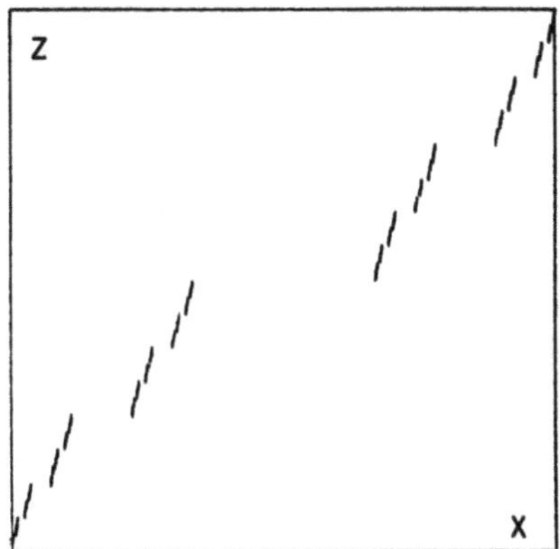

Fig. 3 Section of the filtered baker transformation (eq. (7)) for $\alpha = 0.2$. We have plotted 1000 points, in the variables x (ranging between 0 and 1) and z, rescaled to yield a square picture.

Fig. 4 Asymptotic section of map (7) at the critical point T_2 ($\alpha = 1/2$), obtained from the expansion (12). Notice that, joining all adjacent points with (horizontal) segments, we obtain a complete Devil's staircase.

$$\underline{v} = (0, 2-\alpha, 4/5) \quad .$$

Of course, the linearity of the unstable manifold enables us to exploit every iteration of the map, to extrapolate points on the chosen section y=0. For this model, the two transition points correspond to $\alpha = \alpha_1 = 1/3$ and $\alpha = \alpha_2 = 1/2$. In Fig. 3, we have plotted the section in the variables x,z for $\alpha = 0.2$ ($< 1/3$), which lies in region c. The asymptotic structure of the attractor is unravelled by means of an analytic investigation. The linearity of eq. (7) allows, in fact, its formal resolution as

$$z_{n+1} = \sum_{j=0}^{\infty} w_{n+1-j}\, \alpha^j = \sum_{j=0}^{\infty} (x_{n-j} + 4y_{n-j})\, \alpha^j/5 \quad . \tag{10}$$

We then expand x_n and y_n in the bases of 3 and 2, respectively

$$x_n = 2 \sum_{i=1}^{\infty} a_{n-i}\, 3^{-i} \quad , \qquad y_n = \sum_{i=1}^{\infty} a_{n-1+i}\, 2^{-i} \quad , \tag{11}$$

where a_k ($-\infty < k < +\infty$) is a doubly-infinite sequence of random bits 0,1. Clearly, the action of the baker transformation amounts to a shift by one unit of the index in the sequence a_k. Substituting the relations (11) into eq. (10), after some algebraic manipulations, we obtain an explicit expansion for z_{n+1}. A further specialization to the section $y_n = 0$ requires $a_k = 0$ ($k > n$). The final result is, then

$$z_{n+1} = \frac{2}{5} \sum_{j=1}^{\infty} a_{n-j} \left\{ \alpha^j \left[\frac{1}{\alpha(1-\alpha/2)} - \frac{1}{1-3\alpha} \right] + \frac{3^{-j}}{1-3\alpha} \right\} \quad . \tag{12}$$

Being the attractor composed of many identical subsets, we focus on a small region around the origin: this is tantamount to choosing the first j_0 bits in the sum (12) equal to 0. As a consequence, either the first or the second term in the r.h.s. dominates for j_0 large enough, according to whether α is larger or smaller than 1/3, respectively. In the latter case the variable z is proportional to x: this im-

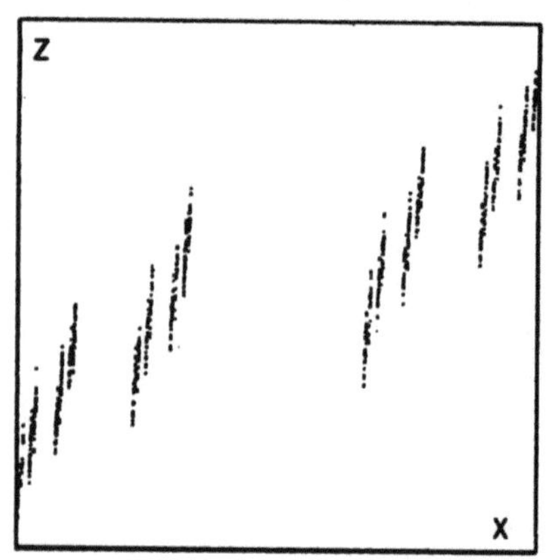

Fig. 5 Same as in Fig. 4 for α =0.7 Notice that the correspondence between z and the points on the section is no more biunivocal.

plies the existence of an asymptotic self-similar structure in agreement with that suggested by Fig. 3. In the former case, every subset (labelled by the index j) can be rescaled to the previous level, provided that z is multiplied by $1/\alpha$ and x by 3: in other words, the set is asymptotically self-affine . The transition at the critical point α = 1/3 can be, therefore, recognized as a typical crossover effect characterizing the start of dimension increase in the filtered signal. The analysis of this phenomenon in a continuous case is presented in the Appendix. It can be easily found that the KY relation yields the correct result for the information dimension D_1 (which coincides with the fractal dimension D_0 , being the set uniform). At α = 1/2, the second transition occurs when the projection of the set onto the z-axis starts being continuous. The value α = 1/2 is found by requiring that the uppermost point of a generic subset has the same ordinate z as the lowermost point of the right--contiguous subset, at the same scale. This condition is simultaneously met at all length scales. The attractor's structure is exemplified in Fig. 4, obtained from eq. (12), neglecting the last term in the sum. The picture can be recognized as the set of all discontinuity points of the complete Devil's staircase [4].
A further increase in the value of α brings to situations like the one depicted in Fig. 5, corresponding to α = 0.7, where the projection onto the z-axis is no more single-valued.

4   Conclusions

In this paper, we were restricted to the particularly simple example of a single-pole low-pass filter. Of course, in real experiments, more complicated kinds of arrangements occur, involving many poles, in order to achieve sharper cut-offs. However, the implications of this can be easily taken into account, as already discussed in Section 2. The main effect arising from our investigation is that a dimension increase can be theoretically predicted, contrary to the intuitive idea that filtering reduces the dimensionality of the signal.

As a second result, the validity of the KY conjecture is theoretically confirmed, for a uniform attractor (baker-map), even in the case of a self-affine structure. In this respect, it is interesting to compare our map with that of Shtern. Indeed, he has a self-affine, uniform set which is the Cartesian product of two Cantor sets: in his case, however, there is the additional freedom of disposing the various subsets in any order, changing the dimension without affecting the Lyapunov exponents.

Differently, nothing conclusive can be said in the nonuniform case (Duffing equation), where we only presented numerical evidence of a nonuniformity increase for small η-values.

Acknowledgements

The authors wish to thank Prof. B.B. Mandelbrot for an enlightening discussion on self-affine sets. Dr. G.L. Oppo is acknowledged for useful exchanges of ideas.

The numerical calculations have been performed on the VAX 11/780 of the Astrophysics Observatory of Arcetri.

Appendix

The filter's equation (2) can be formally solved, yielding

$$z(t,\underline{X}) = \int_{-\infty}^{t} e^{-\eta(t-\tau)} x(\tau,\underline{X}(t))\, d\tau, \tag{A.1}$$

where, we recall, x is the filtered variable, while the N-dimensional vector $\underline{X}(t)$ represents all the coordinates needed to generate the attractor's dynamics. Hence, eq. (A.1) expresses the "new" variable z as a function of the "old" ones $\underline{X}$, or, in other words, it defines the hypersurface in the $(z,\underline{X})$ space, the attractor lies onto.

It is easy to convince oneself that, as far as eq. (A.1) represents a smooth surface, filtering simply corresponds to embedding the set $\left\{\underline{X}(t)\right\}$ in an (N+1)-dimensional space, without affecting the fractal dimension. It is, hence, useful to compute the effect of a slight perturbation of the "final condition" $\underline{X}(t)$ along a generic direction $\underline{v}$ on $z(t,\underline{X})$, that is, introducing the gradient,

$$\underline{\nabla} z \cdot \underline{v} = e^{-\eta t} \int_{-t}^{\infty} e^{-\eta\tau} \underline{\nabla}x \cdot \underline{v}\, d\tau , \tag{A.2}$$

where generic is to be referred to the directions locally spanned by the attractor. As a consequence, the backward evolution of $\underline{\nabla} x \cdot \underline{v}$ is dominated by the most negative Lyapunov exponent entering the KY formula. Referring, for simplicity, to a 3-dimensional flow, we obtain the relation

$$\underline{\nabla} z \cdot \underline{v} \propto e^{-\eta t} \int_{-t}^{\infty} e^{-(\eta-|\lambda_3|)\tau} d\tau , \tag{A.3}$$

which converges only for $\eta > |\lambda_3|$, as expected from eqs. (4). The divergence of the integral, instead, for $\eta < |\lambda_3|$, indicates a strong stretching of points in the z-direction, thus affecting the fractal dimension of the attractor.

References

1 F. Takens, in <u>Lecture Notes in Mathematics</u> 898, D.A. Rand and L.S. Young eds. (Springer, Berlin 1981)

2 J.L. Kaplan and J.A. Yorke, in <u>Lecture Notes in Mathematics</u> 730, H.O. Peitgen and H.O. Walther eds. (Springer, Berlin 1979)

3 Y. Aizawa, this volume

4 B.B. Mandelbrot, <u>The Fractal Geometry of Nature</u>, (Freeman, San Francisco 1982)

5 R. Badii and A. Politi, Phys. Rev. Lett. <u>52</u>, 1661 (1984)

6 R. Badii and A. Politi, J. Stat. Phys. <u>40</u>, 725 (1985)

7 F. Ledrappier, Commun. Math. Phys. <u>81</u>, 229 (1981)

8 L.S. Young, J. Ergodic Theory and Dynm. Syst. <u>2</u>, 109 (1982)

9 V.N. Shtern, Phys. Lett. <u>99A</u>, 268 (1983)

Efficient Algorithms for Computing Fractal Dimensions

F. Hunt and F. Sullivan

Center for Applied Mathematics, National Bureau of Standards,
Gaithersburg, MD 20899, USA

1. Introduction

Our purpose is to describe a new class of methods for computing the "capacity dimension" and related quantities for point-sets. The techniques presented here build on existing work which has been described in the literature. The novelty of our methods lies first in the approach taken to the definition of computation of dimension (namely, via Monte Carlo calculation of the volume of an ε-cover of the point-set), and second in the use of data structures which result in extremely efficient codes for vector computers such as the Cyber 205 (the computation is reduced to the sorting and searching of one-dimensional arrays so that a calculation employing one million points requires less than 2 minutes).

The definition of "capacity dimension" and the related "information dimension" and "correlation dimension" have been given elsewhere. We use these terms here in quotation marks in part because we will be dealing with finite point sets for which, strictly speaking, all of these dimensions are zero. However, as a practical matter, one tries to extrapolate results for small ε and large point sets to the $\varepsilon=0$ limit. This leads to various numerical difficulties which have been reported by many authors and most of which we have encountered in our own calculations. Two advantages of our algorithms are the fact that, because of the Monte Carlo integration, error estimates are available; and since the methods are <u>extremely efficient</u>, very large computations can be performed.

An interesting observation which can be made about the calculations is this: two quantities under our control are N, the number of points in the set for which the dimension is to be computed, and ε, the size of the cover of the set. The ε-cover of the set can be thought of as a collection of boxes with side length ε, and the number of these is denoted by $N(\varepsilon)$. The definition of capacity dimension, d, says that for small ε, $N(\varepsilon) \sim \varepsilon^{-d}$. It seems that the smallest ε one should use in a calculation is one which puts each point in its own box. Smaller ε lead to underestimates of d. By looking at cases for which the answer is known in advance, we find that the result is found near <u>saturation</u>, that is to say, when ε is small enough so $N=N(\varepsilon)$. Call this value ε_s. This gives the saturation expression:

$$\frac{\ln N}{d} \sim \ln\left(\frac{1}{\varepsilon_s}\right).$$

Roughly speaking, the accuracy of results depends on the size of ε_s -- the smaller ε_s, the smaller the relative error in the answer. Notice that if $d>1$ the saturation expression says that in order to halve ε_s (= improve the accuracy by one binary bit) the size, N, of the point set must be doubled. To add one decimal digit of accuracy N must be increased by a factor of 10. (This gives some indication of why evaluation of capacity dimension has been computationally intractable.)

In the next section we outline the theoretical basis for our algorithms -- namely the connection between the volume of an ε-cover and capacity dimension and the use of Monte Carlo integration to compute this volume. Section 3 is concerned with data structures. We first adapt an idea of F. Varosi [7] for representing point sets as trees and discuss how such trees can be used for the computation of capacity dimension, information dimension and correlation dimension by box counting. We next

show how to combine Monte Carlo integration with the tree structure. Finally we
describe a method for collapsing the tree to a one-dimensional array, which leads to
efficient vector algorithms. The one-dimensional array also gives a very convenient
method for calculating the "dimension function" which was described by Badii and
Politi [1]. In the last section we report some numerical results and comment further
on the saturation phenomenon mentioned earlier and its relation to the ideas of Badii
and Politi.

2. Monte Carlo Integration

Assume that $\mathcal{Ol}$ is a compact point set contained in the unit cube of Euclidean
space of dimension D. For each $\varepsilon > 0$ let $N(\varepsilon)$ denote the minimal number of
D-dimensional cubes of side ε needed to cover $\mathcal{Ol}$. The capacity dimension of
is the number

$$d = \lim_{\varepsilon \to 0} \sup \frac{\ln N(\varepsilon)}{\ln(\frac{1}{\varepsilon})}$$

If $\mathcal{Ol}$ is finite $d=0$; in all cases $d<D$ and for the familiar sets of Euclidean
geometry, d is the topological dimension. The interesting cases are those for which
d is not an integer, i.e., $\mathcal{Ol}$ has fractional dimension.

For $y \varepsilon R^D$, $dist(y, \mathcal{Ol})$ denotes the distance from y to $\mathcal{Ol}$ i.e., the infimum of
distances $||y-x||$ for $x \varepsilon \mathcal{Ol}$. Now, for $\varepsilon > 0$, $A(\varepsilon)$ will be the volume of an ε-cover
of $\mathcal{Ol}$

$$A(\varepsilon) \equiv volume \ \{y \,|\, dist(y, \mathcal{Ol}) \leq \varepsilon\}$$

There is an elementary argument found in [6] from which it follows that

$$- \lim_{\varepsilon \to 0} \frac{\ln A(\varepsilon)}{\ln(\frac{1}{\varepsilon})} + \lim_{\varepsilon \to 0} \frac{\ln N(\varepsilon)}{\ln(\frac{1}{\varepsilon})} \sim D. \tag{1}$$

The idea is that for $\varepsilon > 0$, $N(\varepsilon)$ cubes are required to completely cover $\mathcal{Ol}$ and each y
with $dist(y, \mathcal{Ol}) \leq \varepsilon$ is in one of these cubes or in an ε-cube adjacent to one of these
$N(\varepsilon)$ cubes. Therefore,

$$A(\varepsilon) \leq 3^D \ \varepsilon^D \ N(\varepsilon).$$

On the other hand, if $\mathcal{Ol}$ is covered by cubes of size $\varepsilon / \sqrt{D}$, then every y in these
cubes is within ε of $\mathcal{Ol}$. Hence,

$$A(\varepsilon) \geq (\varepsilon / \sqrt{D})^D \ N(\varepsilon / \sqrt{D}).$$

Taking logarithms and limits gives (1).

Equation (1) says that an accurate calculation of $A(\varepsilon)$ could be used to
evaluate d. Of course, for a finite $\mathcal{Ol}$, $\lim A(\varepsilon)=0$. However, the inequalities
which give equation (1) can be thought of as saying that $\frac{\ln A(\varepsilon)}{\ln(\frac{1}{\varepsilon})}$ behaves like

$\frac{\ln N(\varepsilon)}{\ln(\frac{1}{\varepsilon})} - D$ for small ε. In other words, as a method to estimate d, computing $A(\varepsilon)$

is no worse than computing $N(\varepsilon)$. In practice Monte Carlo integration gives an
accurate $A(\varepsilon)$, while box counting on regular grids is extremely time consuming and
seems to understimate $N(\varepsilon)$.

Assume now that $\mathcal{Ol}$ is a subset of the unit cube in R^D. To compute $A(\varepsilon)$ using
Monte Carlo integration, one would generate D-tuples of uniform random numbers $\{y(i):$
$i=1, .., M\}$ and for each $y(i)$ evaluate

$$f_\epsilon(i) = \begin{cases} 1 & \text{if dist } (y(i), \mathcal{Q}) \leq \epsilon \\ 0 & \text{otherwise.} \end{cases}$$

The average $(\Sigma f_\epsilon(i))/M$ is the Monte Carlo estimate of $A(\epsilon)$.

The difficult part of this calculation is the determination of the $f_\epsilon(i)$. If $\mathcal{Q}$ has many points, it is not practical to test $||y(i)-x||$ for every $x \epsilon \, \mathcal{Q}$.

In the next section we discuss data structures which lead to efficient methods for determining $f_\epsilon(i)$. For the moment we assume that this can be accomplished, and concentrate on determining how many evaluations are needed for an accurate estimate of $A(\epsilon)$.

Assume $f_\epsilon(i)$ are drawn (computed) from a distribution with mean $A(\epsilon)$ and variance σ^2. Thus, $E\{f_\epsilon(i)\} = A(\epsilon)$ and $\Sigma f_\epsilon(i)/M$ is the Monte Carlo estimate of $A(\epsilon)$.

$$\begin{aligned} \sigma^2 &= E\{[f_\epsilon(i) - A(\epsilon)]^2\} \\ &= E\{f^2 - 2fA + A^2\} \\ &= E\{f^2\} - A^2 \\ &\simeq A - A^2 \end{aligned}$$

Since $A(\epsilon)$ is small, we approximate σ^2 by $A(\epsilon)$ below.

We want to choose M (the number of random D-tuples) large enough to guarantee that the standard deviation of the mean, $\sigma\phi(M)$, is sufficiently small relative to the mean. Specifically, for given $\delta > 0$, we want

$$\frac{\sigma(M)}{A(\epsilon)} > \frac{\delta}{2} \quad .$$

Of course, $\sigma^2(M) = \sigma^2/M \approx A(\epsilon)/M$, so to satisfy the inequality, we must have that

$$M > \frac{4}{\delta^2 A(\epsilon)} \tag{2}$$

Inequality (2) is important in case adaptive Monte Carlo integration [5] is used, because it gives an estimate of the relative error for M evaluations. Rearranging (1) gives that $A(\epsilon) \sim \epsilon^{D-d}$, and hence

$$M \sim \frac{1}{\epsilon^{D-d}} \quad .$$

This is an interesting connection between M, the "amount" of computation and d, the result. Inequality (2) says that keeping the relative variance small is equivalent to requiring that

$$\Sigma \, f_\epsilon(i) > \frac{4}{\delta^2} \quad ,$$

i.e., to increase the accuracy the number of those $y(i)$ which "hit", i.e., for which $f_\epsilon(i) = 1$, must increase. In case of adaptive integration this means that more Monte Carlo samples are placed in regions where there are more points of $\mathcal{Q}$, so that the probability distribution associated with the point set can be estimated. Alternatively, one could simply choose samples from the probability distribution associated with the point set $\mathcal{Q}$. This leads to a computation of the various "moments" introduced by Badii and Politi, as we shall discuss in the next section.

3. Data Structures

For purposes of exposition we take D=1 and we assume that $\mathcal{A} \subset [0,1]$. The interval $[0,1]$ can be associated with a binary tree. On the first level is the entire interval. Level one has two branches for the half intervals; level two has four branches, etc.

Each point x of $\mathcal{A}$ is associated with a path $p(x)$ in this tree determined by the binary expansion of x. Since computers have finite precision, the tree has only finitely many levels. If $\mathcal{A}$ is a set of uniformly distributed random numbers, all paths in the finite tree are equally likely. In fact, it is easy to see that if p has ℓ levels, then the number of elements of $\mathcal{A}$ following path $p(x)$ is $|\mathcal{A}|/2^\ell$. In case $\mathcal{A}$ is not uniformly distributed the situation is quite different. Paths do not have equal weight, and in fact, some paths never occur. From the definition of the capacity dimension, one would expect that as ℓ gets large the number of occupied nodes at level ℓ is approximately equal to $(2^\ell)^d$, where d is the capacity dimension of $\mathcal{A}$. In other words, the paths which are distinct at level ℓ is the same as the number of intervals of size $1/2^\ell$ which are occupied by points of $\mathcal{A}$. This is $N(\varepsilon)$ for the case $\varepsilon = 1/2^\ell$.

A method based on this idea has been developed by F. Varosi [7]. Begin with an empty tree T. For each $x \in \mathcal{A}$ create a path p in T by adding nodes and left or right branches as needed, according to the binary expansion of x. As new nodes are added, record their levels. Early in the computation when only a few elements of $\mathcal{A}$ have been added to the tree, most branches will call for the creation of new nodes. Later many nodes will be already occupied. As the calculation proceeds it is easy to keep a record of the total number of paths which have passed through a given node. After the tree has been constructed statistics can be gathered.

Assume that NODE is an integer indexing the nodes, LEVEL is the array of levels for the nodes, and C is a count of the number of paths that have passed through a given node. All arrays are initialized to zero; MA is $|\mathcal{A}|$ and MAX is the total number of nodes generated in the calculation. The following simple loop accumulates statistics about the set .

```
          DO  10 NODE = 1, MAX
             L = LEVEL (NODE)
             N(L) = N(L) +1
             P = C(NODE)/MA
             INFO(L) = INFO(L) - P*LOG(P)
             COR(L) = COR(L) + P*P
      10  CONTINUE
```

Upon completion of this loop the $N(L)$ array gives $N(\varepsilon)$ for $\varepsilon_\ell = 1/2^\ell$. It is clear that P is the approximate probability of finding a point of $\mathcal{A}$ in the box associated with the path leading to NODE. Hence INFO(L) approximates the following sum, for $\varepsilon_\ell = 1/2^\ell$

$$I(\varepsilon_\ell) = -\Sigma\, P_i\, \log(P_i)$$

and COR(L) approximates

$$COR(\varepsilon_\ell) = \frac{1}{|\mathcal{A}|^2} \Sigma_{i;j}\, \delta(|x_i - x_j| - \varepsilon_\ell)$$

Here $\delta(|x_i-x_j| - \varepsilon) = 1$ if $|x_i-x_j| < \varepsilon$ and zero otherwise. For attractors studied in the literature [4], $I(\varepsilon_\ell) \sim I_0 - \sigma\ell$ and $COR(\varepsilon_\ell) \sim (\varepsilon_\ell)^{-\nu}$. The quantities σ and ν are called the information dimension and correlation dimension respectively. These dimensions are close to the capacity dimension, and in [4] it is shown that in certain cases $\nu \le \sigma \le d$.

It is not hard to adapt the Monte Carlo integration procedure to this data structure. Assume that T is the tree associated with the point set $\mathcal{A}$, and let r be a random number in $[0,1]$. The binary expansion of r specifies a path $p(r)$, which is to

be compared with T; each bit in r specifies a left or right branch in T. If, at
level ℓ, the specified branch exists, we set $f_\epsilon(i)=1$ where $y(i)=r$ and $\epsilon = \dfrac{1}{2^\ell}$.
This is because to reach level ℓ , p(r) must agree with some path of T for ℓ levels
and, therefore, r must agree with some element of for ℓ bits, i.e.,

$$\text{dist}\ (r,\ \mathcal{Q}\ ) \leq \frac{1}{2^\ell}\ \ .$$

In case the branch specified by r does not exist, we quit and start again with a new
random number and so on until all M random numbers have been generated and tested.

Notice that this method generates all $A(\epsilon_\ell)$ for $\ell =1,\ 2,\ ..$ up to some maximum
level. This is useful when extrapolation is used to determine $\lim_{\epsilon \to 0}(\ln N(\epsilon))$. Notice
also that if the quantities $I(\epsilon)$ and $COR(\epsilon)$ are not required, counts need not be
accumulated at the nodes of T. All that is needed is information about the structure
of T. Unfortunately, the tree data structure uses a great deal of storage, since it
can be expected to have on the order of $\left(\dfrac{1}{\epsilon}\right)^d$ nodes. While the tree is being built
nodes are added as required and so, in the one dimensional case, the index of a
parent node may differ from a child by as much as N. In the two dimensional case
the situation is even worse because the tree is four-fold and, in cases of interest,
d is larger than one.

Direct implementation of the tree leads to an inefficient, non-vectorizable code
for the Cyber 205. A way out of such difficulties is to "map" a significant part of
the method to an algorithm which is known to vectorize, in this case the Diamondsort
method [2].

For the purposes of discussion, assume that $\mathcal{Q} \subseteq R^2$ Each point in is a pair (x,y)
of real numbers. Denote the binary expansion of x by $b_1 b_2 \ldots b_k \ldots$ and the expansion
of y by $\overline{b_1}\overline{b_2} \ldots \overline{b_k} \ldots$. A unique base-4 number q with expansion $q_1 q_2 \ldots q_k \ldots$
can now be generated according to the prescription

$$q_k = b_k + 2 \cdot \overline{b_k}\ \ . \tag{3}$$

The q array is then sorted using a vector method. To do the Monte Carlo integration,
generate pairs of random numbers $(r,\overline{r})$. These are combined as in formula (3) to
give a base-4 number s . Next use a fast search method to locate s in the q array.
The result is an index k such that

$$q(k) < s < q(k+1)\ \ \ .$$

Suppose that $s-q(k) < 1/4^j$; for some j . It is easy to see that this means
$f_\epsilon(s) = 1$ for all $\epsilon = 1/4$, $1/4^2$, $..$, $1/4^j$; . In this way the information
required for the Monte Carlo averages is accumulated directly from the sorted list.
Notice that since the square is not homeomorphic to an interval, the one dimensional
representation can generate some inaccuracies. If $\mathcal{Q}$ contains points whose
compressed representation starts 03333...and a Monte Carlo sample point has
representation 30000, the program will omit this sample point from the count.
However, this additional error is not a serious difficulty for large $\mathcal{Q}$, or when
adaptive methods of sampling are used.

As has been mentioned, if d is known for a set with $|\mathcal{Q}| = N$, then the
saturation level can be predicted from the formula

$$\frac{\ln N}{d}\ =\ \ln\left(\frac{1}{\epsilon_s}\right)$$

In calculations, $\varepsilon_s = 1/2^k$ for some k and so the formula suggests where to stop halving the box size. Of course, in general one does not know d in advance. There is a related idea which does not require prior information about d .

For each $x\varepsilon$ let $\delta(x)$ denote the distance from x to its nearest neighbor, i.e., dist $(x,\mathcal{Q}\setminus(x))$. Various mean values of the function $\delta(x)$ and related dimensions have been of interest. In particular, it is known that the information dimension can be estimated by the expression

$$D(0) = \frac{-\ln N}{\langle \ln(\delta) \rangle}$$

Here the denominator is the mean of the log of the nearest distance function i.e.

$$\langle \ln(\delta) \rangle = \int_0^\infty \ln(\delta) \, P(\delta,N) d\delta$$

where $P(\delta,N)$ is the probability distribution associated with the function δ .

The number $D(0)$ is one value of the dimension function $D(\gamma)$, discussed by Badii and Politi [1]. In general each real γ defines a dimension

$$D(\gamma) = \lim_{N \to \infty} \frac{\gamma \ln N}{\ln(M_\gamma(\delta,N))}$$

where

$$M_\gamma(\delta,N) = \langle \delta^\gamma \rangle \equiv \int_0^\infty \delta^\gamma P(\delta,N) d\delta \quad .$$

In their paper [1] Badii and Politi report results based on approximating the mean $\langle \delta^\gamma \rangle$ by an average

$$\frac{1}{M} \Sigma \, \delta^\gamma(x_i)$$

where the x_i are M points chosen at random from $\mathcal{Q}$ (= chosen from the probability distribution associated with $\mathcal{Q}$). The Monte Carlo technique can be modified as follows, so that $P(\delta,N)$ is approximated:

1. Generate a set of M points of $\mathcal{Q}$ to be used as samples.
2. Generate an additional N points of $\mathcal{Q}$(M<<N).
3. Use the sorting procedure and the M sample points to compute volumes $A(\delta_\ell)$, where $\delta_\ell = 1/2^\ell$.
4. Approximate $P(\delta_\ell,N)$ by $A(\delta_\ell) - A(\delta_{\ell+1})$.

It is not hard to see that $A(\delta_\ell)$ is approximately the probability that some point of the sample agrees with a point of $\mathcal{Q}$ to ℓ bits. Hence, $A(\delta_\ell) - A(\delta_{\ell+1})$ is the probability that some x has

$$\delta_{\ell+1} < \delta(x) \leq \delta_\ell$$

and therefore it approximates $P(\delta_\ell,N)$.

The integrals M_γ can now be estimated by weighted sums

$$\sum_{\ell=1}^{\ell=L} w_\ell \delta_\ell^\gamma \, P(\delta_\ell,N)$$

Here the w_ℓ are weights for a quadrature formula and L is the number of levels used.

4. Results

Results are reported for four cases: the Baker's transformation, the Henon map, the Zaslavskij map and the Lorenz attractor. In the 2-D cases, $D(0)$, $D(1)$ and the fixed point, $D(\overline{\gamma})=\overline{\gamma}$ were computed. Badii and Politi discuss cases for which the fixed point is the value of the capacity dimension. In all cases an estimate based on box counting was also computed, and Monte Carlo calculations were performed for the Baker's transformation, the Henon map, and the Lorenz attractor. The fixed point $\overline{\gamma}$ was used to determine the saturation level from the expression $\log_2(N)/\overline{\gamma}$. The integer part of this was used to determine the level at which to calculate the ratio $-\log N(\varepsilon)/\log(\varepsilon)$. All calculations are for $N=10^6$.

For the Baker's transformation the agreement with the known value, $1+\log(2)/\log(3)=1.6309...$, is quite good. The values for the Henon map are similar to those reported by Badii and Politi for $D(0)$ and $D(1)$. Our value of $\overline{\gamma}$ is higher than their value of $D(2)$. This may be due to the difficulty in computing $\sum \delta_j^\gamma$ for $\gamma>1$ and small δ_j. In any case, the computed values for the Henon attractor do not violate the Lyapunov inequality. Our values for the Zaslavskij map are larger than those reported by Bqdii and Politi, but similar to results reported elsewhere. The Lorenz attractor results are also larger than some reports, but consistent among themselves.

	$D(0)$	$D(1)$	$D(\gamma)=\gamma$	Monte Carlo	Cell Count
Baker	1.572	1.563	1.590	1.635	1.603
Henon	1.248	1.283	1.409	1.383	1.304
Zaslavskij	1.792	1.758	1.786	–	1.762
Lorenz	2.25		2.33	2.21	2.04

The following figures show $D(\gamma)$ and $P(\delta,n)$ for the three 2-D attractors that were studied.

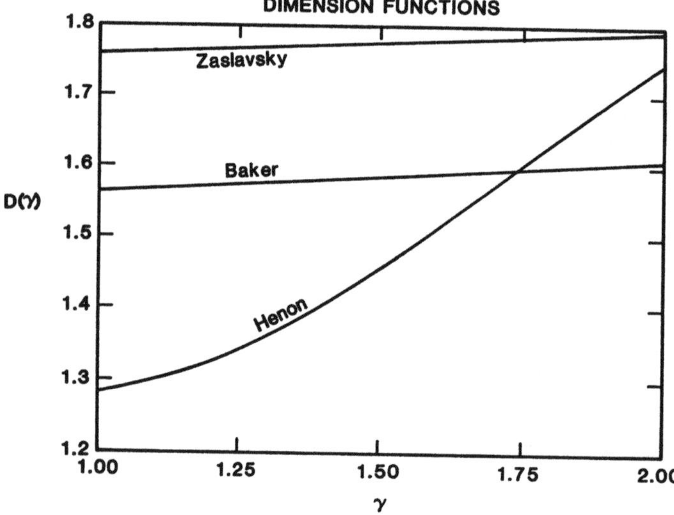

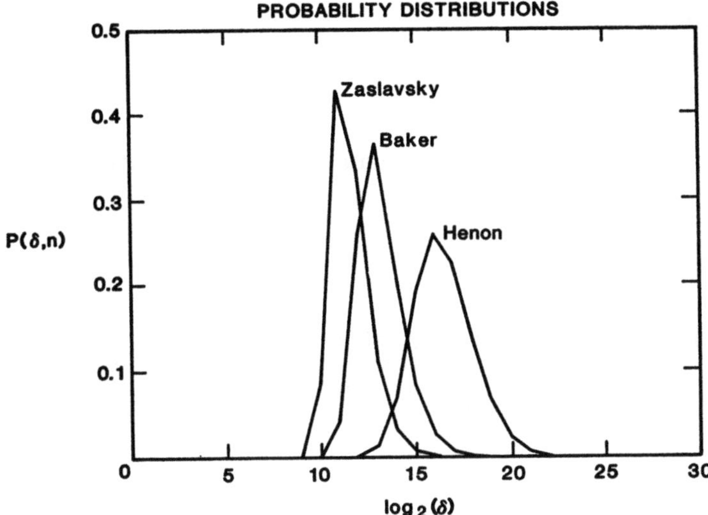

References

1. R. Badii and A. Politi, Statistical Description of Chaotic Attractors, J. of
    Statistical Physics, 40, 1985 (725-750).

2. B. Brooks, H. Brock, F. Sullivan, Diamond: A Sorting Method for Vector
    Machines, BIT, 21, 1981 (142-152).

3. J. D. Farmer, E. Ott, J. A. Yorke, The Dimension of Chaotic Attractors, Physica
    7D, 1983 (153-180).

4. P. Grassberger and I. Procaccia, Phys. Rev. Lett. 50, 1983 (346).

5. S. Haber, private communication.

6. E. Ott, E. D. Yorke, J. A. Yorke, A Scaling Law: How An Attractor's Volume
    Depends on Noise Level, preprint, 1984.

7. F. Varosi, private communication.

Using Mutual Information to Estimate Metric Entropy

A.M. Fraser

Physics Department and the Center for Nonlinear Dynamics, The University of Texas, Austin, TX 78712, USA

A technique for deriving the metric entropy of strange attractors from estimates of the mutual information in scalar time series is presented and applied to experimental and model data. The results are accurate enough to determine if a system is chaotic.

I Introduction

We have developed an algorithm that estimates information theoretic characteristics of systems from time series and a procedure for calculating the metric entropy h_μ from these estimates. The metric entropy, which we will define exactly in section III, was introduced by KOLMOGOROV [1] and refined by SINAI [2] as a classification tool for ergodic theory. Traditionally, ergodic theory[1] consists of taking simply described systems and characterizing the statistics of their trajectories. Long trajectories are not usually calculated and the work can be done analytically. The current interest in dimensions and entropies arises from an attempt to go through this process in reverse. Statistics of actual trajectories are calculated in an attempt to make statements about the simplicity of the underlying systems. This work can only be done with computers.

In this paper we will present our method of estimating h_μ and the results that the method provides for two systems. Before we describe our method we review some basic definitions from information theory and ergodic theory, and we briefly touch on some of the previous approaches to estimating h_μ.

II Information Theory

Information theory [4] provides quantitative measures of information and its transmission. The theory works in terms of the probabilities of messages. We represent general messages by lower case variables such as s or q. The variables can be almost anything from multidimensional vectors to bistable indicators of whether a coin came up heads or tails. We indicate the probability distribution of a variable q by P_q. The probability at a particular value of q, say 5, is denoted $P_q(5)$. While the variables are lower case, we use the upper case S or Q to represent the domain of the variable and its associated probability distribution P_s or P_q. If the messages are discrete and independently identically distributed, the average information contained in a message is the entropy

$$H(Q) = -\sum_q P_q(q) \log[P_q(q)] . \tag{1}$$

---

[1] PETERSON [3] is a helpful reference on ergodic theory.

We will always take logs to the base 2 so that our results will be in bits. We indicate that $H(Q)$ is a functional of P_q, not a function of q, by using capital letters as the arguments of H. If the space of messages is continuous, the expression for H becomes

$$H(Q) = -\int P_q(q)\log[P_q(q)]dq. \tag{2}$$

Again H can be calculated for almost any kind of distribution. For the distribution of an ordered pair (q,s) we get

$$H(Q,S) = -\int P_{q,s}(q,s)\log[P_{q,s}(q,s)]dqds. \tag{3}$$

For continuous variables the value of H depends on the coordinates chosen. If $s=F(q)$ and F is invertible,

$$H(S) = H(Q) -\int P_q(q)\log[|dq/ds|]dq, \tag{4}$$

where $|dq/ds|$ is the Jacobian determinant of F.

A communication system is described in terms of the messages sent and received. We denote the messages sent by s and the messages received by q. The whole system is characterized by the joint distribution $P_{s,q}$. From the $P_{s,q}$ the conditional distributions $P_{s|q}$ and $P_{q|s}$ can be derived. If one knows that the message sent was say 5, then the entropy of the received message is

$$H(Q|5) = -\int P_{q|s}(q,5)\log[P_{q|s}(q,5)]dq. \tag{5}$$

If this is averaged over all possible messages sent, one obtains

$$H(Q|S) = -\int P_{q,s}(q,s)\log[P_{q|s}(q,s)]dqds$$
$$= H(Q,S) - H(S), \tag{6}$$

which is called the <u>conditional entropy</u>. Up to this point we have not considered connections between successive messages. If these connections are considered, the entropy of an information source is defined in terms of conditional entropy. We denote by $\underline{S}_k(t)$ the space of sequences of messages of length k beginning at time t, i.e., $\{s(t),s(t+1),\ldots,s(t+k-1)\}$. If a source sends a sequence of messages and the statistics of the sequence are stationary, i.e., the origin of time makes no difference to any statistic, the <u>entropy of the source</u> is

$$\mathbf{H}(S) = \lim_{k\to\infty} H(\underline{S}_1(t+k)|\underline{S}_k(t)), \tag{7}$$

where the bold $\mathbf{H}$ indicates that connections along the sequence are being considered. There is a theorem due to SHANNON [4] which justifies using $\mathbf{H}$ to measure information. If a sequence of messages is sent from a source, then binary codes can be constructed that represent the sequence, and the average length of the code per message must be at least $\mathbf{H}$ bits per message, but the code length is not bounded above $\mathbf{H}$ by any finite amount. Of course, if the messages are independent of each other $H = \mathbf{H}$.

$H(Q|S)$ is the average uncertainty in the received message given that the sent message is known. Thus the amount that knowing the sent message reduces the uncertainty in the received message is

$$I(Q;S) = H(Q) - H(Q|S) = H(Q) + H(S) - H(Q,S) = I(S;Q)$$
$$= \int P_{q,s}(q,s)\log\{P_{q,s}(q,s)/[P_q(q)P_s(s)]\}dqds. \tag{8}$$

$I(Q;S)$ is called <u>the mutual information</u>. Shannon called it the rate because it is the rate (in bits per symbol) that information goes across a channel. It is important to notice that unlike H, I is independent of the coordinates chosen.

A channel is characterized by the conditional distribution $P_{q|s}$. Given $P_{q|s}$ and an input distribution P_s, the joint distribution is given by

$$P_{q,s}(q,s) = P_{q|s}(q,s)P_s(s). \tag{9}$$

Thus, given a channel, one can choose the distribution of input messages and thereby change the rate. The channel capacity is the maximum rate that one can obtain for a given $P_{q|s}$ by varying P_s. Shannon proved that codes exist that enable one to use any channel at a rate arbitrarily close to the channel capacity with arbitrarily small error rates, and that if one tries to use a rate higher than the channel capacity one will get faulty communication. Shannon's result justifies the definition of channel capacity and shows that mutual information is the most fundamental measure of fidelity (how closely linked an output signal is to an input signal). In the final section of [4], Shannon compares the mutual information to the more familiar R.M.S. measure of fidelity, and quantifies how much more fundamental the former is.

A generalization of mutual information which retains the property of coordinate independence but can be defined for any number of variables is the redundancy

$$R(Q_1,Q_2,\dots,Q_n) = \sum_i H(Q_i) - H(Q_1,Q_2,\dots,Q_n). \tag{10}$$

We have developed an algorithm that estimates the redundancy in joint distributions of scalar variables from a large number of samples. This is useful because it enables us to calculate several other quantities. For example the mutual information between ordered pairs (v,w) and triples (x,y,z) is

$$I(V,W;X,Y,Z) = H(V,W) + H(X,Y,Z) - H(V,W,X,Y,Z)$$

$$= R(V,W,X,Y,Z) - R(V,W) - R(X,Y,Z). \tag{11}$$

Similarly

$$I(V;W,X,Y,Z) = R(V,W,X,Y,Z) - R(W,X,Y,Z). \tag{12}$$

III <u>Metric Entropy</u>

In 1958 and 1959 Kolmogrov[1] and Sinai[2] introduced an invariant now called the K.S. entropy, or the metric entropy h_μ and used it to show that certain dynamical systems are not isomorphic to each other. The metric entropy of a system is its information production rate. To understand its definition we have to develop a few preliminary concepts. We will consider an ergodic dynamical system that moves in a continuous phase space S, the space is mapped into itself by a map M, and we suppose that there is an attractor in S that has an invariant measure that we label P_s. A discrete version of the continuous space can be obtained by applying a <u>partition</u>

$$\alpha = \{a_0,a_1,a_2,\dots,a_k\}, \tag{13}$$

where the a's are regions or <u>elements</u> which are disjoint and together cover all of S. If there is a second partition β and none of its elements b cross the borders of any of the a's, β is said to be a refinement of α. Given two partitions α and β their <u>least common refinement</u>

$$\chi = \alpha \vee \beta = \{c: c=a\cap b, \; a\in\alpha \text{ and } b\in\beta\} \tag{14}$$

is a partition whose elements are the intersections of the elements of α and β. We define the preimage of a partition α under the map M

$$M^{-1}\alpha = \{M^{-1}a_1, M^{-1}a_2, \ldots, M^{-1}a_k\} \tag{15}$$

and the least common refinement of α and its preimage

$$\alpha^2 = \alpha \vee (M^{-1}\alpha). \tag{16}$$

α^k is defined inductively as

$$\alpha^k = \alpha^{k-1} \vee (M^{1-k}\alpha). \tag{17}$$

If α is a <u>generating partition</u> then α^k becomes arbitrarily fine everywhere as $k\to\infty$.

We are now in a position to discuss entropy in dynamical systems. The <u>entropy of a partition</u> $H(\alpha)$ is

$$H(\alpha) = -\sum_a P_a(a) \log[P_a(a)]. \tag{18}$$

and the <u>entropy of the transformation M with respect to the partition α</u> is

$$h(\alpha,M) = \lim_{k\to\infty} (1/k)[H(\alpha^k)]. \tag{19}$$

Instead of thinking in terms of an ever-refined partition, one could consider the entropy of a string of messages of ever-increasing length. If we let a(t) denote the message that tells which element of the partition α the system was in after t iterations of the map M then

$$H(\alpha^k) = H(A(0), A(1), \ldots, A(k-1)) = H(A_k(0)), \tag{20}$$

where as before $A_k(t)$ denotes the space of strings of length k beginning at time t. $h(\alpha,M)$ can be expressed in the same form as (7)

$$h(\alpha,M) = \lim_{k\to\infty} H(A_1(t+k) \,|\, A_k(t)). \tag{21}$$

The value of $h(\alpha,M)$ depends on the partition α chosen, but it can never be larger than when α is a generating partition. This fact is called the Kolmogorov-Sinai theorem [1,2], and it assures that the definition of <u>metric entropy</u>

$$h_\mu(M) = \sup_\alpha h(\alpha,M) = \sup_\alpha \lim_{k\to\infty} (1/k) H(\alpha^k) \tag{22}$$

makes sense. The metric entropy "is a measure of our average uncertainty about where M moves the points of S" [3] or it is "the ratio of 'paths' to 'states'" [5]. If the sup were taken before the lim, the definition would not make sense because $H(\alpha^k)$ increases without bound as the resolution of α becomes finer.

SHAW [5] has suggested using mutual information to calculate h_μ. Notice that if α is a generating partition for the map M, then α^T is a generating partition for M^T,

$$h(\alpha^T, M^T) = Th(\alpha, M), \text{ and} \tag{23}$$

$$h_\mu(M^T) = Th_\mu(M). \tag{24}$$

If we let m, T and U be integers with $m \geq T > U$, and let $A^T_j(t)$ denote the space of message strings of length j beginning at time t which result from the map M^T and the partition α^m, and let $A^U_k(u)$ denote the space of message strings of length k beginning at time u which result from the map $M^{(U)}$ and the same α^m, then

$$h_\mu = \lim_{k \to \infty} [1/(T-U)] [H(A^T_1(t+k)|A^T_k(t)) - H(A^U_1(u+k)|A^U_k(u))]. \tag{25}$$

Recall $I(S;Q) = H(S) - H(S|Q)$. Since the partitions are the same, $A^T_1(t) = A^{(U)}_1(u)$ for all t and u, and $H(A^T_1(t)) = H(A^U_1(u))$. So

$$h_\mu = -\lim_{k \to \infty} [1/(T-U)] [I(A^T_1(t+k);A^T_k(t)) - I(A^U_1(u+k);A^U_k(u))]. \tag{26}$$

While (26) may seem inferior to (22) because the former requires a priori specification of a generating partition and the latter "finds one by itself", computationally the feature of "finding one by itself" is worthless because it takes forever.

IV   <u>Techniques for Estimating Metric Entropy from Experimental Data</u>

The other papers in this volume describe a variety of the statistics that are used to measure the simplicity of systems which can be calculated from experimental data, such as the dimension, the metric entropy, and the Lyapunov exponents. Most of the procedures for calculating the statistics are based on similar experimental situations or models. It is assumed that the system studied lies on an ergodic attractor in a multidimensional phase space S. Measuring consists of projecting S onto a scalar observable O, and recording a value x which is the value of the observable o plus a stochastic instrumental noise term. A partition of O induces a partition of S, and it is hoped that this induced partition is generating. If it is generating, all the information about a trajectory in S is available in a time series of o.[1] Most of the procedures take the measured time series and calculate statistics on them that are equivalent to estimates of the probability of particular strings of messages or symbols from particular partitions of O using histograms.

Before any other choices regarding technique are made, it is clear that there are two constraining factors in any set of experimental data. Instrumental noise will spread histograms out, and the fact that data sets have finite length will force a trade off between partition resolution and statistical sampling errors.

The metric entropy is defined in terms of the probabilities of all sequences of messages that could come from all partitions, unless a generating partition is known, in which case only the probabilities of all sequences from it are required. KRIEGER [8] has shown that

---

[1] It is this fact that lies at the heart of reconstruction of phase space by the method of delays [6,7].

generating partitions exist, but there is not a general procedure for finding them.

CRUTCHFIELD and PACKARD [9] have calculated h_μ of one-dimensional maps from time series data. They observe that (21) converges more quickly than (19) and so use the former. They have done a careful study of the dependence of $h(\alpha,M)$ on noise and choice of partition. In the presence of noise $h(\alpha,M)$ diverges as α becomes finer. So the choice of partition is critical and there seems to be no easy criterion for its choice.

SHAW [5] has calculated h_μ for one-dimensional maps and for experimental data using (26), but setting k=1. He observes that a key advantage of I over H is that in the presence of noise I does not diverge as the partition is made finer. This means that the only requirement of a partition is that its elements be as small as the scale of the noise. With such a fine partition one needs a lot of data to prevent sampling errors.

We have generalized Shaw's approach by using larger values of k in (26). Using the algorithm outlined in section V, we obtain estimates of $R(X^T_j(t))$ from which we derive estimates of $I((X^T_1(t+k);(X^T_k(t))$ by manipulations like (12). We introduce the abbreviation

$$R'_k(T) = I((X^T_1(t+k);(X^T_k(t)), \qquad (27)$$

and call $R'_k(T)$ the __marginal redundancy__ because of the way (12) is used to derive it from the redundancy. In words that are perhaps more familiar, k is the embedding dimension, $T\Delta t$ is the delay time, and Δt is the sampling time. Given a time series of scalar measurements $x(t)$, $R'_k(T)$ is the number of bits that can be predicted about $x(t+kT\Delta t)$ from a knowledge of $\{x(t),x(t+T\Delta t),x(t+2T\Delta t),...,x(t+(k-1)T\Delta t)\}$. With this notation (26) becomes

$$h_\mu = -\lim_{k\to\infty} \frac{R'_k(T_1) - R'_k(T_2)}{T_1 - T_2}. \qquad (28)$$

To estimate h_μ we plot $R'_k(T)$ against T for a range of values of k and T and look at the slopes of the curves. As k increases the slopes should approach h_μ for small values of T. We do not use the same partition for each calculation of $R'_k(T)$ because we assume that we have enough samples to see the structure of the distributions all the way down to the scale of the noise, in which case, our estimate of $R'_k(T)$ is an estimate of the limiting case of a partition with infinite resolution, i.e., an integral like (8). As we noted earlier, the integral in (8) is coordinate independent. It is the coordinate independence of mutual information and its robustness with respect to noise that make it the best information theoretic statistic for characterizing experimental systems.

V __Our Algorithm__

The algorithm we use to calculate redundancies is a generalization of an algorithm we described in a previous paper [10] that only calculates mutual information $I(S;Q)=R(S,Q)$ between two scalar variables S and Q. The generalization to more scalar variables entailed completely rewriting the computer code, but the concepts are almost identical. We will outline the algorithm for the case of two scalar variables here and indicate how it can be generalized.

The principal difficulty in calculating mutual information from experimental data is in estimating P_{sq} from histograms. If a box in the (s,q) plane of size $\Delta s \Delta q$ has N_{sq} points in it, we estimate P_{sq} to be $N_{sq}/(N_{total}\Delta s \Delta q)$ uniformly across the box. Any particular box size has advantages and liabilities. For a given number of data, larger boxes have more points; hence, the estimate of the average probability is more accurate, but the estimates of P_{sq} are too flat, which underestimates $I(S;Q)$. Smaller boxes let one follow changes in P_{sq} over short distances but allow the fluctuations that are due to small sample size to be interpreted as small scale structure in P_{sq}, which overestimates $I(S;Q)$. No single box size is best over the whole (s,q) plane.

The key feature of our algorithm is that it covers the (s,q) plane with boxes whose sizes are <u>tailored to local conditions</u>. It avoids using boxes so small that statistical fluctuations would influence the results. This means that if there are enough samples to fill out the features of the probability distribution well, the algorithm will provide an accurate estimate of the redundancy, but if there are not enough samples the algorithm will smooth out some features of the distribution, consequently underestimating the redundancy. Thus the estimates of R that our algorithm provides are lower bounds.

The first step in our algorithm is to change variables to equiprobable coordinates. Next the (s,q) plane is divided into quarters and the number of observations in each part is counted. If the distribution of counts is not flat at the 20% confidence level, each of the quarters is divided in a similar fashion. This process is continued until a flat distribution is observed. Figure 1 illustrates one such recursive sequence. At the bottom of each such sequence is a box in which the sample distribution cannot be confidently used to show that the parent distribution is not flat. As each such box is found its contribution to the estimate of the integral is calculated.

The generalization from the case of two to n scalar variables consists of subdividing the region into 2^n sub-regions at each step instead of just 4, and modifying the statistical test for flatness to the appropriate number of degrees of freedom.

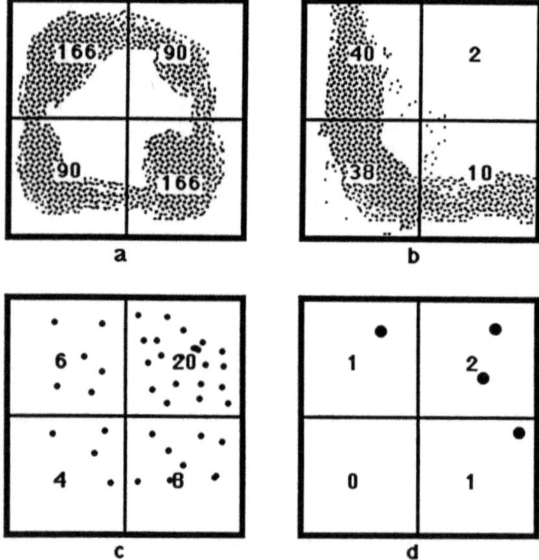

Figure 1. A recursive partitioning of the sample plane (s,q). At each step the lower left quadrant is expanded. Because no substructure is indicated in d, the lower left quadrant of c will be used as an undivided partition element.

VI Results

We have applied our procedure to experimental and numerically generated
time series data. While the results show that there are errors in our
estimates of I, the estimates of h_μ are good enough to determine whether
or not a system is chaotic. The results for experimental data from an
experiment on a Couette cell [11] and for the Rossler system [12] are
shown in Fig.2 and Fig.3, respectively. From BRANDSTATER et al. [11] we
know that the Couette data represent a quasiperiodic state, i.e. h_μ=0,
and from the calculations of WOLF et al.[13] we know that h_μ for the
Rossler system is 0.79 bits/orbit. Figures 2 and 3 correctly indicate
h_μ=0 and h_μ>0 respectively. While Fig. 3 shows a value that is 39% too
large, the more disturbing feature of both figures is that the estimates
of $R'_n(T)$ are not monotonically increasing functions of n as they should
be.

VII Conclusion

We have presented a technique for estimating the metric entropy of a
system from the time series of a scalar observable. By using mutual
information I rather than entropy H we have made our technique more
robust with respect to experimental noise. We hope to improve our
algorithm to eliminate the problems we have estimating $R(X_k)$ and $R'_k(T)$
at large values of n and to understand the limitation of our technique

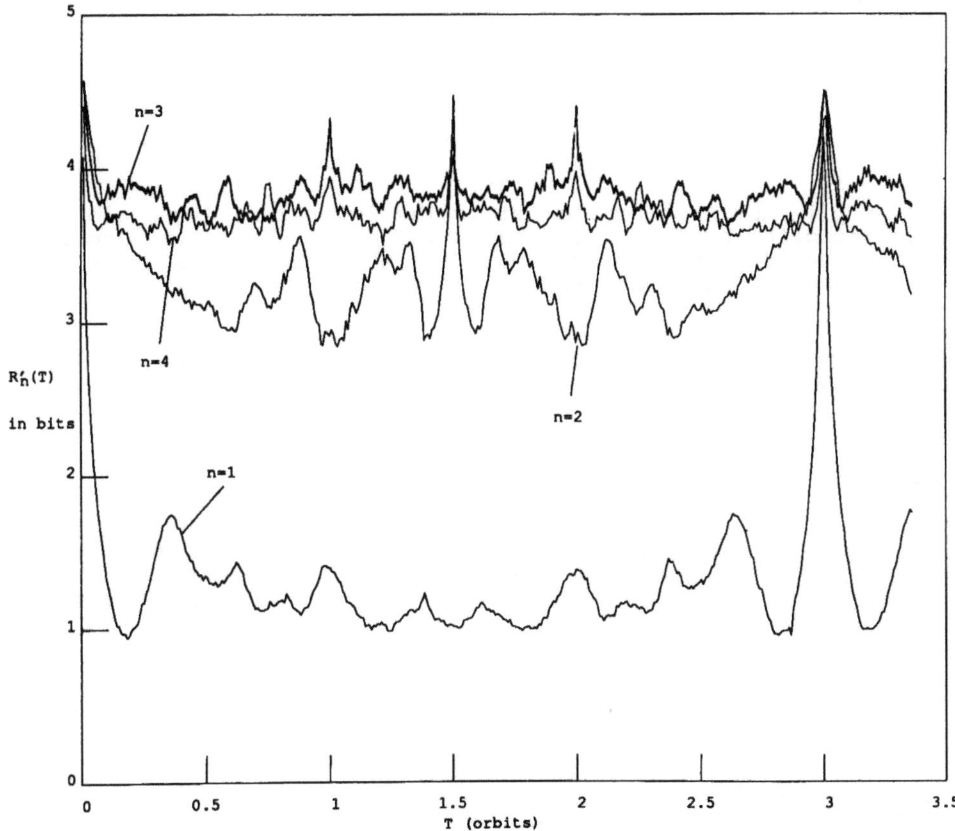

Figure 2. $R'_n(T)$ for quasiperiodic Couette data from BRANDSTATER et al.
[11]. The n=3 and n=4 curves correctly indicate h_μ=0, that is, the
average slope of $R'_n(T)$ is zero.

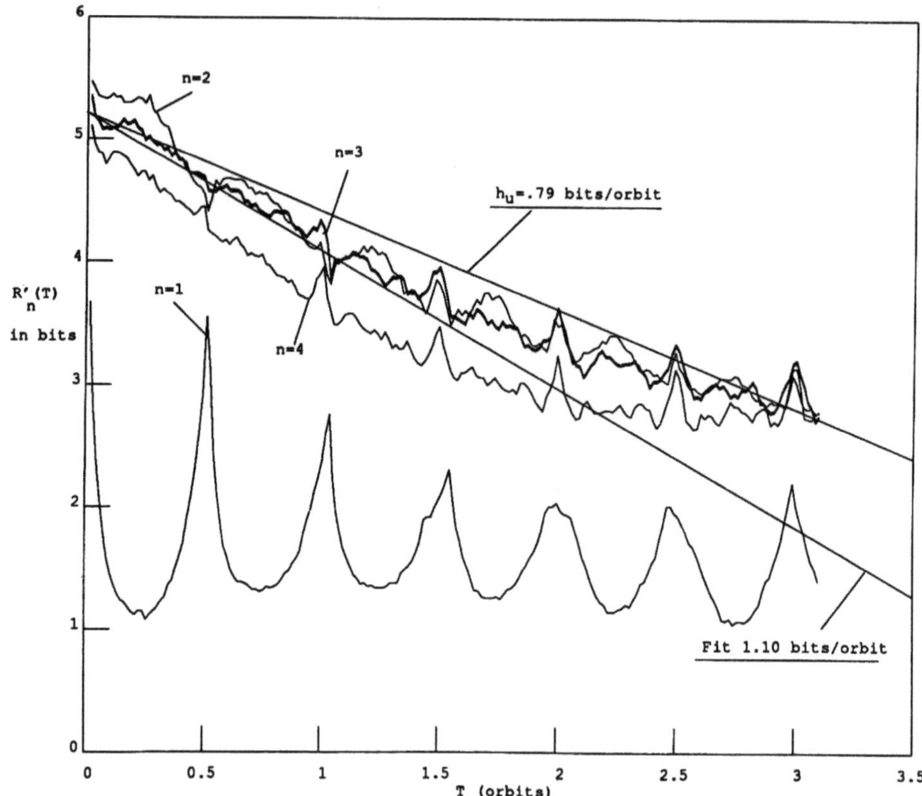

Figure 3. $R'_n(T)$ for Rossler attractor. The lower line drawn along the n=3 curve indicates $h_\mu=1.10$ bits/orbit, while the upper line indicates a correct value of $h_\mu=0.79$ bits/orbit from WOLF et al. [13]. The time series analyzed here is described in [10].

in terms of length of time series required and the sensitivity of the results to experimental noise.

We have demonstrated that the technique is useful as it stands, and we believe that with better estimates of R we can obtain very accurate measurements of h_μ. The promise of the technique justifies the effort required to perfect it.

I would like to thank Harry Swinney for guidance in this work and Anke Brandstater, Eric Kostelitch, and Laurette Tuckerman for helpful discussions. This work was supported by NSF Fluid Mechanics program grant MSM8206889.

References

1 A. N. Kolmogorov, Dokl. Akad. Nauk. SSSR 119, 861 (1958); Dokl. Akad. Nauk. SSSR 124, 754 (1959). English summary in MR 21, 2035 (1960).
2 Y. Sinai, Dokl. Akad. Nauk. SSSR 124, 768 (1959). English summary in MR 21, 2036 (1960).
3 K. Peterson, Ergodic Theory, (Cambridge University Press, Cambridge, 1983).
4 C. E. Shannon and W. Weaver, The Mathematical Theory of Communication, (University of Illinois Press, Urbana, 1949).

5 R. S. Shaw, <u>The Dripping Faucet as a Model Chaotic System</u>,
(Aerial Press, Santa Cruz, CA, 1985).

6 N. H. Packard, J. P. Crutchfield, J. D. Farmer, and R. S. Shaw,
Phys. Rev. Lett. <u>45</u>, 712 (1980).

7 F. Takens, in <u>Dynamical Systems and Turbulence, Warwick, 1980</u>,
Lecture notes in mathematics Vol. 898, ed. by D. A. Rand and L. S. Young
(Springer, Berlin-Heidelberg-New York, 1981), p. 366.

8 W. Krieger, Trans. Amer. Math. Soc. <u>149</u>, 453 (1970).

9 J. P. Crutchfield, N.H. Packard, Physica <u>7D</u>, 153 (1983).

10 A. M. Fraser and H. L. Swinney, Phys. Rev. A (1985).

11 A. Brandstater, J. Swift, H. L. Swinney, and A. Wolf, Phys. Rev.
Lett. <u>51</u>, 1442 (1983).

12 O. E. Rossler, Phys. Lett. <u>57A</u>, 397 (1976).

13 A. Wolf, J. B. Swift, H. L. Swinney and J. A. Vastano, Physica
<u>16D</u>, 285 (1985).

Part IV

Computation of Lyapunov Exponents

Intermediate Length Scale Effects
in Lyapunov Exponent Estimation

A. Wolf[1] and J.A. Vastano[2]

[1] The Cooper Union, School of Engineering, New York, NY 10003, USA
[2] Department of Physics and Center for Nonlinear Dynamics,
University of Texas, Austin, TX 78712, USA

Abstract

Algorithms for estimating Lyapunov exponents from experimental data monitor the divergence of nearby phase space orbits. These algorithms rely on the assumption that the dynamics on intermediate length scales are "close" to the dynamics on infinitesimal length scales. We have studied two one-dimensional maps in which intermediate length scale dynamics may result in inaccurate exponent estimates. This effect is found to be small enough so that the exponent estimates are still good characterizations of the systems. Similar effects are likely to be present whenever a finite quantity of data is used for Lyapunov exponent estimation.

1. Introduction

This conference has demonstrated that fractal dimension calculations can fail to detect experimental chaos. However, a reasonably accurate estimate of the largest Lyapunov exponent not only confirms the presence of chaos but also quantifies it. It is therefore essential to develop methods to accurately estimate the Lyapunov exponents of attractors defined by finite amounts of experimental data. The techniques which have been developed to date for estimating Lyapunov exponents from a single time series [1,2] are based on the assumption that by studying the evolution of finitely separated phase space orbits we are "approximately" monitoring the behavior of infinitesimally separated orbits. We present a calculation of the effect of this approximation on estimates of the largest Lyapunov exponent for two simple chaotic systems.

An n-dimensional dynamical system has n one-dimensional Lyapunov exponents. The exponents measure the long-term average exponential growth or decay rates of infinitesimal perturbations from a phase space trajectory. We give an operational definition of the largest Lyapunov exponent λ_1 for a system whose equations of motion are explicitly known: starting from a point $x(0)$ within the phase space basin of an attractor, and (almost any) initial perturbation $\delta x(0)$ in the tangent space of $x(0)$, find $x(t)$ from the equations of motion and $\delta x(t)$ from the linearized equations of motion. The largest Lyapunov exponent is then defined by

$$\lambda_1 = \lim_{t \to \infty} \frac{1}{t} \log_2 \left[\frac{||\delta x(t)||}{||\delta x(0)||} \right] \tag{1}$$

if the limit exists [3]. In an attractor defined by a finite data set, monitoring orbital divergence requires (at least) two nearby phase space orbits rather than one phase space orbit and infinitesimal perturbations away from it. It is not possible to take the limit of infinite times or probe infinitesimal length scales, so we cannot be certain that orbital divergence rates for the data set accurately reflect the underlying Lyapunov exponents. Our viewpoint is that the orbital divergence rates for a data set provide a useful characterization of the data, and we therefore define an effective Lyapunov exponent, λ_{eff}, as

$$\lambda_{eff}(\varepsilon,t) = \frac{1}{t} < \log_2 \left[\frac{||\Delta x(t)||}{||\Delta x(0)||} \right] > \;.
\tag{2}$$

The average is taken over the ensemble of all possible initial separations $\Delta x(0)$ with length ε. In the limit as ε goes to zero and t goes to infinity, λ_{eff} approaches λ_1.

2. Estimating Lyapunov Exponents

There are currently two techniques for estimating Lyapunov exponents from experimental data for a dynamical system. The first, due to WOLF *et al.* [1,4,5], estimates the exponents directly from the growth of intermediate separations [6] defined by the data. The second, which was suggested by ECKMANN and RUELLE [2; see also 7,8], involves the approximation of the product Jacobian for the system, also employing the evolution of a small number of orbital separations. With each approach there is the hope that nearby orbits - in small data sets - diverge, as predicted by the linearized equations of motion. Intuitively, we expect this to happen as the amount of data grows large.

Unfortunately, many strange attractors contain regions in which nonlinear effects for orbital divergence are important on very small length scales. For example, in the Lorenz attractor [9] such a region is the separatrix of the two lobes of the attractor. Points on opposite sides of the separatrix diverge at an enormous rate, resulting in a large spurious contribution to λ_1. In the Rossler attractor [10], nearby orbits passing through the folding region may emerge much closer than when they entered. As a result the exponent is underestimated. Avoiding the separatrix in the Lorenz attractor is not difficult, but in a noisy experimental data set that requires a high-dimensional embedding, we are not aware of a general approach to detecting and avoiding such troublesome regions.

We have studied these finite length effects for two chaotic one- dimensional maps. The maps are depicted in Fig. 1. The first map is

$$x_{n+1} = 2x_n \ (mod\, 1),
\tag{3}$$

and the second is

$$\begin{aligned}
x_{n+1} &= 2x_n & (x_n \le 0.5) \\
x_{n+1} &= 2 - 2x_n & (x_n > 0.5)
\end{aligned}
\tag{4}$$

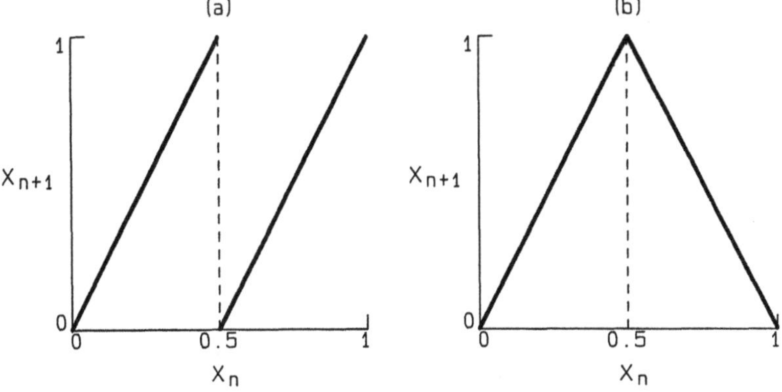

Fig. 1. Maps used in this paper: (a) the map defined by (3); (b) the map defined by (4).

Each map has a Lyapunov exponent of 1 bit/iteration because every infinitesimal separation is doubled upon application of the map. These maps are especially useful because the magnitudes of their slopes are two everywhere (except at x= 0.5) and thus their invariant probability distributions are uniform. This makes taking the average in (2) simple: each element of the ensemble of possible initial separations is equiprobable.

Suppose we iterate these maps a finite number of times and treat the output as experimental data. Most small separations double each iteration, but pairs of points split across x= 0.5 act differently. Note that map (3) contains a feature very similar to the separatrix in the Lorenz attractor. A very small separation which lies across the midpoint will be mapped to a very large separation: if the initial separation is 0.001, it will be mapped to a separation of length 0.998. It is also possible for a very large separation to be mapped to a very small separation. Map (4) mimics the Rossler attractor folding region: a separation of length ε which lies across the midpoint will be mapped to a separation with a length in the range $[0,2\varepsilon]$. Because of these intermediate length scale effects we cannot expect a finite data set from a 1-D map to give us the correct exponent of 1 bit/iteration. We now calculate the effective Lyapunov exponent for these maps.

Our calculation begins with the ensemble of all possible separations of length ε. To evaluate the effective Lyapunov exponent exactly, we need to know both the possible evolved lengths and the fraction of the ensemble which has been mapped to each of these. That is, we need to know the probability distribution of evolved lengths on the interval [0,1]. With the simple structure of our two maps, an exact recursive expression for this distribution is easily obtained. We have solved these expressions numerically.

In the case of map (3), for any number of iterations there will be a finite number of possible evolved lengths. The rule for obtaining the probability distribution can be summarized as follows: starting with an initial ensemble of separations of length ε, after one iteration, if $\varepsilon \leq 0.5$, a fraction $(1-2\varepsilon)/(1-\varepsilon)$ of the initial ensemble maps to separations of length 2ε, and a fraction $\varepsilon/(1-\varepsilon)$ maps to separations of length $(1-2\varepsilon)$. If $\varepsilon > 0.5$, all of the ensemble maps to separations of length $(2\varepsilon-1)$.

In general the initial ensemble is split into two new ensembles, each with a different effective exponential growth rate [11]. Each new ensemble is now treated as an initial ensemble as the map is iterated again. Thus we can find the probability distribution for the evolved lengths for any number of iterations of the map. This process is illustrated in Fig. 2 for two iterations. We compute the effective Lyapunov exponent from (2) by averaging the exponential growth rates weighted by their relative probabilities. Our results are presented in Fig. 3. For one iteration the estimate of the Lyapunov exponent is within 10% of the correct value for λ_1 for a

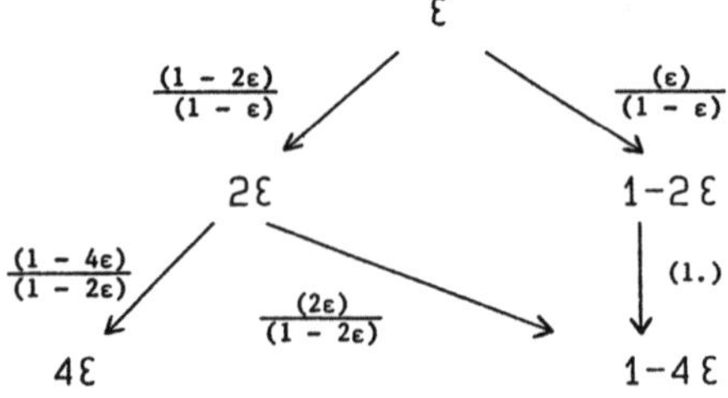

Fig. 2 The distribution of evolved lengths for two iterations of map (3). Transition probabilities for each ensemble are shown in parentheses. The initial value of ε is less than 0.25, so that neither 2ε nor 4ε is greater than 1.

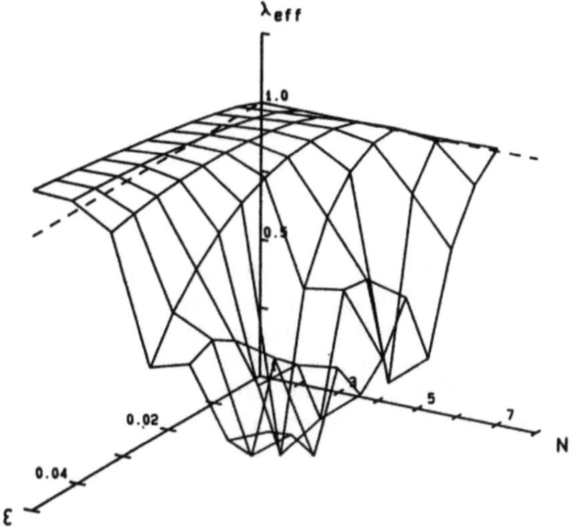

Fig. 3 The estimated exponent λ_{eff} as a function of initial length ε and number of iterations N for map (3). The dashed line indicates the true value of λ_1 is 1.0 bit/iteration.

wide range in ε. For further evolutions, there is only a small range of iterations where a good exponent estimate is obtained. The accuracy of the exponent improves as ε is made very small (in the limit ε goes to zero we must recover λ_1), but we have not shown results for ε less than 1%, since that is very often the smallest accessible (or numerically meaningful) length scale for experimental data.

The results for map (4) are presented in Fig. 4. In this case the rule for determining the probability distribution is much more complicated, since the behavior of a separation straddling the midpoint is determined by the positions of the points which define the separation. The evolution rule for the probability distribution is

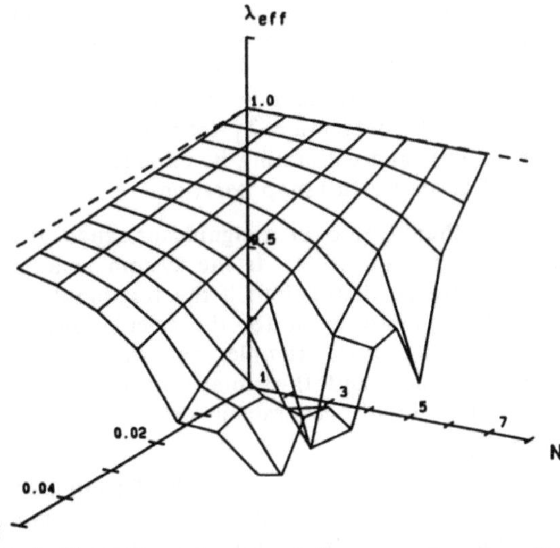

Fig. 4 The estimated exponent λ_{eff} as a function of initial length ε and number of iterations N for map (4). The dashed line indicates the true value of λ_1 is 1.0 bit/iteration.

$$p_{i+1}(x) = \left[\frac{2 - 2x}{2 - x}\right] p_i(\frac{x}{2}) + \int_{\frac{x}{2}}^{1 - \frac{x}{2}} \left[\frac{1}{2 - 2x'}\right] p_i(x')dx'$$

where the initial distribution is

$$p_0(x) = \delta(x - \varepsilon)$$

Rather than solve this equation for the probability distribution, we evolved 500,000 randomly chosen initial separations of length ε to approximate the distribution used to calculate λ_{eff}. The results show that our underestimatation of the exponent increases as ε increases, but the problem is much less severe than was found in map (3). The broad plateau in Fig. 4 means that an accurate estimate of λ_1 for this map, or for a continuous system in 3 (or more) dimensions containing this map, should not be too hard to obtain. The plateaus in Figs. 3 and 4 occur for the intermediate length scales over which the dynamics is a good approximation of dynamics on infinitesimal length scales.

3. Summary and Conclusions

Our experience [1] has been that estimating the first Lyapunov exponent for a Lorenz-like attractor is significantly more difficult than for a Rossler-like attractor. In this work we find that this effect is due to the different degree to which each system's intermediate length scale behavior corresponds to its infinitesimal length scale behavior. In computing such dynamical diagnostics as fractal dimensions, metric entropies, and Lyapunov exponents it is well known [12] that the amount of data required for a fixed level of accuracy rises exponentially fast with the fractal dimension of the data set. The finite length scale effect we have considered here places an additional constraint on the size of experimental data sets.

There are two concluding points we would like to make.

(1) In a one-dimensional map the smallest time step is a single iteration. In a continuous system containing a chaotic map, the corresponding time interval is one (mean) orbital period. Typically, an experimental time series provides us with many points per mean orbital period so that it is possible to "go through the map" much more slowly than one iteration at a time. It is possible that the worst of the erroneous contributions may be avoided this way. Increasing the frequency of replacement may introduce other errors, however [1]. Attempts to determine λ_1 from experimental data by forming the underlying $1-D$ map (if any) and estimating its exponent cannot avoid the problems arising from time steps of at least a full map iteration.

(2) The errors in our estimates were less than 10% over a significant range of ε. This is more accuracy than we need to answer the question: Is the system chaotic? This should be contrasted to estimates of the fractal dimension: if the fractional part of the estimated dimension is larger than the absolute error for the calculation, the result does not prove that the system is chaotic. For example, an attractor with dimension $3.2 \pm 10\%$ may be of integer dimension and thus may not be chaotic. Lyapunov exponent estimates do not have to be highly accurate in order to provide useful information about an experimental system.

We believe that Lyapunov exponent estimation is crucial to the task of confirming and quantifying chaos in experimental data. We also believe that accurate exponent estimation is often possible for experimental systems. It is important, however, to examine the basic assumptions of exponent estimation, and know the limitations of the techniques.

Acknowledgements

The authors would like to thank Gottfried Mayer-Kress and the Center for Nonlinear Studies at Los Alamos for sponsoring the conference "Dimensions and Entropies in Chaotic Systems" at the Pecos River Conference Center, where an ideal work environment brought this calculation into being. The research of John Vastano, an Exxon Fellow at the University of Texas, is partially supported by the Department of Energy Office of Basic Energy Sciences.

References

1. A. Wolf, J. Swift, H. L. Swinney, and J. A. Vastano, *Physica* **16D**, 285 (1985).

2. J. P. Eckmann and D. Ruelle, *Rev. Mod. Phys.* **57**, 617 (1985).

3. More precisely we should say that the long-term growth rate of $\delta x(t)$ is λ_1. For the $1-D$ map these definitions are equivalent, but in higher dimensions the transient behavior of the vector $\delta x(t)$ will almost always depend on other Lyapunov exponents.

4. A. Wolf, in *Nonlinear Science: Theory and Applications*, ed. by A. Holden, (Manchester University Press, 1986).

5. A. Brandstater, J. Swift, H. L. Swinney, A. Wolf, J. D. Farmer, E. Jen, and J. P. Crutchfield, *Phys. Rev. Lett.* **51**, 1442 (1983).

6. By intermediate length scales we mean length scales above the noise level but small compared to the overall size of the attractor. We recommend either imposing a small-distance spatial cutoff throughout exponent calculations, low-pass filtering the data, or repeating calculations with the least significant data bit(s) truncated, to test the stability of exponent values.

7. M. Sano and Y. Sawada, *Phys. Rev. Lett.* **55**, 1082 (1985).

8. J. A. Vastano and E. J. Kostelich, in *Dimensions and Entropies in Chaotic Systems - Quantification of Complex Behavior*, ed. by G. Meyer-Kress (Springer, 1986).

9. E. N. Lorenz, *J. Atmos. Sci.* **20**, 130 (1973).

10. O. E. Rossler, *Phys. Lett.* **57A**, 397 (1976).

11. It would appear that after n iterations there could be 2^n possible ensembles. However, the algebra is such that there are only two possible ensembles after any number of iterations. This is shown in Fig. 2.

12. H. S. Greenside, A. Wolf, J. Swift, and T. Pignataro, *Phys. Rev.* **A25**, 3453 (1982).

Comparison of Algorithms for Determining Lyapunov Exponents from Experimental Data

J.A. Vastano and E.J. Kostelich

Department of Physics and Center for Nonlinear Dynamics,
University of Texas, Austin, TX 78712, USA

Abstract

Two methods for estimating the Lyapunov exponents of attractors reconstructed from a time series are compared. A method due to Wolf *et al.* for computing the largest Lyapunov exponent λ_1 is found to be robust with reasonable changes in input parameters. In contrast, a least-squares method suggested by Eckmann and Ruelle yields estimates for the Lyapunov exponents that vary considerably depending on the embedding dimension of the attractor. It appears that only the Wolf algorithm is suitable for the analysis of experimental data.

1. Introduction

A fundamental problem in the analysis of possibly chaotic systems is the extraction of the Lyapunov exponents of an attractor reconstructed from a time series of experimental measurements. Recently WOLF et al. [1] have described an algorithm to extract the largest one or two Lyapunov exponents from time series data. Another method, proposed by ECKMANN and RUELLE [2], in principle permits the estimation of all the positive Lyapunov exponents. The basic idea in both methods is to measure the rates of separation of points in small ε neighborhoods of the attractor. SANO and SAWADA [3] have reported some numerical results for the Eckmann-Ruelle method and argue that it permits the recovery of "all the Lyapunov exponents with great ease"; we find that this is not the case.

In this paper we are concerned with the robustness of the two methods with respect to changes in parameters that are common to both: the number of data points, the distance ε between nearby points, the time step Δt, and the embedding dimension of the attractor. We sketch the definition and numerical computation of Lyapunov exponents in the remainder of this section. The next section describes the Wolf and Eckmann-Ruelle algorithms in more detail. In the last section we compare the results of the two methods using time series of a single variable generated by the HENON [4] and ROSSLER [5] equations.

The first step towards computing Lyapunov exponents, whichever method is used, is to construct the attractor from the experimental data. TAKENS [6] discusses the theory that underlies this approach. We begin with a time series $\{\xi_t\}$ of measurements, from which we construct a set of points of the form

$$\mathbf{x}_i = (\xi(t_i), \xi(t_i+\tau), \ldots, \xi(t_i+(m-1)\tau)), \tag{1}$$

where τ is the time delay. One supposes that the behavior of the experiment can be described by a finite dimensional attractor whose dynamical properties can be recovered by this method of time-delay reconstruction if the embedding dimension m is large enough. In principle, the choice of τ is arbitrary. Recently, FRASER and SWINNEY [7] developed a criterion to choose τ which we have used in our numerical tests below, so we do not consider the dependence of these methods on τ.

The Lyapunov exponents measure the average rate of separation of nearby points on an attractor. They were defined originally by OSELEDEC [8]; a sketch of the mathematical theory is given in [2]. Let $f : \mathbf{R}^n \to \mathbf{R}^n$ be differentiable, and suppose that A is an attracting limit set for the map $x_{k+1} = f(x_k)$. We let f^k denote the composition of f with itself k times, and $Df^k(x)$ denote the $n \times n$ matrix of partial derivatives of f^k evaluated at x. Let $a_i(k,x)$ be the modulus of the ith eigenvalue of $Df^k(x)$, ordered so that $a_1(k,x) \geq \cdots \geq a_n(k,x)$. We define the ith Lyapunov exponent $\lambda_i(x)$ as

$$\lambda_i = \lim_{k \to \infty} \frac{1}{k} \log_2 a_i(k,x), \quad i = 1, \cdots, n. \tag{2}$$

Under suitable assumptions [8], this limit exists and is the same for typical points x on A. A similar definition can be given for the attracting limit set of an autonomous system of ordinary differential equations $\dot{x} = f(x)$, where the index k is replaced by time t.

The Lyapunov exponents quantify the local dynamical behavior on the attractor, because they measure the sensitivity of trajectories to small changes in initial conditions. To see this, let $B_\varepsilon(x)$ be a ball of radius ε about a point x of A, where ε is very small. It is mapped after k iterations to an ellipsoidal set the length of whose ith semi-axis on the average is approximately $2^{k\lambda_i}\varepsilon$. The ball B_ε may be contained in a subspace of $\mathbf{R}^n$. For example, a line segment of length ε connecting two nearby points on the attractor (a one-dimensional ball) typically is stretched into a curve whose length is approximately $2^{k\lambda_1}\varepsilon$ after k iterations. More generally, an m dimensional volume element grows by an average factor of $\lambda_1 + \cdots + \lambda_m$ at each iteration. If $\lambda_1 > 0$, then the attractor is *chaotic*, that is, very small changes in initial conditions grow exponentially quickly, at least for a short time.

BENETTIN et al. [9] have devised a numerical algorithm to compute the Lyapunov exponents of an attractor produced by a given map. It begins with an orthonormal basis $\{ u_1^{(0)}, \cdots, u_n^{(0)} \}$. At the kth stage one computes

$$\overline{u}_j^{(k+1)} = Df(x_k)u_j^{(k)}, \quad j = 1, 2, \cdots, n. \tag{3}$$

Gram-Schmidt orthonormalization is done as follows. Let

$$\delta_1^{(k+1)} = \| \overline{u}_1^{(k+1)} \| \tag{4}$$

so that

$$u_1^{(k+1)} = \overline{u}_1^{(k+1)} / \delta_1^{(k+1)}. \tag{5}$$

At subsequent steps we let

$$\delta_m = \| \overline{u}_m - \sum_{i=1}^{m-1} (u_i, \overline{u}_m) u_i \|, \tag{6}$$

where (u,v) is the usual dot product (we omit the superscripts $(k+1)$ for clarity). Here δ_m is the norm of the mth vector after orthogonalization but before normalization. The mth Lyapunov exponent is evaluated numerically as

$$\lambda_m = \frac{1}{k} \sum_{i=1}^{k} \log_2 \delta_m^{(i)} \tag{7}$$

for large k. A similar procedure is applied to the flow $\dot{x} = f(x)$, except that the variational equations

$$\dot{u}_i = Df(x)u_i \tag{8}$$

are integrated from t to $t + \Delta t$ beginning with an orthonormal basis $\{ u_i \}$ at t. The orthonormalization procedure is applied as above, but the factor $1/k$ in (7) is replaced by $1/(k\,\Delta t)$.

The Wolf algorithm follows a pair of nearby points on the attractor to estimate δ_1 at each time step, from which the the largest Lyapunov exponent λ_1 can be calculated using (7). In the approach suggested by Eckmann and Ruelle, one follows at least $n+1$ points, from which a least-squares estimate of the Jacobian $Df(x)$ is computed; that is, one attempts to estimate the variational equations (8) at every point. We discuss the implementation of these algorithms in the next section.

2. Algorithm descriptions

Fig. 1 is a schematic illustration of the Wolf procedure to calculate λ_1. One begins with the first data point $y(t_0)$ and its nearest neighbor $z_0(t_0)$, which are a distance L_0 apart. The two points are evolved by time steps Δt until the distance L'_0 between them exceeds some value ε. The evolved first data point $y(t_1)$ is retained, and a new neighbor $z_1(t_1)$ is sought such that the distance

$$L_1 = \| y(t_1) - z_1(t_1) \| \tag{9}$$

is again less than ε and such that $z_1(t_1)$ lies as nearly as possible in the same direction from $y(t_1)$ as $z_0(t_1)$.

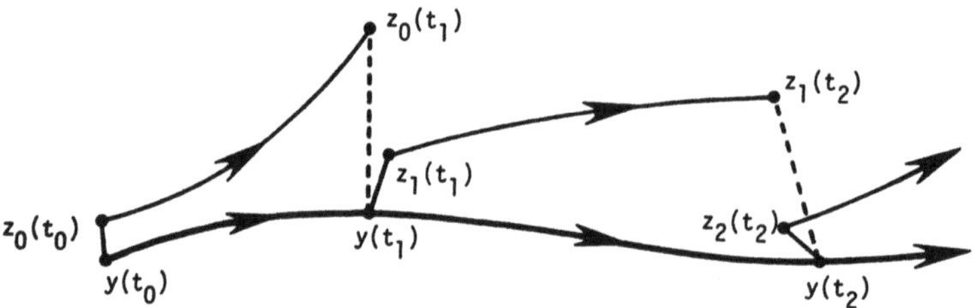

Figure 1. Schematic representation of the Wolf algorithm to compute λ_1.

The procedure continues until the fiducial trajectory y has been followed to the end of the time series. The largest Lyapunov exponent of the attractor is estimated as

$$\lambda_1 = \frac{1}{N\,\Delta t} \sum_{i=0}^{M-1} \log_2 \frac{L'_i}{L_i} \tag{10}$$

where M is the number of replacement steps and N is the total number of time steps that the fiducial trajectory y has been followed.

Every replacement point should lie in the same direction as the old one, but compromises are necessary with a finite data sample. The search for replacement points initially is confined to a cone of angular width θ and height ε about $y(t)$; typically, θ is set to $\pi/9$. The value of θ is increased as necessary until a replacement can be found. If this does not succeed, then we take the nearest neighbor to y, irrespective of ε. It is shown in [1] that the orientation errors generally result in small errors in the estimation of λ_1. Thus, the results usually are insensitive to the choice of θ.

The implementation of the Wolf algorithm is straightforward. As we demonstrate in the next section, it is relatively insensitive to reasonable changes in the search radius ε and the evolution time step Δt before point replacements are attempted. Its data requirements also are modest; only a few thousand attractor points are needed to estimate λ_1 to within 10 percent of the true value when the attractor is less than three dimensional.

Unless λ_1 is very small, the Wolf method is useful for determining whether the observed time series is chaotic. Often, however, one is interested in more than just the first Lyapunov exponent. The algorithm has been extended to measure $\lambda_1 + \lambda_2$, but the implementation is more complex; it must follow triangles of points and preserve orientation with respect to the triangle which choosing replacements [1]. This approach becomes unwieldy for more than two positive Lyapunov exponents.

As we stated in the Introduction, Eckmann and Ruelle have suggested a method to estimate the variational equations, which in principle should give all the positive Lyapunov exponents of the attractor. The algorithm is the same regardless of the number and is illustrated schematically in Fig. 2. One begins with a fiducial trajectory $y(t)$ and at least n additional points $z_i(t)$ which fall within ε of $y(t)$. The solution curves through each point are followed for a time Δt. Since $z_i(t+\Delta_i)$ and $z_i(t)$ are known along with $y(t+\Delta t)$ and $y(t)$, one can make a least-squares estimate of the spatial derivatives $Df(y(t))$ of the flow using $y(t)$ and at least n points z_i. At each time step, one takes *all* the points in the data sample in an ε ball about $y(t)$ to obtain the best possible fit, continuing the process until the data are exhausted. Notice that unlike the Wolf algorithm, it is not necessary to consider the orientation of replacement points. With a least-squares estimate of the Jacobian at $y(t)$, the variational equations (8) can be integrated, using Benettin's algorithm to compute the Lyapunov exponents. In the remainder of this paper, we call this the *Jacobian* method.

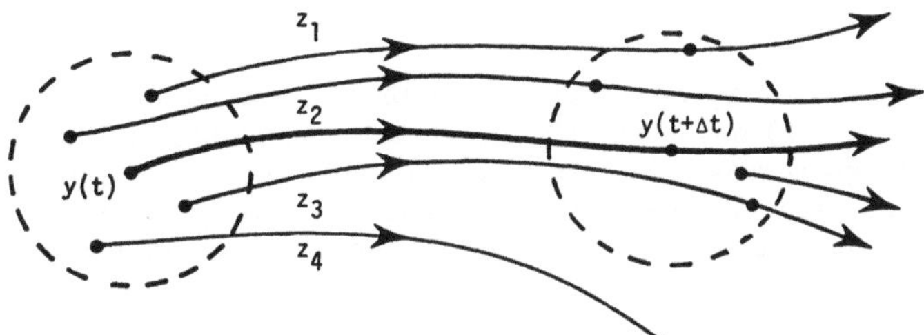

Figure 2. Schematic representation of the Eckmann-Ruelle method.

3. Comparison of methods

In this section we test the robustness of the Wolf and Jacobian methods with respect to changes in the ball size ε, the time step Δt, the number N of attractor points, and the embedding dimension m. Our objective is to determine how reliable they would be in the analysis of experimental data. We have chosen to study the effects of these parameters because they are varied in the analysis of a data set and because they are common to both algorithms.

We tested both methods using time series data generated from the Rossler equations

$$\dot{x} = -(y+z)$$
$$\dot{y} = x + 0.15y$$
$$\dot{z} = 0.2 + z(x-10) \tag{11}$$

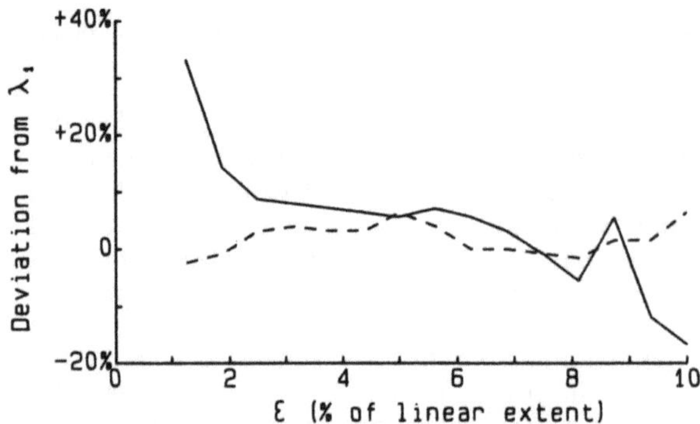

Figure 3. Dependence of the estimates of λ_1 on the ball size ε, expressed as a percentage of the horizontal (x) extent of the reconstructed attractor. The dotted line is the estimate obtained by the Wolf algorithm; the solid line, the Jacobian method. A time series of x coordinate values from the Rossler equations was used; the time step Δt was 0.5. The numerically computed value of λ_1 for the Rossler equations (11) is 0.13 bits/sec.

integrated using a fourth order Runge Kutta method with a fixed time step $\Delta t = 0.01$. The time series consisted of x coordinate values taken every 50 time steps. The attractor was reconstructed from the time series using a delay $\tau = 1.0$, which is about 1/6 of a mean orbital period.

Fig. 3 illustrates the effects of changes in the ball size ε on the estimates of λ_1. A time series of 8192 points was used. The estimates do not change much as ε is varied over a relatively large interval.

Fig. 4 shows how the estimates of λ_1 vary with the number of data points. As before, a time series generated by (11) was used. The results are poor with only 1024 points, but the Wolf algorithm gives reasonable estimates with 4096 points. Similar results are found for the Henon map. It appears that the data requirements of both algorithms are generally modest for these low-dimensional examples; the estimates of λ_1 are accurate to within 10 percent with only a few thousand data points.

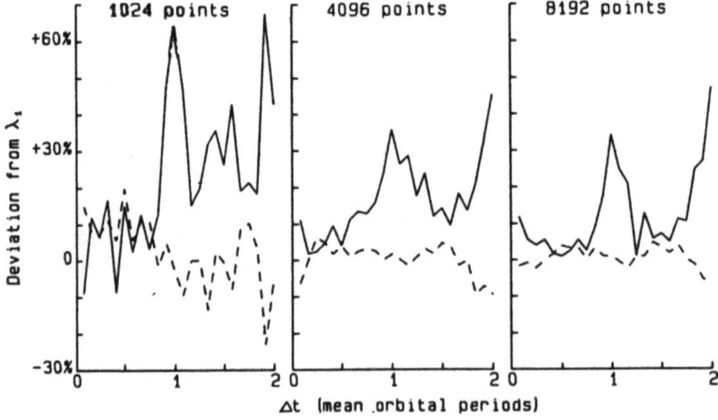

Figure 4. Dependence of the estimates of λ_1 on the number of data points and time step Δt. Subsets of the time series in Fig. 3 were used. The ball size ε was fixed at 4% of the x-axis extent of the reconstructed attractor.

Based on these tests, the algorithms appear to give similar results. In fact, the Jacobian method seems preferable because of its simplicity and its potential to recover more than one positive Lyapunov exponent. However, this is not the end of the story, for we must compare the estimates of λ_1 as a function of the embedding dimension of the attractor. This is a crucial test, because the embedding dimension is not known *a priori* in the analysis of experimental data; as we noted in the introduction, we assume that the behavior of the experiment is governed by a finite dimensional attractor whose Lyapunov exponents can be recovered if the embedding dimension m is large enough. If these assumptions are valid, then the estimates of λ_1 from the Wolf and Jacobian methods should approach a limit as a function of m.

Using a time series of 65,536 points generated as above, we reconstructed the Rossler attractor in $m=3$ and $m=6$ dimensions and computed estimates λ_i^{est} of the nonnegative Lyapunov exponents. The results for the Jacobian technique are illustrated in Fig. 5. The value of λ_i^{est} changes dramatically with the embedding dimension. Although the Rossler attractor has only one positive Lyapunov exponent, $\lambda_1 \approx 0.13$ bits/sec, the estimates of the Lyapunov exponents computed by the Jacobian method include *two* positive values. In fact, $\lambda_1^{est} > \lambda_1$. We obtain similar results whenever $m > 3$; only in the three-dimensional reconstruction does the method produce an accurate estimate of λ_1. The Wolf algorithm yields an accurate estimate of λ_1 in each reconstruction.

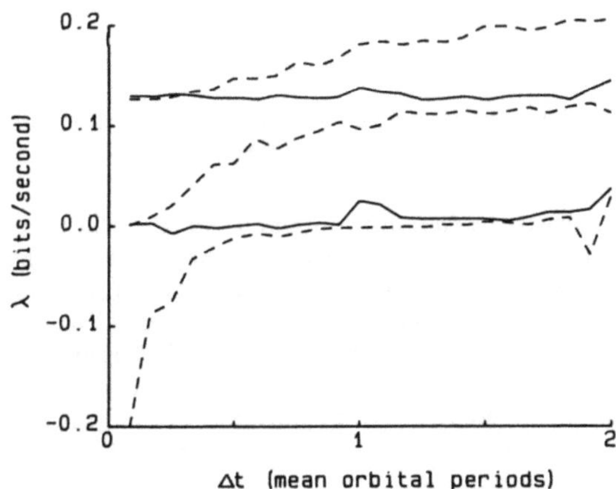

Figure 5. Embedding dimension dependence of the Jacobian method for estimating the Lyapunov exponents of the Rossler attractor. The attractor was reconstructed from a time series of 65,536 points. The solid lines are for the 3 dimensional reconstruction; dotted lines, 6 dimensional. The ball size ε was 2.5% of the x extent of the original Rossler attractor (so $\varepsilon \approx 0.8$).

The same thing happens when the attractor of the Henon map

$$x_{n+1} = 1 - 1.4x_n^2 + 0.3y_n$$
$$y_{n+1} = x_n \tag{12}$$

is reconstructed from an x-coordinate time series consisting of 32,768 points, as shown in Table 1. (For completeness, all the estimated Lyapunov exponents are shown.) The ε ball size was 2% of the x extent of the original attractor (so $\varepsilon \approx 0.04$). In this case, a two-dimensional reconstruction produces the same attractor as the original map, and both algorithms give good results for λ_1. However, as the embedding dimension is increased, the Jacobian method finds two positive Lyapunov exponents where $\lambda_1^{est} > \lambda_1$ as before. Once again, the Wolf algorithm gives an accurate estimate of λ_1 in all the reconstructions.

These results suggest that there are serious numerical difficulties associated with the Jacobian method. We conjecture that they are due to the fact that the vectors $\{z_i - y\}$ typically span only a subspace of the entire space. Suppose that the attractor has p nonnegative Lyapunov exponents, and the reconstruction is in an m dimensional space where $m > p$. The points in a small neighborhood of the reconstructed attractor tend to lie on a p dimensional plane which is tangent to the unstable manifold; hence, the Jacobian matrix is not well defined. ECKMANN and RUELLE [2] have recognized this possibility, but they have conjectured that the method nevertheless ought to be able to measure the rate of expansion in the direction of the unstable manifold (that is, the positive Lyapunov exponents should be recoverable). However, we believe that the numerical estimates of the $m \times m$ Jacobian matrix are vulnerable to roundoff errors and global effects like the curvature of the embedded attractor.

m	λ_1^{est}	λ_2^{est}	λ_3^{est}	λ_4^{est}	λ_5^{est}	λ_1^{Wolf}
2	0.61	-2.30				0.60
3	1.18	0.56	-2.39			0.60
4	1.24	0.57	-1.90	-2.46		0.61
5	1.24	0.57	-0.93	-1.66	-2.48	0.61

Table 1. Embedding dimension dependence for estimates of the Lyapunov exponents of the Henon attractor. The actual values are $\lambda_1 \approx 0.603$ bits/iteration, $\lambda_2 \approx -2.34$ bits/iteration.

In our numerical tests the most negative exponent obtained by the Jacobian method is approximately the same as the most negative Lyapunov exponent of the original attractor. As we stated above, however, one expects the Jacobian method to be able to estimate only the positive Lyapunov exponents of the attractor. Yet, we find that the estimate of the most positive exponent becomes considerably larger than the most positive exponent of the original attractor as the embedding dimension m is increased. The other values λ_i^{est} appear to fall somewhere in between the largest and smallest Lyapunov exponents of the original attractor. The relationship between the estimated and true exponents is unclear.

4. Conclusion

Two methods have been proposed to calculate the positive Lyapunov exponents from experimental data. An algorithm due to Wolf *et al.* follows the separation between nearby pairs of points on the attractor to estimate the largest Lyapunov exponent λ_1. Another approach, suggested by Eckmann and Ruelle, is to follow groups of nearby points to compute a least-squares estimate of the Jacobian at each point, which is then used to integrate the variational equations, from which the Lyapunov exponents are calculated. The results of the Jacobian method depend strongly on the embedding dimension of the reconstructed attractor. Unless a criterion can be developed to choose the embedding dimension, the Jacobian method does not appear useful for the analysis of experimental data.

Acknowledgment

The authors thank Harry L. Swinney and Alan Wolf for helpful discussions. This research was supported by the Department of Energy Office of Basic Energy Sciences. J. Vastano acknowledges the support of an Exxon Fellowship.

References

1. A. Wolf, J. Swift, H. L. Swinney and J. Vastano, *Physica* **16D**, 285 (1985).

2. J. P. Eckmann and D. Ruelle, *Rev. Mod. Phys.* **57**, 617 (1985).

3. M. Sano and Y. Sawada, *Phys. Rev. Lett.* **55**, 1082 (1985).

4. M. Henon, *Comm. Math. Phys.* **50**, 69 (1976).

5. O. E. Rossler, *Phys. Lett.* **57A**, 397 (1976).

6. F. Takens, in *Proc. on Dynamical Systems and Turbulence*, Springer Lecture Notes in Mathematics, no. 898 (1980).

7. A. Fraser and H. L. Swinney, *Phys. Rev. A*, to appear.

8. V. I. Oseledec, *Trudy Mosk. Mat. Obsc.* **19**, 179 [Moscow Math. Soc. **19**, 197 (1968)].

A Measure of Chaos
for Open Flows

R.J. Deissler[1] *and K. Kaneko*[2]

Center for Nonlinear Studies, MS B258, Los Alamos National Laboratory,
Los Alamos, NM 87545, USA

Since the subject matter of the presentation given at the conference is (or
will be) well represented elsewhere [1-3], here we just give a brief account.
Two systems were studied:

1) The time-dependent generalized Ginzburg-Landau equation [2,3],

$$\frac{\partial \psi}{\partial t} = a\,\psi - v_g\,\frac{\partial \psi}{\partial x} + b\frac{\partial^2 \psi}{\partial x^2} - c\,\mid \psi \mid^2 \psi \tag{1}$$

where the dependent variable $\psi(x,t)$ is in general complex; a, b, and c are
constants which are in general complex; and v_g is the group velocity. The
term with the first order spatial derivative is a convective term which is
responsible for the "mean flow".

2) A system of coupled logistic maps [4],

$$X_{n+1}^{(i)} = (1-d)f\,(X_n^{(i)}) + d\,\{\alpha f\,(X_n^{(i-1)}) + (1-\alpha)f\,(X_n^{(i+1)})\} \tag{2}$$

where $f(X)=1-aX^2$ is the logistic map and $i=1,2,...,N$. n and i are
integers representing discrete time and space variables resp.

Both these systems were studied in the presence of low-level external noise
under conditions when the equilibrium or fixed point solutions were convectively (ie. spatially) unstable (eg. see [2]). Under these conditions and with a
fixed boundary condition at the left boundary, perturbations are amplified and
convected downstream (ie. in the $+x$ direction). Therefore the noise near the
left boundary is spatially amplified, causing the fluctuations to be larger for
larger values of x. Because of the nonlinearity, the fluctuations saturate at
some spatial point producing a structure. In a region sufficiently far downstream, the structure changes in a chaotic fashion with time.

---

[1] Also at: Physics Department, 307 Nat Sci II, University of California at Santa Cruz,
Santa Cruz, CA 95064
[2] Permanent address: Institute of Physics, College of Arts and Sciences, University of
Tokyo, Komaba, Meguro, Tokyo 153, Japan

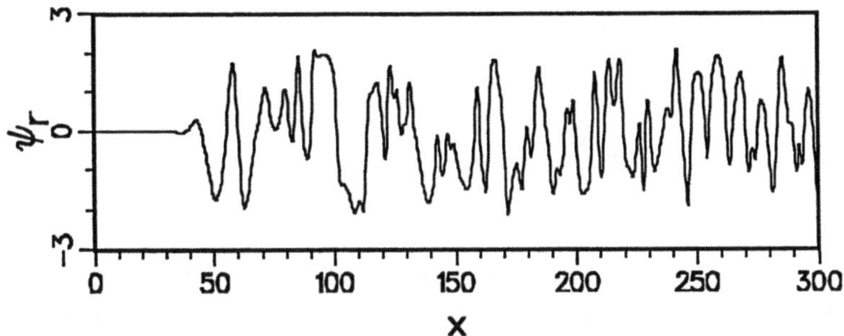

Figure 1 Plot of ψ_r [for Ginzburg-Landau equation (1)] as a function of x for a given t ($t=400$) after the system has reached a statistically steady state. The initial state was $\psi=0$ and noise is introduced into the system by adding, at each time step, random numbers uniformly distributed between $-r$ and r to ψ_r and ψ_i at all grid points except the boundary points. Second order Runge-Kutta is used in the time differencing (with $\Delta t = .01$) and fourth order differencing is used in the space differencing (with $\Delta x = .3$) except at the grid points adjacent to the boundaries where second order differencing is used. The parameter values are $a=2$, $v=6$, $b_r=1$, $b_i=-1$, $c_r=.5$, and $c_i=1$. The noise level is $r=10^{-7}$.

For the Ginzburg-Landau equation (1) this behavior may be seen in fig. 1. The noise near the left boundary is spatially and selectively amplified forming spatially growing waves, which saturate at some spatial point forming the observed structure. This structure changes in a chaotic fashion with time. If the usual Liapunov exponents are calculated, we find that there are no positive Liapunov exponents even though the flow appears to be chaotic. This occurs since perturbations are convected downstream out through the right boundary. Similar behavior occurs for the coupled map (2).

The question is: How do we define a measure for this chaos? In ref. [1] a measure for this chaos -- a velocity-dependent Liapunov exponent -- was defined. For the Ginzburg-Landau equation (1) the velocity-dependent Liapunov exponent may be defined as follows:

$$\lambda(v;x_1,x_2) = \lim_{t\to\infty} \frac{1}{t}\ln\left[\frac{\varsigma(v,x_1,x_2,t)}{\varsigma(v,x_1,x_2,0)}\right] \tag{3}$$

where

$$\varsigma(v,x_1,x_2,t) = \left[\int_{x_1+vt}^{x_2+vt} |\delta\psi(x,t)|^2 dx\right]^{1/2}$$

Here $\delta\psi$ is an infinitesimal perturbation about the state ψ initially localized within the region $\{x_1,x_2\}$ and v refers to the velocity of the frame of reference from which the system is observed. As t increases the region $\{x_1+vt,x_2+vt\}$ moves downstream with velocity v. $|x_2-x_1|$ is usually taken sufficiently large such that $\lambda(v)$ is independent of x_1 and x_2. For $v>0$

the system must be extended in the $+x$ direction. The perturbation $\delta\psi$ satisfies the following equation:

$$\frac{\partial \delta\psi}{\partial t} = a\,\delta\psi - v_g\frac{\partial \delta\psi}{\partial x} + b\frac{\partial^2 \delta\psi}{\partial x^2} - 2c\,|\psi|^2\delta\psi - c\,\psi^2\delta\psi* \qquad (4)$$

Equation (3) essentially says the following: Instead of calculating the Euclidean distance between two nearby trajectories for the whole system, calculate the Euclidean distance between two nearby trajectories for the given spatial region $\{x_1+vt,x_2+vt\}$. If the region moves at a small velocity, the perturbation initially localized within this region will "outrun" the region and the velocity-dependent Liapunov exponent $\lambda(v)$ will be negative. If the region moves at a large velocity, the region will "outrun" the perturbation and $\lambda(v)$ will again be negative. However, for some intermediate velocity, the region will move with the growing perturbation and thus $\lambda(v)$ will be positive. Let v_m be that velocity which gives a maximum value for $\lambda(v)$. Then we define $\lambda(v_m)$ as a measure of chaos for the fully developed portion of the flow.

For the Ginzburg-Landau equation it is not practical to directly calculate (3) since the system would have to extend very far in the $+x$ direction in order to get an accurate value for $\lambda(v)$. In ref. [1] this difficulty was circumvented by transforming (1) and (4) into a frame of reference moving at $v=v_g$ and approximating open boundaries at both boundaries (ie. $\partial^2\psi/\partial x^2=0$). We then found $\lambda(v_m)=\lambda(v_g)=.466\pm.004$ as a measure for the chaos. For comparison,the value for the usual Liapunov exponent is $-2.55\pm.02$.

For the coupled map (2) the definition (3) must be modified for discrete time and space (ie. $t\rightarrow n,x\rightarrow i,vt\rightarrow[vn],\delta\psi(x,t)\rightarrow\delta X_n^{(i)}$, and $\int\rightarrow\sum$). Here the brackets mean "the integer part of". Due to the ease of iterating (2) the velocity-dependent Liapunov exponent was calculated directly for a range of velocities by taking a very long system and following a region in a moving frame of reference. The expected behavior was found (ie. $\lambda(v)$ is negative for $v=0$, increases to a positive maximum as v is increased, and then decreases until it again becomes negative as v is further increased).

Under convectively unstable conditions these systems represent "open flow" systems. The ideas presented in ref. [1-4] therefore have relevance to open-flow fluid systems such as fluid flow in a pipe, channel flow, and fluid flow over a flat plate.

To summarize, we found that even though two nearby trajectories may exponentially converge on the average in the stationary frame of reference [corresponding to a negative value for $\lambda(0)$], a moving frame of reference may exist in which two nearby trajectories exponentially diverge on the average [corresponding to a positive value for $\lambda(v_m)$]. For a more complete set of references the reader is referred to refs. [1-4] and the references contained therein.

This work was partially supported by the Air Force Office of Scientific Research under AFOSR grant #ISSA-84-00017.

References:

1. R. J. Deissler and K. Kaneko: "Velocity-Dependent Liapunov Exponents as a Measure of Chaos for Open-Flow Systems", Los Alamos Preprint LA-UR-85-3249 and submitted to Phys. Rev. Lett.
2. R. J. Deissler: J. Stat. Phys. *40* Nos. 3/4, 371 (1985)
3. R. J. Deissler: "Spatially-Growing Waves, Intermittency, and Convective Chaos in an Open-Flow System", Los Alamos Preprint LA-UR-85-4211 and submitted to Physica D
4. K. Kaneko: "Spatial Period Doubling in Open Flow", to appear in Phys. Lett. A; Also, for another convectively unstable map lattice, see R. J. Deissler: Physics Letters *100A* ,451 (1984); For stability conditions for coupled map lattices see appendix 3 in ref. [3].

Part V

**Reliability, Accuracy
and Date-Requirements of
Different Algorithms**

An Approach to Error-Estimation
in the Application of Dimension Algorithms

J. Holzfuss and G. Mayer-Kress*

Center for Nonlinear Studies, MS B258, Los Alamos National Laboratory,
Los Alamos, NM 87545, USA

Three different methods for calculating the dimension of attractors are analyzed. An approach
to error-estimation is presented and is used on various data sets. In some cases it is shown that
the errors can become very large.

1. Introduction

The dimension of attractors reconstructed from a time series [1] is of great physical interest
especially in experimental situations [2,3]. It is a measure for the number of active modes
modulating a physical process, and therefore a measure of complexity. Many different methods
[4,5,6] of calculating the dimension of attractors have been introduced. Very important questions are, how far these methods are reliable and how large the uncertainty of a calculated
dimension is. Most of the algorithms used for dimension measurements of attractors, reconstructed from numerical and experimental data, average over certain variables, such as the
number of nearest neighbors, the mass of a cube of a certain sidelength or pointwise dimensions from different reference points on the attractor. All these averages must be taken into
account, if one wishes to determine a realistic error estimate of a fractal dimension. In the
literature, error estimates are mainly calculated by just averaging over some values of the scaling exponents obtained in different length scales (least squares fit). This method can truly
under estimate existing errors. The errors can make it useless for experimentalists to deal with
smaller/larger relationships between different definitions of fractal dimensions. The largest
error source is the limitation in the number of data available to reconstruct an attractor from a
time series. The number of data points necessary for filling this subset of a phase space with
points to get the same probability measure as given by the attractor of the physical process
might be very large, and even increases exponentially with the dimension of the attractor. This
also gives rise to the question of whether an analysis of high dimensional attractors is possible.

We first recall several methods for calculating the dimension (II), and then introduce our
approach to determine the error (III). After a short description of the computer programs and
some ideas for automation of the dimension calculation (IV), we analyze different data sets,
including a 5-torus, gaussian noise and the Lorenz attractor (V).

2. Methods for Calculating the Dimension of an Attractor

To reconstruct an attractor from a time series of a single probe, we use the now classical
method of time-delay coordinates. In this method a vector $\vec{X}(t_k)$ in an n-dimensional phase
space is constructed by taking delayed samples of the time series $x(t_k)$ as coordinates [1], such
that

$$\vec{X}(t_k) = \left(x(t_k),\, x(t_k+T),\, x(t_k+2T),\, \cdots,\, x(t_k+(n-1)T) \right) \tag{1}$$

---

* Permanent address: Drittes Physikalisches Institut, Universitaet
   Goettingen, D-3400 Goettingen, Fed. Rep. of Germany

where t_k is the discrete time with k running from 1 to the number of data points and T is an 'arbitrary', but fixed time delay. The embedding dimension n is the number of coordinates of the embedding space. If T is chosen to be equal to the time delay, where there is a minimum in the mutual information between two measurements [7,8], a D-dimensional attractor is constructed best by taking the embedding dimension larger than $2D + 1$. For signals with strong periodic contents,this time delay is approximately equal to the first zero-crossing of the auto-correlation function.

We used three different methods for calculating the "fractal" dimension of an attractor.
1. The pointwise dimension (mass dimension) [4,9] can be defined as

$$ D_x = \lim_{n_{data} \to \infty} \lim_{r \to 0} \frac{\log \dfrac{1}{n_{data}} N_{X_0}(r)}{\log r} \tag{2} $$

and it consists of counting the number of data points $N_{X_0}(r)$ within a cube of sidelength r centered at a point $\vec{X}_0$ on the attractor. Due to the fact that one does not have an infinite amount of data points and also no infinite precision, as required in the definition, one has to average over several reference points $\vec{X}_0$. Fig. 1 shows the scaling behavior of $N_{X_0}(r)$ of different reference points. We calculate the pointwise dimensions for 200 reference points and take the average, which yields a good estimate for the dimension of the attractor.

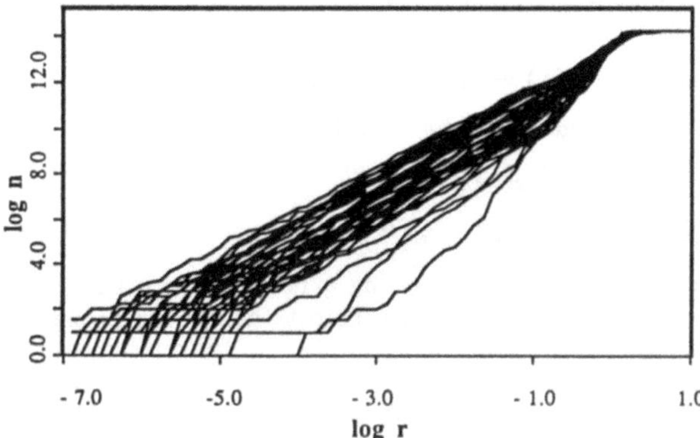

Fig. 1: Scaling behavior of $N_{X_0}(r)$ at 50 reference points $\vec{X}_0$ on the Lorenz attractor. The embedding dimension is 5.

2. Another method we used was the determination of the correlation dimension D_2 with the algorithm proposed by GRASSBERGER and PROCACCIA [5]. They showed the scaling of the correlation integral $C(r)$ for small r

$$ C(r) \sim r^{D_2} \tag{3} $$

with

$$ C(r) = \lim_{n_{data} \to \infty} \frac{1}{n_{ref}} \sum_{j=1}^{n_{ref}} \frac{1}{n_{data}} \sum_{i=1}^{n_{data}} \Theta(r - |\vec{X}_i - \vec{X}_j|) \tag{4} $$

and Θ equal to 1 for positive and 0 for negative arguments.
Also

$$C(r) = \lim_{n_{data} \to \infty} \frac{1}{n_{ref}} \sum_{j=1}^{n_{ref}} \frac{1}{n_{data}} N_{\vec{X}_j}(r) \tag{5}$$

with $N_{\vec{X}_j}$ equal to the rightmost sum in (4). $C(r)$ counts the number of points $N_{\vec{X}_j}(r)$ in a cube of fixed sidelength r, averages over all the cubes that are centered at different reference points $\vec{X}_j$, and normalizes. An example shows how the averaging is done.

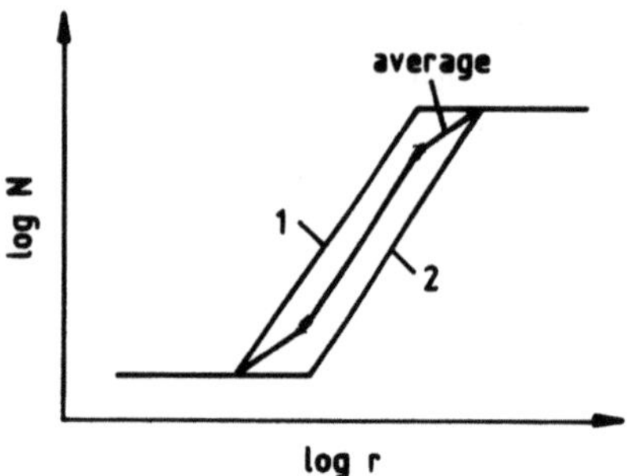

Fig. 2: sketch of the scaling behavior for 2 reference points with different scaling regions $[r_{min}, r_{max}]$ and their average (Grassberger and Procaccia method).

The scaling of the contents of the cubes with respect to r in two different regions on the attractor (fig. 2) is discribed by

$$N_1(r) = (\alpha_1 r)^D \qquad \text{and} \qquad N_2(r) = (\alpha_2 r)^D \qquad \alpha_1, \alpha_2 = const. \tag{6}$$

The Grassberger and Procaccia algorithm averages over the N-values, while r is fixed:

$$C(r) = (N_1(r) + N_2(r))/2 = 1/2 \tag{7}$$

$$\overline{N}(r) = r^D ((\alpha_1{}^D + \alpha_2{}^D)/2) \tag{8}$$

For infinitely expanded scaling behavior we get the same scaling exponent D for the averaged values. Large differences in the α's and small scaling regions distort the scaling properties of $C(r)$.

3. The third method considered here, is the one proposed by TERMONIA and ALEXANDROWICZ [6]. It consists of averaging over the different radii of cubes, which contain a fixed number of data points. They showed that N, the 'number of nearest neighbors', behaves like

$$N \sim \overline{r}(N)^{D_F{}'}, \tag{9}$$

where $\overline{r}(n)$ is the average radius of the cubes containing N data points and $D_F{}'$ the "fractal" dimension of the attractor. As an example we consider again the scaling behavior for two reference points, this time keeping N fixed and r variable.

$$N = (\alpha_1 r_1)^D \qquad \text{and} \qquad N = (\alpha_2 r_2)^D \qquad \alpha_1, \alpha_2 = const. \tag{10}$$

Then

$$N^{1/D}/\alpha_1 = r_1(N) \qquad \text{and} \qquad N^{1/D}/\alpha_2 = r_2(N) \tag{11}$$

and

$$N^{1/D}(1/2(1/\alpha_1 + 1/\alpha_2)) = [r_1(N) + r_2(N)]/2 = \overline{r}(N) \tag{12}$$

gives

$$N = \overline{r}(N)^D (1/2(1/\alpha_1 + 1/\alpha_2))^{-D} , \tag{13}$$

which results in the same scaling exponent D for the averaged values as for the single reference points.

All three methods use a different kind of averaging. The last two methods have one thing in common: they average over an ensemble of single values (N in case 2, r in case 3) and don't care about the orientation of the lines in fig. 1, i.e., the scaling behavior of the attractor at the different reference points. This may in some cases lead to a misinterpretation of the results if the averaged values obey a scaling behavior, that the values obtained from different reference points may have never had. In method 1 the additional information of the reference points is used when the averaged pointwise dimension is calculated.

3. Error Estimation

Each value of the pointwise dimension D_x was obtained by calculating the slope of a fitted straight line to each curve in the log r / log N plot using least squares fit. In order to determine the spread of values of the D_x we consider their standard deviation, given by

$$\Delta D = \sqrt{\overline{D^2} - \overline{D}^2} \tag{14}$$

which says, that 68.3 % of all values of a normal distribution lie in the interval $[D - \Delta D, D + \Delta D]$. In case 2 and 3 straight lines are also fitted to the values in the doubly logarithmic plot using least squares fit. The least squares fit consists of finding a straight line

$$y_{ij} = a + bx_{ij} + z_{ij} \tag{15}$$

such that $\sum(z_{ij})^2$, the sum of squares of the errors z_{ij} is minimized. The index j denotes the different reference points and the index i their average value. It is also possible to use weighted least squares fit, which minimizes

$$\sum w_i z_i^2 = \sum w_i (y_i - a - bx_i)^2 \qquad \text{with} \tag{16}$$

$$w_i = 1/(\Delta y_i)^2 \qquad \text{and} \tag{17}$$

$$y_i = 1/n_{ref} \sum_{j=1}^{n_{ref}} y_{ij} \tag{18}$$

z_i and x_i have the same definition. Using weighted least squares fit reduces the weight of values, if their variance $(\Delta y_i)^2$ is large. However, because of the finite size of the attractor, the variances become very small for large values of log r (saturation region) and therefore could give rise to a false fit. Therefore we have to exclude this possibility by e.g. imposing a lower bound for the slope which should be fitted, or by restricting the possible scaling ranges.

In terms of the Grassberger and Procaccia algorithm y_i is the average of the log N_{ij} values of a fixed radius r_i over all reference points $\vec{X}_j$ (fig. 3 a). In the Termonia and Alexan-

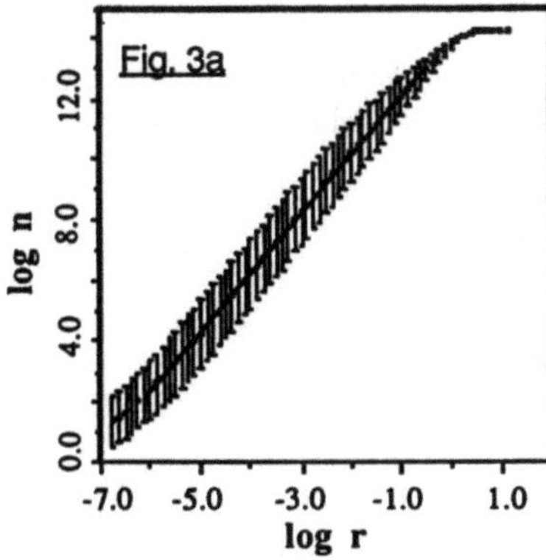

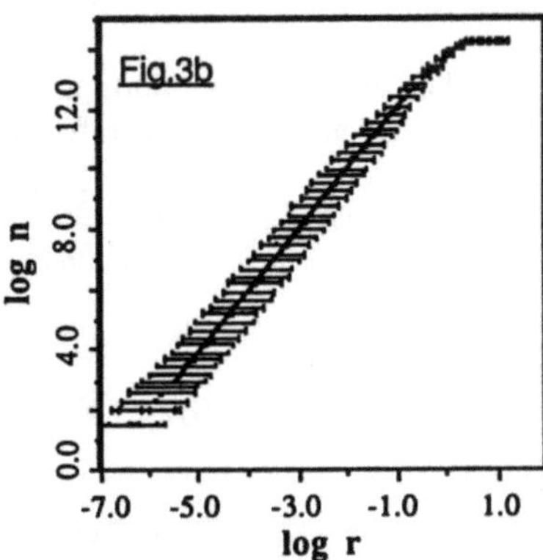

Fig. 3 a,b: Plot of number of points N in a cube versus cubesize r for the Lorenz attractor. The error bars show the standard deviations of the $\log N(r)$ values (3a) and the standard deviations of the $\log r(N)$ values (3b). The embedding dimension is 5.

drowicz method y_i would be the average of the logarithms of the radii $\log r_{ij}$ with a fixed number of nearest neighbors N_i (fig. 3b).

We use the average over the logarithmic values in determining $\Delta^2 y_i$ and minimize the errors z_{ij} of the straight line in the $\log r$ / $\log N$ plot. This gives a good approximation of the fully consistent way of averaging over the nonlogarithmic values of r or N, which would require a minimizing of

$$\sum_i (y_{ij} - ax_{ij}^b)^2 \tag{19}$$

In order to get the standard deviation of the slope, we have $\Delta y + \Delta z = (b \, \Delta x) + (\Delta a + \Delta b \, \Delta x)$. We exclude possible parallel shiftings of the lines obtained at the different reference points to be able to evaluate the largest possible error of the slope, because methods 2 and 3 yield no information about the orientation of these lines. This is done by setting Δa to zero. Now we get for the standard deviation of the slope

$$\Delta b = \Delta z / \Delta x \tag{20}$$

Δz is the standard deviation of the differences of the actual y-values to the y-values of the fitted line. Δx is the standard deviation of the x-values. From fig.4a,b we see, that Δb is the standard deviation of all the possible values of the slope b .

In case. 2 the errors are symmetric around the average slope. In case 3, where we calculate $r(N)$ and therefore get an error for $1/D$, they get asymmetric when solving for the slope D.

To modify the definition into a more computable form we have

$$(\Delta b)^2 = \frac{(\Delta z)^2}{(\Delta x)^2} = \frac{\overline{z^2} - \overline{z}^2}{\overline{x^2} - \overline{x}^2} \tag{21}$$

Due to the least squares fit $\overline{z} \approx 0$. Taking the average between the differences of all the $\log r$ or $\log N$ values and the values of the straight line, which is fitted to the logarithms of the averages of the nonlogarithmic r or N-values implies a very small correction $\overline{z} \neq 0$, which can be neglected. Therefore we get

$$(\Delta b)^2 = \frac{\displaystyle\sum_{i=1}^{n_{rad}} \sum_{j=1}^{n_{ref}} (y_{ij} - a - bx_{ij})^2}{\displaystyle\sum_{i=1}^{n_{rad}} \sum_{j=1}^{n_{ref}} (x_{ij} - \overline{x})^2} \tag{22}$$

where n_{ref} is the number of reference points and n_{rad} is the number of the averaged values of distances r sub i. With the average over all reference points (17), its variance is given by:

$$(\Delta y_i)^2 = \sum_{j=1}^{n_{ref}} \frac{(y_{ij})^2}{n_{ref}} - \left(\sum_{j=1}^{n_{ref}} \frac{y_{ij}}{n_{ref}} \right)^2 \tag{23}$$

and with

$$\sum_{i=1}^{n_{rad}} \sum_{j=1}^{n_{ref}} (x_{ij} - \overline{x})^2 = n_{ref} \sum_{i=1}^{n_{rad}} (x_i - \overline{x})^2 \tag{24}$$

where: $x_{ij} = x_i$ for all $j = 1,...,n_{ref}$ we get:

$$(\Delta b)^2 = \frac{\sum\limits_{i=1}^{n_{rad}} (\Delta y_i)^2 + \sum\limits_{i=1}^{n_{rad}} (y_i - a - bx_i)^2}{\sum\limits_{i=1}^{n_{rad}} (x_i - \bar{x})^2} \tag{25}$$

This expression allows to compute the variances of all the y_i's first and to get the variance of the slope when fitting over a certain number of y_i's. Expression (22) can also be modified by using

$$a = \bar{y} - b\bar{x} \tag{26}$$

and

$$b = \frac{\sum\limits_{i=1}^{n_{rad}} \sum\limits_{j=1}^{n_{ref}} \dfrac{x_{ij}\, y_{ij}}{n_{rad} * n_{ref}} - \bar{x}\,\bar{y}}{(\Delta x)^2} \tag{27}$$

with $\bar{x}$ and $\bar{y}$ equal to the overall mean into

$$(\Delta b)^2 = (\Delta y)^2/(\Delta x)^2 - b^2 \tag{28}$$

The variances are taken over all $n_{rad} * n_{ref}$ points.

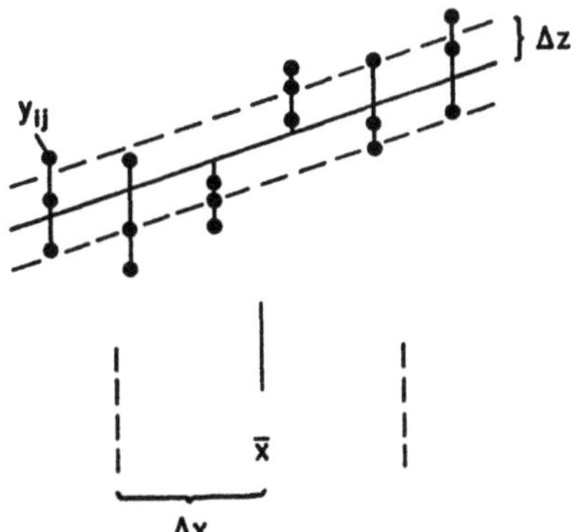

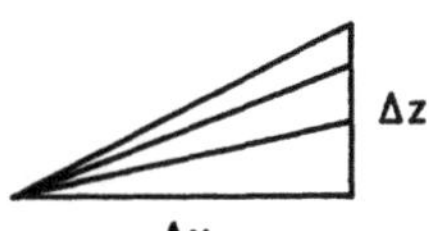

Fig. 4a : standard deviation of x-values
and of the errors z of the y-values

Fig. 4b : possible ranges for the slope b

4. Description of the Programs

The programs we use are designed for a fully automatic analysis of the data sets. First of all the program has to find the scaling region. This is done by looking for the interval of a given length in the $\log r$ / $\log N$ plot, where the root-mean-square error σ of the deviation from a fitted straight line, which is defined as (see also (16))

$$\sigma = \sqrt{\frac{1}{n_{rad}} \sum_{i=1}^{n_{rad}} w_i\, z_i{}^2} \quad , \tag{29}$$

becomes minimal. In case 1, where we consider the averaged pointwise dimension, the weights w_i are all 1. Here the program calculates the slope of a fitted line at each reference point. Then it averages over just 20% of the obtained values of the pointwise dimensions, neglecting all the curves that don't show a scaling behavior over the entire range of r-values. After calculating the average and the standard deviation it repeats these steps in a different embedding dimension. In case 2 and 3 the procedure is about the same, except there is just one line to be fitted. Also the weights are set to their respective values (17). The procedure also features a "self-blowup" of the length of the fitted line: If the total length of the scaling region is unknown, the program starts with a given short interval length and finds the scaling region by minimizing the RMS error σ (29). Then it repeats this step with an enlarged interval until a certain threshold value of the σ is reached, thus indicating that the fitted interval length exceeds the length of the scaling region. In all cases we find that 0.05 is a "good" threshold value.

5. Analysis of Some Typical Datasets

In the analysis the total number of data points is always 20,000. We average over 200 reference points. The time delay T for the reconstruction of the attractor is chosen according to [7,8] by calculating the mutual information. As the first example we consider a 5-torus, constructed from a time series with a Fourier spectrum of 5 incommensurate frequencies. We analyze the data with all three methods (fig. 5a,b,c). In all three figures we see that the calculated dimension of the attractor converges with increasing embedding dimension, and it has the value $D = 5$ from about $2D + 1 = 11$. The standard deviations of the attractor dimension, described by the error bars, converge also to some fixed values. For the averaged pointwise dimension we get 5 ± 0.3, for the correlation dimension 5 ± 0.5 and for the dimension obtained by the TERMONIA/ALEXANDROWICZ [6] method $5 + 0.8 / - 0.7$.

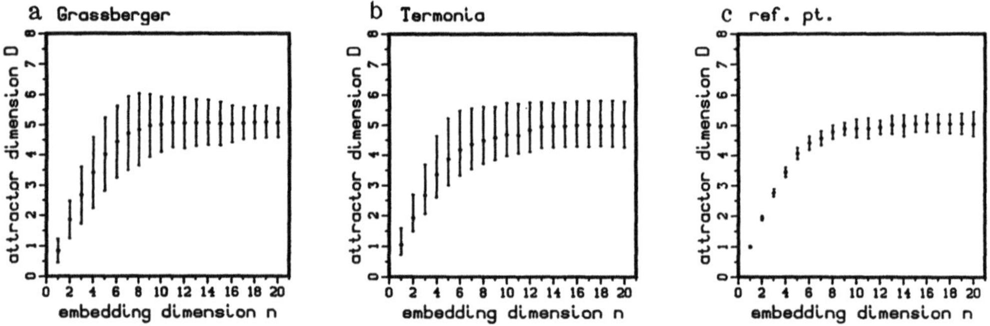

Fig. 5 a,b,c: Graph of the attractor dimension D versus the embedding dimension for a 5-torus. 20,000 data points and 200 reference points were used to calculate the dimension. The used methods are Grassberger/Procaccia (a), Termonia/Alexandrowicz (b) and the averaged point-wise dimension (c).

The calculation of the dimension of the Lorenz attractor (fig 6a,b,c) shows also good convergence at about $2D + 1$. The values for D obtained by the different methods were 2 ± 0.15 in the case of the averaged pointwise dimension, 2 ± 0.6 in case 2 and $2 + 0.6 / - 0.4$ in case 3. The growing of the standard deviation in case 2 and 3 is due to successive shortening of the scaling region in higher embedding dimensions because of geometrical effects. Smaller scaling regions with constant variance at each average value in the log r / log N diagram result in a larger variance of the slope of the fitted line. The average of the pointwise dimension is not affected by this, when the length of the fitted line equals the length of the scaling region.

In the third example we analyzed gaussian noise (fig. 7a,b,c). Noise is considered to be high dimensional and spacefilling, i.e. each phase space of every embedding dimension is filled. None of the three methods were able to produce this result. This is truly seen in the deviation from the 45° line. The deviation is due to the increasing amount of data necessary for calculating

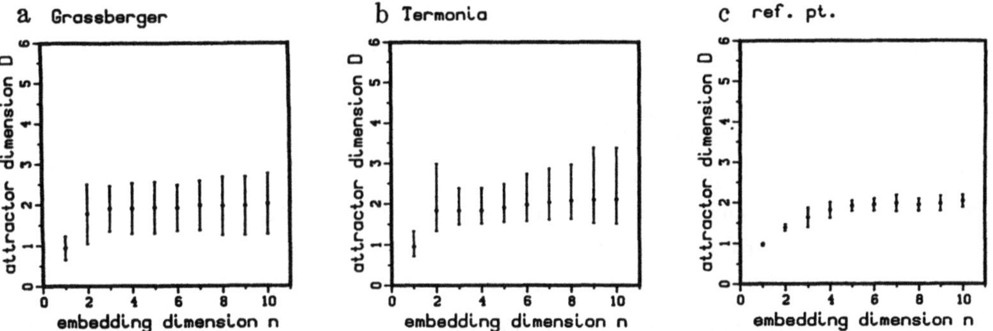

Fig. 6 a,b,c: Graph of the attractor dimension D versus the embedding dimension for the Lorenz attractor. (description see fig. 5).

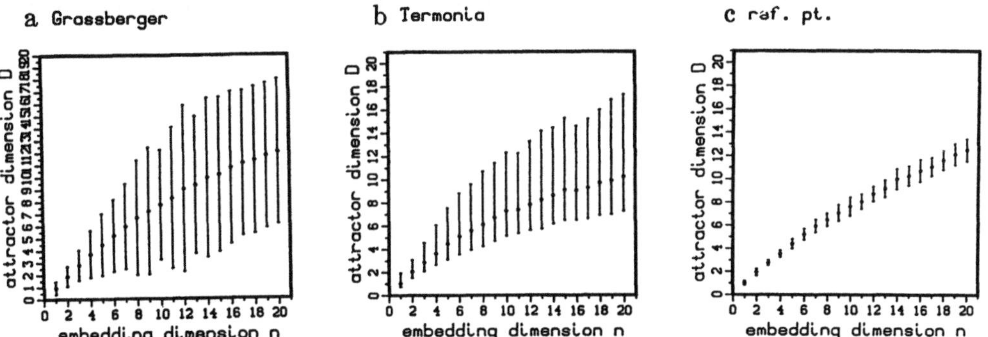

Fig. 7 a,b,c: Graph of the attractor dimension D versus the embedding dimension for gaussian noise (description see fig. 5).

higher dimensions. Also it can be extracted that the error increases almost linearly with the attractor dimension. Compared to the averaged pointwise dimension, the dimension calculated with the two other methods showed a very large possible error.

6. Conclusions

Three different methods of calculating the dimension of attractors were analyzed. To each of those an approach to estimate the error was presented, which was based on calculating standard deviations for certain variables. Examples of analyzed data sets showed that the averaged pointwise dimension provided the smallest possible error in the calculated dimension. It also seemed to be closer to the real value. It was shown that the possible errors in the methods of GRASSBERGER/ PROCACCIA [5] and TERMONIA/ ALEXANDROWICZ [6] can be very large. High-dimensional analysis is shown to be very difficult, because of the linear growth of the errors with the embedding dimension.

Acknowledgements

J.H. wants to thank the Center for Nonlinear Studies for the hospitality and financial support and W. Lauterborn for very useful discussions. We are deeply indepted to Erica Jen for permission of using versions of her codes which are the basis of parts of our numerical calculations. We also appreciate very helpful discussions with Erica Jen and J. Doyne Farmer. All computations were done in Los Alamos on CRAY 1 computers.

References:

1   Packard, N. H.,Crutchfield, J. P., Farmer, J. D., Shaw, R. S. , Phys. Rev. Lett. **45** 9 (1980) 712

2   Brandstaetter, A., Swift, J.,Swinney, H. L., Wolf, A., Farmer, J. D., Jen, E., Crutchfield, P. J. , Phys. Rev. Lett. **51** 16 (1983) 144

3   Lauterborn, W., Holzfuss, J. , Preprint Goettingen

4   Farmer, J. D., Ott, E. and Yorke, J. A. , Physica **7D** (1983) 153-180

5   Grassberger, P. and Procaccia, I., Phys. Rev. Lett. **50** 5 (1983) 346

6   Termonia, Y. and Alexandrowicz, Z. , Phys. Rev. Lett. **51** 14 (1983) 1265

7   R.S. Shaw, *The Dripping Faucet as a model Chaotic System* (Aerial Press, Santa Cruz, CA, 1985)

8   A. Frazer, this volume

9   Mandelbrot, B.: *The Fractal Geometry of Nature* (W. H. Freeman and Co., San Francisco, 1982).

Invisible Errors in Dimension Calculations: Geometric and Systematic Effects

W.E. Caswell and J.A. Yorke[1]

Naval Surface Weapons Center, White Oak, Silver Spring, MD 20910, USA
and [1]University of Maryland, College Park, MD 20742, USA

We use box-counting methods to attempt to reliably calculate the generalized dimensions (including box-counting dimension, i.e., capacity dimension, and information dimension) for the Henon attractor (a = 1.4, b = 0.3). In order to investigate possible errors arising in more general situations, we have analyzed the asymptotic behavior of the cover of the attractor as the number of iterates considered approaches infinity. The error in estimating the box-counting dimension depends in part on the geometric shape of the "boxes" used, and we give a heuristic derivation of the rate of approach. We introduce the use of disks rather than squares to minimize errors in estimates of the number of "boxes" required. The resulting dimension estimates have very small fitting errors: the points in a log-log plot are quite well fit by a straight line. However, what would happen for even smaller box sizes cannot be estimated.

We conclude that it is safest to include the binary range of box sizes tested (as powers of one-half down from the attractor size) when giving dimension estimates. For the Henon attractor the information dimension $D[7,13] = 1.254 \pm 0.006$, (that is the attractor dimension estimate using box sizes from 2^{-7} down to 2^{-13}) while $D[9,13] = 1.271 \pm 0.001$. Obviously, these standard error estimates are inconsistent. The first value is in agreement with the rigorous upper-bound D(Lyapunov) = 1.25826 ± 0.00006, while the second and presumably more accurate (due to the smaller error) value violates the bound.

1. Introduction

Searching for (and sometimes finding) fractal dimensions has become a favorite pursuit shared by physicists and mathematicians. Both groups wish to find measureable aspects of systems which have non-integer dimensions, which they can use to characterize and quantify the systems. Our understanding of the mathematical objects involved, which we will generically refer to as chaotic attractors, is only in its infancy. A variety of properties intrinsic to the attractors have been proposed as worth measuring and remembering, the most natural being the box-counting dimension of the attractor. Reviews of the current status of the measurement of dimensions of attractors are available [1]. A continuum of other dimensions, which includes the box-counting and the information dimensions, has been proposed [2]. While quite easy to define, the practical measurement of these dimensions has been the subject of much research, to which we now add.

Any dimension of a system tells something about how much information it takes to locate a point on an attractor. We will investigate box-counting algorithms, where typically the attractor is covered with boxes of a particular size and shape, and the variation of the number and/or probabilities of the boxes is analyzed. With such an approach it is rather easy to average over all of the points available, for a moderate range of box sizes (down to, for the Henon attractor, in two dimensions, a box perhaps 10^{-4} the size of the attractor). The practical limit is computer-dependent, typically several hundred thousand non-empty boxes. This has a practical value for the physicist, who normally will have a limited amount of data, and is interested in using all of it (averaging over the attractor, for example) in analyzing his experiment. Further, methods which require many decades of scale size will not be very useful to the physicist, who is limited by the noise effects to a relatively small range of signal size.

We plan to pursue the idea of obtaining the best estimates of box-counting dimensions by extrapolating the relevant quantities (the generalized entropies) from finite to an infinite number of iterates on the attractor. This procedure has been advocated by GRASSBERGER [3] and we apply it here to dimensions other than the box-counting dimension. We also gain some understanding of the source of the striking linear extrapolation found by Grassberger, which is more a reflection of the geometric shape of the boxes used, interacting with the one-dimensional bands of points of an attractor in two dimensions (the Henon attractor appears to be made up of lines), than it is of an intrinsic property of the attractor.

We find that, even with much effort, the dimensions found are not known nearly as well as the standard error estimate coming from the theory of least-squares fits would have us believe. We further believe that as a standard procedure **dimension estimates should be stated with a range of distances** (or box sizes) **used in arriving at the estimate.** The reader can do his or her own estimation of what would happen outside the range, and the author does not have to be in the position of claiming more than the data implies.

2. Box-counting

One naive approach to calculating the dimension of an attractor is to cover the attractor with an array of "boxes" (which we put in quotes because we will also consider disks), and then to study how the number of boxes scales as we change the box-size: this is the box-counting dimension. Attempting to cover the attractor may be unrealistic, as often extremely low probability events may be missed, and a variety of other dimensions, such as the information dimension, have been introduced. Recently, many of these dimensions have been placed into a continuum of correlation exponents based on moments of the probability distribution of boxes [2].

First, we must choose a system on which to calculate. We have chosen the venerable Henon attractor, as much effort has been expended on it in the past (so we have many comparisons available). The Henon attractor is generated by the Henon map [4],

$$x_{n+1} = 1 + y_n - a\,x_n^2$$

$$y_{n+1} = b\,x_n.$$

We choose the standard values a = 1.4 and b = 0.3.

We start by covering the region occupied by the attractor with a grid of "boxes". We then iterate the Henon map many times, obtaining a picture like that in Fig. 1.

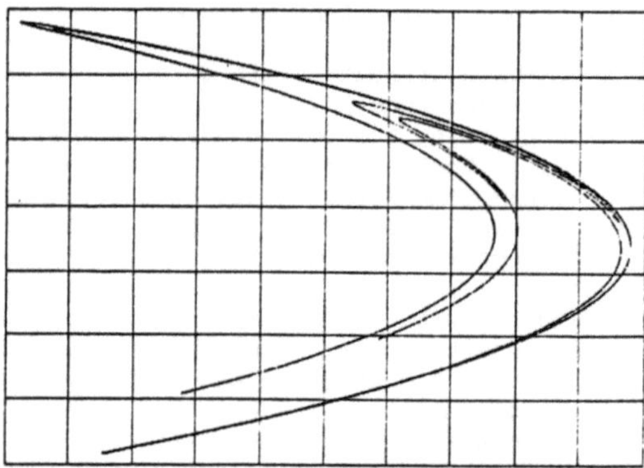

Fig. 1. Here we show the Henon attractor covered with a grid of boxes (which are about a thousand times larger than the smallest boxes which we have utilized). Box-counting methods require that the attractor be covered with a grid. The grid is typically made up of identical boxes, equally for computer and for conceptual simplicity. The probabilistic methods require that we not·only record whether the attractor hits a box, but also the probability that we are in the box (fraction of the total number of iterates in the box)

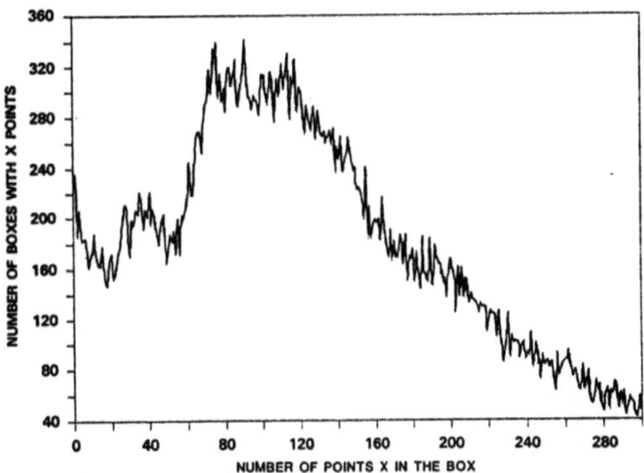

Fig. 2. Density distribution for the Henon attractor (a = 1.4, b = 0.3), with box-size 0.000125 square. Plotted are the number of points in a box (divide by 4.4 $\times 10^7$ to get the probability contained in the box) versus the number of boxes. There are many low-probability boxes. This is a geometric effect, not a third peak

If we keep track of the number of times the iterated point falls in each box, we will be able to calculate the probability of each box by dividing its contents by the number of iterates. We obtain a density distribution (number of boxes of each possible probability) such as that shown in Fig. 2. Averages over such distributions will give us the necessary information (generalized entropies) to calculate box-counting, information, and other dimensions.

To calculate the box-counting dimension we ignore the probability, simply counting each occupied box as 1, and consider how the number of boxes scales as we change the box size (which we will call ε). Noting that 1 is p^0, we are instructed to calculate

$$N(\varepsilon) \ = \ \sum 1 \ = \ \sum p^0,$$

where the sum is over the boxes containing the attractor. The dimension is defined by how $N(\varepsilon)$ scales as ε goes to zero. If we assume a scaling with the box-size ε of the form $N_{boxes}(\varepsilon) = C\,\varepsilon^{-D}$, then we can find D by fitting this functional form.

For other dimensions, we generalize [2] the p^0 to p^q, q>0. The precise formula comes from the definition of the generalized entropy:

$$I_q \ = \ \log(\sum p^q) \ / \ (1-q)$$

where $\sum$ again goes over all boxes which have been hit.

It is possible to try other box shapes than the usual square (or cube). We have studied covering the attractor with disks: the dimensions measured are expected to be the same as for hypercubes. We have, for computational simplicity, chosen disks that are concentric (Fig. 3) with the grid of boxes discussed above, though a more adventurous soul might try close-packing hyperspheres in n dimensions.

We replace the squares covering the attractor by the inscribed disks, i.e. the largest disk that can be inscribed in each square. Notice that the attractor is not covered, since points which fall near the corners will be in no disk. None-the-less **the number of disks hit by the trajectory can be expected to scale in exactly the same way as the number of squares.** The difficulty with squares is that if the attractor just clips the edge of the square, many iterates may be needed to detect the fact that the square should be included. Such an effect is of course possible with disks but it occurs less often. Therefore, given a limited number of points (and the number of points is always limited), **the number of disks hit can be determined more accurately than the number of squares.**

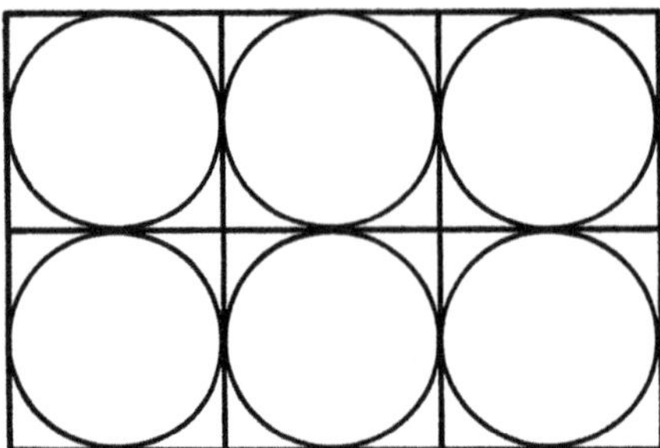

Fig. 3. We may cover the attractor with other geometric figures. Disks have the
particularly nice property that they have no corners, and we have studied the dimensions found using disks. We do not cover the whole attractor (disks do not tile
the plane), and for simplicity we have chosen to simply inscribe the disks inside
corresponding boxes, as shown.   Using disks should not change the generalized
dimensions

3. Heuristic Derivation of Grassberger's Power Falloff

Imagine that the attractor is made up of one (long and twisty) line whose apparent
length increases at a rate (as a function of the box size ε) adequate to account for
the box-counting dimension D of the attractor, i.e.

$$L = L_o \, \varepsilon^{-(D-1)}.$$

Then, up to a geometric factor, the number of boxes covering the line will be

$$N(\varepsilon) = L/\varepsilon = L_o \, \varepsilon^{-D}.$$

where we have also assumed that a box covers one line segment at a time.   We are
interested in doing a probability analysis, so we define the probability density per
unit length of the line

$$d = 1/L = (1/L_o) \, \varepsilon^{(D-1)}.$$

If the box (disk) intercepts the line in a chord of length $w\varepsilon$ (we scale the chord
length by the size of the box), then the probability of a given iterate falling in
the box is

$$prob = w\varepsilon d.$$

How likely is it that a line, dropped randomly on the plane, will intercept a box in
a chord of length w, given that it intersects the box? This is a simple probability
calculation which we discuss in the appendix. For a single line the probability is,
for small w (the line barely clipping the box)

$$p(w) \; = \; c \, w^q \quad \text{where} \quad \begin{cases} q = 0 \ (\text{box}), \\[2ex] q = 1 \ (\text{disk}). \end{cases}$$

Note that q is not really a known constant, because we may in fact intersect more
than one line at a time and, worse, the lines of the attractor have varying densities and are made up of lots of little lines.
  We are interested in small probabilities, so we assume that the above form of $p(w)$
is true in general. What is the probability P for a **single** box of the $N(\varepsilon)$ boxes
available to contain probability p? It is just the integral over all w (though

126

only small w are important) constrained to give the right probability:

$$P(p) = \int dw\ p(w)\ \delta(w\epsilon d - p)$$

$$= \int dw\ c\ w^q\ \delta(w\epsilon d - p)$$

$$= c\ p^q\ (\epsilon\ d)^{-(q+1)}.$$

where $\delta(x)$ is the Dirac delta function, $\int f(x)f(x-x_o)dx = f(x_o)$. The total number of boxes with probability $P(p)$ is just the above multiplied by the number of boxes $N(\epsilon)$, i.e.

$$N(p) = c\ L_o\ p^q\ d^{-q-1}\ \epsilon^{-q-D-1}.$$

For fixed n (number of iterations) and ϵ, this behaves as p^q, and q may be read off the probability plots of the number of boxes, in reasonable agreement with the above expectations.

Now we are in a position to calculate dN/dn, the expected number of new boxes per iteration. A box containing probability p will be new if it is found on the nth iteration (probability p), but was not found on any of the previous iterations probability $(1-p)^{n-1}$, though we drop 1 relative to n and use $(1-p)^n$). dN/dn is just this value summed over all boxes, so

$$dN/dn = \sum (1-p)^n\ p$$

$$= \int dp\ N(p)(1-p)^n\ p$$

$$= c\ L_o d^{-q-1}\ \epsilon^{-q-D-1} \int dp\ e^{-np}\ p^{q+1}$$

$$= C\ \epsilon^{-(q+1)(D-1)}\ \epsilon^{-q-D-1}\ n^{-2-q}$$

$$= C\ \epsilon^{-D(2+q)} n^{-(2+q)}.$$

Grassberger [3] reported observing that dN/dn has the form $\epsilon^{-\alpha}\ n^{-\beta-1}$. Identifying Grassberger's α and β ($\epsilon^{-\alpha}\ n^{-\beta-1}$), we find:

$$\alpha = D(2 + q)$$

$$\beta = 1 + q\ .$$

We have studied the asymptotic behavior as the number of iterations tends to ∞ for both boxes and disks. The curves found are presented in Fig. 4 (boxes) and Fig. 5 (disks). Fits of the power falloff regions give the values shown in Table I. The values obtained are (taking D = 1.28 to get the heuristic estimate):

Table I

	box (q=0)	Grass.[3]	Fit	disk (q=1)	Fit
a	2.56	2.42 ± 0.15	2.3 ± 0.3	3.84	3.2 ± 0.1
b	1	0.89 ± 0.03	0.8 ± 0.2	2	1.3 ± 0.1

The fit is a nonlinear least-squares fit to the asymptotic data in Figs. 4 and 5. The fit is sensitive to which data is selected as asymptotic (close enough to falling on a straight line). This introduces an unknown systematic difference into the fit values of α and of β due to different criteria for asymptopia of different inves-

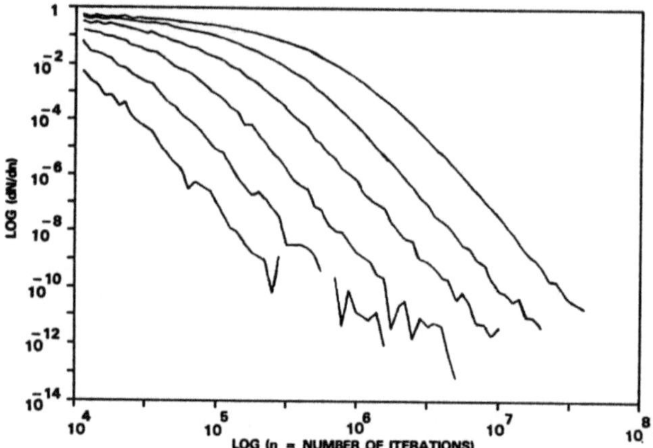

Fig. 4. As we iterate the map more times, the number of new boxes (δN) per iteration (dn) falls off with approximately a power of n. This figure shows the falloff for box-size ε = 0.004 (left side) to 0.000125 (right). Jaggedness at the bottom of the graph comes from statistical fluctuations in the number of new boxes found

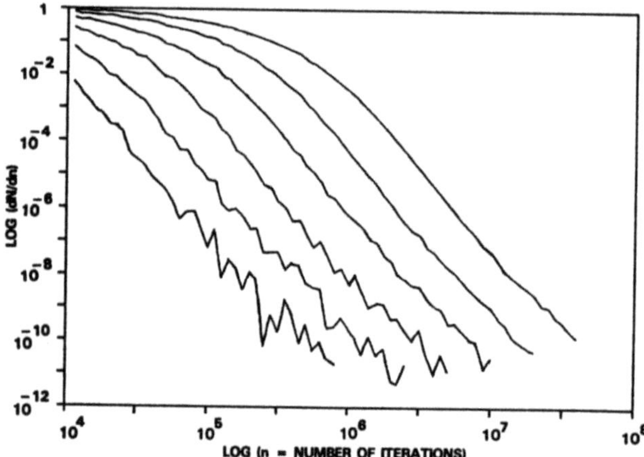

Fig. 5. A similar falloff of (δN)/dn occurs if we use disks to cover the attractor. Comparison with the previous figure shows that the falloff is somewhat faster with disks

tigators. Further, we tried fitting the data for two cases--boxes (resp. disks) un-shifted (a box has its center at (½ ε, ½ ε)) and shifted boxes (a box has its center at (0,0)). Asymptotically, both shifted and unshifted boxes should give us the same fit values for α and β. These values have large errors to reflect the large varia-tion in the fit between shifted and unshifted boxes. Note that we can form a combina-tion of α and β which is independent of q:

$$D = \frac{\alpha}{1+\beta} .$$

It is interesting to see (Table I) that the above formula is well satisfied. How-ever, where the linear region in a log-log plot starts and stops is very subjective. Fewer disks than squares have very low probability regions, so fewer disks are missed due to insufficient data. However, a little "fuzziness" of the "lines" making up the attractor will affect the number of disks found more than the number of boxes found. It is not surprising that our results fit the naive formula better for the boxes than for the disks.

128

The number of boxes yet to be found, after n iterations, may be calculated by integrating the above form for dN/dn out to ∞. The result is (expressed in terms of $(dN/dn)_O$ at the maximum iteration reached, n_O), is:

$$N(\infty) - N(n_O) = n_O(dN/dn)_O/\beta.$$

Since β is approximately one, this means that a reasonable first-cut estimate of the number of boxes left to be found after n_O iterations is simply the product of the current rate at which boxes are being found and the number of iterations made.

4. Systematic Effects on the Extrapolation of Generalized Entropies

We would like to extrapolate the generalized entropies which are defined by:

$$\sum_i p_i^q.$$

We have available a finite estimate based on n iterations and we want an estimate of the limit value as $n \to \infty$ since this is what is needed in the definitions of dimensions. There will of course be the small probability effects discussed in the previous section, but for $q > 0$ these decrease faster than $1/n$. (See the end of this section for a discussion). There are statistical effects, coming from the approach to an asymptotic probability distribution, that give an easily calculable $1/n$ effect. We will calculate this term for two cases, $q = 2$ (the correlation exponent), and for $q = 1$ (the information dimension). The method is exact in the large-n limit, assuming only that the probabilities of the boxes have a well-defined asymptotic distribution.

For a given value of ε (which we will not suppress, we hope for clarity) the number of boxes with probability p will be written $N(\varepsilon,p)$. The total number of boxes is (in all sums the subscript i runs over all the non-empty boxes):

$$\sum_p N(\varepsilon,p) = N(\varepsilon) = \sum_i 1.$$

We are interested in finding the value of

$$\sum f(p)$$

for various functions $f(p)$. If we have made n iterations, the expected number of iterates in box i is $p_i n$. We use the fundamental result of hand-waving statistics that, for large enough n, the deviation from this expected value follows a Gaussian distribution with mean $p_i n$ and standard deviation $(p_i n)^{\frac{1}{2}}$. Calling x the observed number of hits in the box, the expected contribution to our sum from box i is:

$$\int_0^\infty dx\, f(x/n)(2\pi p_i n)^{-\frac{1}{2}}\, \exp(-(x-p_i n)^2/(2p_i n))$$

We may change variables to $y = x/n - p_i$. The integral becomes:

$$\int dy\, f(p_i+y)(n/(2\pi p_i))^{\frac{1}{2}}\, \exp(-y^2/(2p_i/n)).$$

For large n, the Gaussian is very narrow and centered about $y = 0$. We may therefore expand f in a Taylor series about p_i:

$$f(p_i + y) = f(p_i) + y\, f'(p_i) + (y^2/2)f''(p_i) + O(y^3).$$

The integral may now be done for each term in the series, as long as we extend the range of integration to the whole real line--this makes an error which vanishes exponentially with n as n goes to ∞. The odd terms in y then integrate to 0, and we obtain

$$f(p_i) + p_i/(2n)\, f''(p_i) + O(1/n^2).$$

We may now sum this over all boxes. The leading term gives the naive (asymptotic)
generalized entropy, while the term down by $1/n$ is easily calculable. For the cases
of interest ($q = 1$ and $q = 2$), we find that the result is already available:

$$\sum p_i \ln(p_i) = \left\{\sum p_i \ln(p_i)\right\}_{asympt} + (\sum 1)/(2n).$$

The sum is just the total number of boxes, $N(\varepsilon)$:

$$\sum p_i \ln(p_i) = \left\{\sum p_i \ln(p_i)\right\}_{asympt} + N(\varepsilon)/(2n) + \text{H.O.T.}$$

(H.O.T. denotes higher order terms). For $q = 2$, we find (the sum here is just the
probability normalization):

$$\sum p_i^2 = \left\{\sum p_i^2\right\}_{asympt} + 1/n + O(1/n^2) \cdot$$

These results have been checked to several percent using data from the Henon attractor. Similar corrections have been found by Erica Jen (unpublished).

We have found systematic n-dependent corrections to generalized entropies calculated using n iterations. If these terms are not explicitly removed, they will bias
the estimate of the corresponding dimensions for any finite n. To obtain the most
accurate result with a finite number of iterations, these terms should be removed
from the generalized entropy before estimating the dimension. The terms we calculate
here are the dominant correction terms for $q > 0$.

There are also terms analogous to the scaling found by Grassberger, and discussed
in the previous section. For the qth moment, this gives a correction term due to
missed low-probability boxes proportional to $1/n^{1+q}$ for boxes, and to $1/n^{2+q}$ for
disks. $q = 0$ gives the effect Grassberger found (for the box-counting dimension),
with $\beta = 1$ for boxes and 2 for disks. For $q = 1$, this gives a $1/n^2$ correction
(boxes) and a $1/n^3$ correction (disks), and may be neglected relative to the $1/n$
correction we found above.

5. Dimension Calculations

We have performed a variety of dimension measurements on the Henon attractor, with
long runs and extrapolating to an infinite number of iterations: a sample of the
data obtained are presented in Table II, and our results are presented in Table III.
In general, shifting the position of the boxes only slightly affects the measured
dimension (for an exact calculation it is not expected to affect the dimension at
all). Uncertainties from that source can be neglected here.

Table II. Sample Data

boxes					
ε	hits	$N(\varepsilon)$	Extrap.	$\sum -p\ln(p)$	Extrap.
0.000125	39810720	362835	364247	12.4582	12.4628
0.00025	19952624	146902	147372	11.5695	11.5732
disks					
ε	hits	$N(\varepsilon)$	Extrap.	$\sum -p\ln(p)$	Extrap.
0.000125	31267010	328315	328745	12.3608	12.7030
0.00025	15676917	133522	133657	11.4780	11.4842

We include a sample of the data we have been fitting, for unshifted boxes and disks.
The extrapolations of the number of boxes have been performed using the α and β
obtained by fitting the data of Figs. 4 and 5.

Table III. Dimension Results

	Boxes, n.s.	Boxes, sh.	Grassberger	Disks, sh.	Disks, n.s.
Box-counting Dimension					
7 points	1.268 ± 0.006	1.268 ± 0.006		1.271 ± 0.006	1.272 ± 0.006
5 points	1.285 ± 0.003	1.287 ± 0.003		1.289 ± 0.002	1.289 ± 0.002
3 points	1.283 ± 0.008			1.287 ± 0.004	
2 points	1.305	1.306	1.30	1.298	1.300
Information Dimension					
7 points	1.255 ± 0.006	1.252 ± 0.006		1.256 ± 0.006	1.254 ± 0.006
5 points	1.272 ± 0.002	1.271 ± 0.002		1.272 ± 0.002	1.271 ± 0.001
q = 2 Dimension					
7 points	1.223 ± 0.006	1.214 ± 0.008		1.222 ± 0.007	1.216 ± 0.008
5 points	1.242 ± 0.002	1.239 ± 0.003		1.243 ± 0.003	1.238 ± 0.003

Comparison: Lyapunov Dimension Bounds

Box-counting Dimension   1.272 ± 0.006
Information Dimension   1.25826 ± 0.00006
q = 2 Dimension   1.224 ± 0.006

Dimensions coming from fits of Henon attractor data over a
variety of ranges in box-size. The data come from coverings
with boxes and with disks, both unshifted and shifted (by
0.5 in each direction). The Lyapunov dimension bounds are
from GRASSBERGER and PROCACCIA [6].

The estimate of the dimension is the slope of the line in a log-log plot, such as
in Fig. 6. The box-counting dimension D_0 results depend strongly on the range of
box sizes we have looked at, with the smaller scales giving a larger box-counting
dimension when we take the box sizes in pairs. We have also fit the data over a
range of box sizes (this improves the statistical accuracy--there are more points--
but may be more sensitive to transient effects). We have performed least-squares
fits on the data of Fig. 6. The errors come from the usual least-squares analysis
where deviations from the straight line are assumed to be gaussian-distributed random variables.

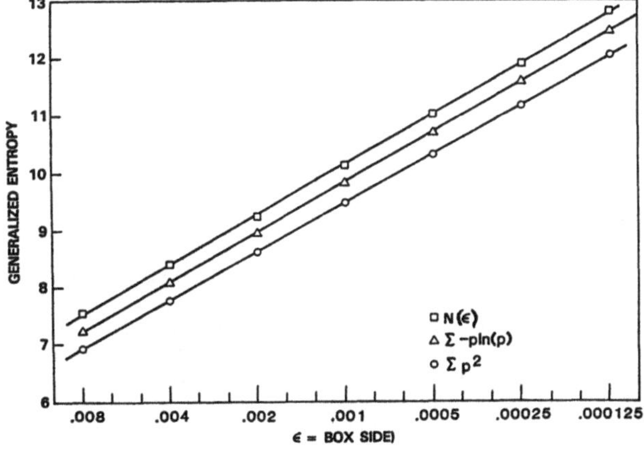

Fig. 6. Least-squares fit of the number of boxes to a power law ϵ^{-D} (top curve) for
the Henon attractor. The bottom two lines are the corresponding fits for the information dimension and to the q = 2 dimension

Perhaps surprisingly, the fit over 5 binary ranges gives $D_o[9,13] = 1.288 \pm 0.002$ while the fit over 7 binary ranges gives $D_o[7,13] = 1.270 \pm 0.006$, quite a difference for the same data. This large difference in estimates of D_o (three standard deviations) is about as large a difference in standard deviation units as is possible (we show this at the end of this section).

There is an upper bound on the information dimension to which we can compare: the information dimension is bounded above by the Lyapunov dimension [5], and often (typically) this dimension may equal the Lyapunov dimension. The Lyapunov dimension can be accurately calculated [6]. For the Henon attractor $D(\text{Lyapunov}) = 1.25826 \pm 0.00006$. This dimension is in good agreement with the 7-point fit $D_o[7,13] = 1.254 \pm 0.006$, but is in clear disagreement with the 5-point fit $D_o[9,13] = 1.271 \pm 0.001$. The 5-point fit, while apparently more accurate due to its smaller error, badly violates the rigorous bound. We could live with a violation of our expectation that the dimensions agree, but this violates a theorem, which is quite another matter. Our only recourse is to plead that we are not yet near the hoped for asymptotic regime where our dimension calculations are justified. These results force us to examine the meaning of the error estimates we have obtained.

The "standard deviations for slopes of the best-fit lines" which are calculated using standard least-squares methods **are not aimed at questions of interest to people making dimension calculations.** The error estimate certainly gives no indication of how much the dimension estimate (slope of the best fit line) will change if points are added or deleted. The following example will illustrate the problem.

Suppose we wish to fit 7 points to a straight line (say x = 1,2,...,7) and we wish to compare the best fit slope and error estimate of that slope with what we obtain using only 5 points: x = 1,2,...,5. To make this example simple, put the first five points on a straight line. There is no harm in choosing that line to be horizontal. Hence we start with

$$(1,0),(2,0),(3,0),(4,0),(5,0)$$

which gives a line of slope 0, with standard deviation 0 of that slope. We continue with two points off that line

$$(6,1),(7,a)$$

where a is to be chosen. Using all seven points, the best slope is

$$m(a) = (2+3a)/28$$

with the standard deviation error estimate of that slope being

$$\sigma(a) = (1 - a + 3a^2/4)^{1/2} /14.$$

The maximum of $m(a)/\sigma(a)$ comes at a = 4/3 yielding a slope m = 3/14 and an error estimate of that slope $\sigma = 1/14$, so for a = 4/3, the slope of 3/14 is exactly 3 standard deviations away from the 0 slope line obtained using 5 points. This is almost identical to the case we found in our data. So, **what does the standard least-square estimate mean to us? The answer is: not much.**

6. Discontinuity in the Box-Counting Dimension

The box-counting dimension is clearly very dependent on the low-probability boxes. As a parameter is varied, an interior crisis can result in a sudden jump in the size of the attractor, and the box-counting dimension can also jump. The addition of a spread-out (high box-counting dimension) region having very low probability p to the attractor can make the box-counting dimension jump while the information dimension must change continuously as p increases from 0. For example, consider a system in which the iterates randomly fill an area A with probability p, and a length of line L with probability 1-p. It is easy to see that the dimension D_q = 2 (for q < 1), 1+p (for q = 1) and 1 (for q > 1). In general, if two "attractors" are combined, visiting one with probability p and the other with probability 1-p, we can easily show that the information dimensions must combine linearly:

$$D_I = p\, D_{I1} + (1-p)\, D_{I2}.$$

We have studied the slow increase in the probability of hitting the low-density attractor after an interior crisis, and conclude that the information dimension will change slowly as we pass through the crisis point.

We have done some numerical tests to see if the box-counting dimension does in fact show a discontinuous jump at an interior crisis. We looked at the box-counting and Lyapunov dimensions just below and just above the crisis in the Henon attractor near a = 1.30868 and b = 0.3. For a = 1.30867 (before the crisis), the box-counting dimension equals $D_O[6,8]$ = 0.6 ± .1. For a = 1.30868 (after the crisis) the box-counting dimension has the value $D_O[6,8]$ = 1.20 ± 0.02. Both values are probably underestimates of the true value, as the dimension is increasing rapidly over the range available. For a below the crisis, the attractor is too dense to elucidate structure. The value D_O is less than 1, also demonstrating that we are below the true dimension. Above the crisis, many low-probability boxes are just being discovered, and long enough runs to do proper extrapolations were not possible especially when the box size is small. However, our point is that the box-counting dimension can show a dramatic increase when going through a crisis, at least when it is studied over comparable scales.

For comparison, the Lyapunov dimension for a = 1.30867 equals 1.08 ± 0.01 and for a = 1.30868 equals 1.085 ± 0.010, i.e. it changes slowly as we pass through the crisis. Above the crisis, it rapidly increases, and is about 1.157 at a = 1.31, and 1.195 at a = 1.32.

7. Conclusions

On the positive side, it is possible to extrapolate estimates of the generalized entropies to many iterations (infinitely long time series). This can be particularly useful for the information dimension, which does not have the extreme sensitivity to low-probability boxes that the box-counting dimension has. The extrapolation for the box-counting dimension (the number of boxes) is less exact, as the low-probability boxes are not under control. Using boxes exacerbates this problem by giving a ready source of low-probability boxes, while using disks allows this region to be seen more clearly. Furthermore, the extrapolations beyond the known data with the disks gives smaller changes, about a third the size of the extrapolation for boxes, for the cases we considered.

On the negative side, having done a careful extrapolation we find that **the dimensions we get are precise but not accurate.** They have very small least-squares errors, but the actual error is at least 13 times larger. We can calculate them, obtaining very small least-squares errors, but the values we obtain appear not to have settled down: the dimensions are still rising as we go to smaller box-size. **We feel that it is important to include the ranges of box-sizes used as part of the dimension reported.** Choosing box-sizes differing from each other by a factor of two suggests that these be called binary ranges, so that our most precise result is $D_I[9,13]$ = 1.271 ± 0.001 --which unfortunately violates the Lyapunov dimension upper bound by 0.013, or thirteen times the supposed standard error obtained from the least-squares method.

8. Appendix: Attractors Intersecting Squares and Disks

We wish to study the geometric reasons why disks are preferable to boxes, why the number of disks converges faster as n → ∞ than does the number of boxes. Boxes and disks have very different α and β in the large-iteration scaling (α and β are defined in section 3). In this appendix we will calculate the probability that a box or disk which intersects the attractor will do so in a segment (chord) of length w.

In the text we argue that for the low-probability boxes, all that is important is the low-w (edge) region. We first give a qualitative argument. We will treat the attractor locally as a straight line, with an arbitrary angle and displacement in the plane. We will consider a box (or disk) which is only constrained in that it must intersect the line (the attractor). A line which makes a tangential hit (e.g.

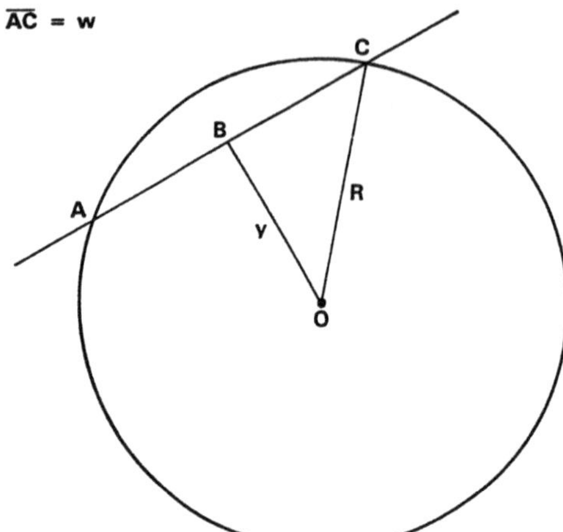

Fig. 7.   Line (AC) hitting a disk. The line can be taken to have a uniform distribution in y and in θ. The intersected segment length (AC) = w has a probability distribution shown in the next figure

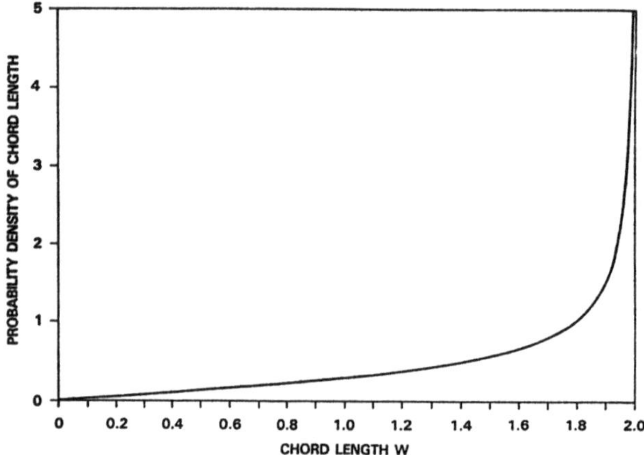

Fig. 8.   Probability density for obtaining an intersected segment of length w, when placing a line down arbitrarily (in translation and rotation) in the plane, such that it intersects a disk of radius 1

line AC in Figs. 7 and 9) will have length w. Displacements in the y-direction are all equally likely (the crucial assumption: the probability distribution is flat in y), and the question is how w changes as y increases to its limit. For the box (Fig. 9), y is related to w by some angular factors which do not change as the line is shifted laterally in y, and so the tangential hits have the same flat probability distribution. For the disk, w vanishes as the square of the difference of y from its limiting value, R. Thus a flat distribution in y will result in a vanishing distribution in w.

We calculate the exact distribution for the disk, and then quote the results for a box. The probability distribution in y is assumed to be flat, $p(y) = \theta(R-y)/R$. For a disk, the angular variable can be ignored, by symmetry. The Pythagorean theorem tells us that $w = 2(R^2-y^2)^{1/2}$. A simple change of variables from y to w gives us the distribution:

$$p(w) \;=\; p(y)dy/dw \;=\; (w/D)(D^2-w^2)^{-1/2}\,\theta(D-w)$$

134

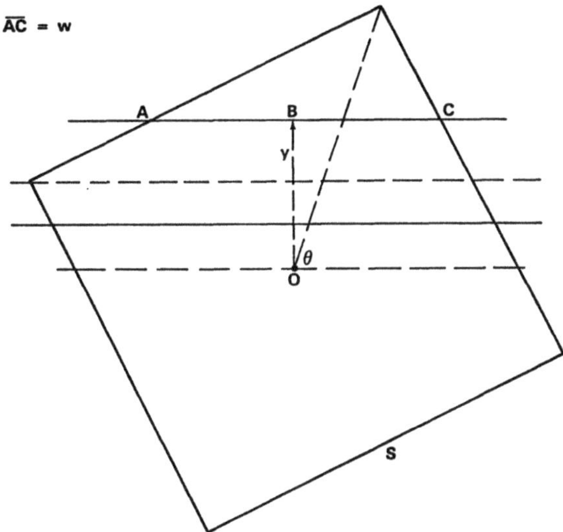

Fig. 9. Line (AC) hitting a square. The line can be taken to have a uniform distribution in y and in θ. The intersected segment length (AC) = w has the probability
distribution shown in the next figure. All lines between the dashed lines have the
same length, giving a δ-function distribution in y for fixed θ

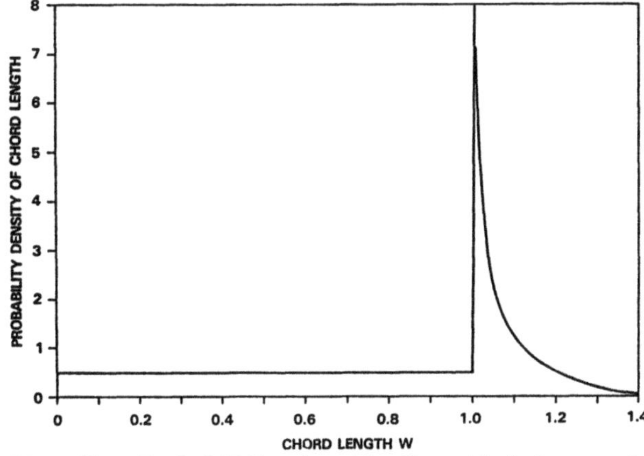

Fig. 10. Probability density for obtaining an intersected segment of length w; when
placing a line down arbitrarily (in translation and rotation) in the plane, such that
it intersects a square of side 1

where $D = 2R$ is the diameter of the disk. For small w, this vanishes as w/D^2. The
resultant distribution is shown in Fig. 8: note also the square root singularity as
$w \to D$.

The calculation is similar for the box, except that now the angular coordinates
must be averaged over. Also, when the line is between the two dashed lines (Fig. 9),
the length w is independent of y, and care must be taken in performing the variable
transform (the δ function goes away in the angular averaging). First, scale out
the side S of the square (set it equal to 1). If we define an angle φ by:

$$\sin(\phi) = (1/w + (1-1/w^2)^{1/2})/\sqrt{2}$$

$$\cos(\phi) = (1/w - (1-1/w^2)^{1/2})/\sqrt{2}$$

135

then the probability distribution in w becomes:

$$\rho(w) = (4/\pi)\,(1 + \ln(\sqrt{2}-1)2^{-\frac{1}{2}})\quad \text{for} \quad w < 1 \quad \text{and}$$

$$p(w) = (4/\pi)(w^{-2})/(1-w^{-2})^{\frac{1}{2}}\cot(\phi)$$

$$+ \sqrt{2}\,\cos(\phi) + \ln(\tan(\phi/2)/\sqrt{2}\,)$$

$$\text{for} \quad \sqrt{2} > w > 1.$$

The distribution is shown in Fig. 10: almost half of the total probability is in the flat region below w = 1.

Acknowledgments

This research was supported in part by the Independent Research Fund at the Naval Surface Weapons Center, by grants from the Office of Naval Research, the Air Force Office of Scientific Research, the National Science Foundation, and the Department of Energy. We wish to thank Robert Cawley and Frank Varosi for helpful conversations.

References

1. Farmer, J. D., E. Ott and J. A. Yorke, Physica 7D (1983) 153. See also J. P. Eckmann, D. Ruelle, Reviews of Modern Physics, 54, 617-656 (1985).
2. Grassberger, P., Phys. Lett. 97A, 227 (1983). Caswell, William E., NSWC Technical Note NLD-1, describes a computer program to calculate these dimensions.
3. Grassberger, P., Phys. Lett. 97A, 224 (1983).
4. Henon, M., Commun. Math. Phys. 50, 69 (1976).
5. F. Ledrappier, L. S. Young, Bulletin of the American Mathematical Society, 11, 343-346 (1984).
6. Grassberger, P. and I. Procaccia, Phys. Rev. 13D, 34 (1984).

Methods for Estimating
the Intrinsic Dimsnionality of High-Dimensional Point Sets

R.L. Somorjai

Division of Chemistry, National Research Council Canada, Ottawa,
Ontario, Canada, K1A OR6

1. Introduction

In recent years the characterization of fractals, strange attractors, dynamical chaos
has received much attention. A particularly attractive measure is the __fractal
dimensionality__ of the dynamical system trajectory. This is a (generally noninteger)
number that characterizes the trajectory (viewed as a collection of causally connected discrete points that are embedded in a D-dimensional space, such that d<D).
Depending on its definition, it may reflect purely geometrical aspects (capacity),
but also probabilistic, information-type features of the system. For practical purposes the Balatoni-Rényi generalized dimension [1] d_q, q>0, subsumes all currently
used definitions of d (see however, [2]). A number of algorithms have been proposed
to determine d [3-10]; they are reasonably successful, but only for d<3.

The purpose of this article is twofold. First, it draws attention to the fact
that methods to determine the __intrinsic dimensionality__ (ID) of point sets have long
been known in a wide range of disciplines (e.g. psychometry, taxonomy, artificial
intelligence, engineering, etc.) under various guises and for varied goals. The
basics of these methods will be reviewed and their scope and limitation discussed.
They will be contrasted with and compared to the algorithms that were specifically
designed for (chaotic) dynamical systems.

In addition, we shall develop the unifying notion that all current d-estimators
for dynamical systems can be usefully considered in the context of __probability
density estimation__ (PDE): in particular, the majority of these algorithms uses im-
plicitly the concept of __k-nearest-neighbor__ (k-NN) estimation. We shall elucidate
this connection explicitly by sketching a particular ID-estimator of the k-NN type
[11]. This will allow us to introduce an idea that seems essential if one wants
accurate estimates of the intrinsic dimensionality d when d is (much) greater than 3
and the data are limited. (E.g. recent experiments in turbulence etc. indicate
d>5.)

Finally, some other possible approaches to d-estimation will be outlined.

2. Extraction of Intrinsic Dimensionality and Dimensionality Reduction

Generally, these two concepts are closely linked. They are the essential goal of a
number of disciplines where one has to classify and interpret __patterns__. Thus methods
of Pattern Recognition and Classification, Cluster Analysis, Factor Analysis, etc.,
have to address this problem. Typically, data are presented either as a __moderate
sized set__ (<1000) of __high-dimensional points__, or as a __matrix of proximities__. In the
former case, the points can be viewed as D-dimensional vectors in some metric space.
The __components__ of the vectors (the "attributes") characterize the points ("features")
which constitute the __pattern__. __Dimension reduction__ is demanded because of the difficulty in interpreting high-dimensional patterns. Thus a new vector is sought whose
dimension is much lower than that of the original, while it still preserves the former's essential characteristics. This new dimension is the __intrinsic dimension__ of the
pattern and represents the minimum number of components required for its faithful
description. Note that the ID is integer-valued. The lower-dimensional pattern has
components that are combinations (linear or nonlinear) of the original attributes.

In certain disciplines (psychology, social sciences, etc.) one does not even know
the embedding dimension, the space is frequently non-metric and the data ordinal,
presented as a matrix of proximities. We shall discuss the ramifications in connection with the specific methods that were designed to deal with these kind of data.

In summary, the salient features of these problems are:

a) moderate sized data base,
b) high dimensionality of the embedding space,
c) ID integer-valued,
d) no causal relationship between points.

Contrast this with the case of strange attractors and other chaotic dynamical
systems where one has

a) large set of points (discretized trajectory),
b) embedding space is low-dimensional,
c) ID is generally non-integer valued,
d) causal link between points.

The ID is one measure of the information content of the trajectory, independently of
the embedding dimension (which is always integer-valued).

3. Methods for Extracting the ID

A. Karhunen-Loève Expansion (Factor Analysis)[12]

The K-L expansion (and the mathematically closely related Factor Analysis (FA)) can
be characterized as the method that finds the best coordinate system for information
compression. The K-L expansion minimizes the average error committed by taking only
a finite number of terms in the infinite series of an expansion when a given set of
functions is expressed in terms of some complete set of orthogonal functions. (It
also minimizes the entropy defined in terms of the average squared coefficients of an
expansion.) FA attempts to reduce many correlated (redundant) random variables to
fewer redundancy-free hidden common variables on which the original variables are
linearly dependent:

$$\{x_i\}, i=1,\ldots,D \;\rightarrow\; \{y_j\}, j=1,\ldots,d \;, \qquad y_j = \sum_{i=1}^{D} a_{ji} x_i \;, \qquad d \ll D \;.$$

The actual procedure is to form the covariance matrix R from the N D-dimensional
points and find its eigenvalues and eigenvectors. The d eigenvectors corresponding
to the d largest eigenvalues of R define a linear transformation from D-space to
d-space, $d < D$; in the latter the original patterns are uncorrelated. The ratio

$$r = \sum_{i=1}^{d} \lambda_i \Big/ \sum_{i=1}^{D} \lambda_i$$

gives the fraction of total variance that is retained in d-space (the λ's are the
eigenvalues of R). Typically, the d that gives $r > 0.9$ is identified with intrinsic
linear dimension of the set.

Note that there are dangers in using a covariance matrix to characterize data
[13]. Possibly because the K-L ID is a linear dimensionality, this method does not
work too well, especially for when that data are distributed in nonlinear subspaces.

A variant of the K-L algorithm was developed by FUKUNAGA and OLSEN [14]. This is
based on local K-L expansions, using subsets of the data that are contained in small
hyperspherical regions. Local ID's are determined from which a global ID may be
obtained.

The other known algorithms for ID can be classified into two categories. They are
called "static", if ID is estimated directly from the original data set. The
"dynamic" algorithms flatten, unfold or linearize the patterns either prior to, or
during the process of ID determination.

<u>Dynamic Approaches</u>

B. <u>The B-C-A Algorithm</u>

This algorithm, proposed by BENNETT [15] and modified by CHEN and ANDREWS [16] is
based on the fact [17] that for two uniformly distributed point sets the configuration with a <u>larger</u> interpoint distance <u>variance</u> will have a <u>lower</u> dimensionality.
The sets are in D-dimensional hyperspheres of radius R and the probability density
function $P_D(r)$ for the normalized interpoint distance $r=r_{ij}/2R$ is

$$P_D(r) = 2^D Dr^{D-1} I_\delta [(D+1)/2, \ 1/2],$$

where $\delta = 1-D^2$, $I_z(p,q)$ is the incomplete beta-function and r_{ij} is the distance from
point x_i to point x_j in the hypersphere. The dimensionality is reduced by iterations
on the interpoint distances which stretch or shrink them to increase the variance.
This "unfolding" is carried out under the constraint that the rank order of the
cluster is maintained [15]. However, this is generally too restrictive, hence the
use of a <u>cost function</u> [16] which places a higher penalty on misranking the order in
local regions than in nonlocal ones. The unfolded pattern is still in the space of
dimensionality D and the reduction to the proper ID is generally carried out via the
K-L expansion. Thus, the method is basically a pre-processing of the original data.

C. <u>The S-V Algorithm</u>

Assume that the D-dimensional points are <u>nodes</u> of a graph, and their interpoint distances are its <u>edges</u>. The main idea of the algorithm is to unfold the original data
cluster, using the notion of the <u>minimum spanning tree</u> (MST) as the information
invariant that represents the data [18]. The MST configuration is "linearized" via a
barycentric transformation.

<u>Definition</u>: The MST of N points in a metric space is a tree (a graph with no
closed-loop paths) which spans all N points and whose total length is minimum.

Its main properties are:

a) any point is connected to at least one of its nearest neighbors
b) any subtree is connected to at least one nearest neighbor by the shortest
   available path
c) the MST minimizes all increasing symmetric functions (maximizes all decreasing
   symmetric functions) of the interpoint distances r_{ij}
d) The MST is invariant under similarity transformations (translation, rotation,
   reflection) and in general under all r_{ij}-order-preserving transformations.

Efficient algorithms are available to generate the MST rapidly [19,20].

Note that the original N points with N(N-1)/2 interpoint distances, after replacement with the MST, reduce to N points with N-1 distances, and the configuration is
now flexible.

The <u>barycentric transformation</u> (BCT) "straightens" the original MST. Given N
points in E_D and its computed MST, the BCT replaces every point x_i , i=1,..,N by the
barycenter of the cluster formed by x_i <u>and all the points connected to it in the MST</u>.
After substitution, the MST is recomputed and the original MST is restored by uniform
scaling. (This prevents the MST from collapsing to a single point.) The stopping
criteria for iterating the above procedure is either that the MST breaks up or that
the interpoint distance variance stabilizes (at a maximum). Thus the latter is identical to the B-C-A criterion for unfolding. The ID is determined by a K-L
expansion.

139

A potential problem with the method arises when the variance of the lengths of the MST branches is large; then long branches are straightened faster than short ones.

D. Multidimensional Scaling (MDS)

The two previous dynamic approaches can be characterized as nonlinear transformations of the original D-dimensional point sets. However, in certain disciplines only a set of proximities are available as input. A proximity is a number which indicates how similar or how different two objects are. Multidimensional scaling is a class of techniques which, given a set of proximities, produces a spatial representation, consisting of a geometric configuration of points. Each of these points corresponds to one of the objects. The configuration reflects the "hidden structure" in the data, and often helps comprehension and interpretation. The essential requirement for good MDS is that the larger the dissimilarity (or the smaller the similarity) between two objects, as shown by their proximity value, the further apart they should be in the configuration space.

Given the proximity matrix $\{\delta_{ij}\}$, a practitioner of MDS has to select the dimensionality D of the embedding space, the metric of the space and an initial configuration. The central comcept of MDS is to find the best monotonic correspondence between the proximities δ_{ij} and the distances r_{ij} in the chosen D-dimensional space: $f(\underline{\delta})=\underline{r}$. Typically, a quantity, called the stress F is minimized in the least squares sense:

$$F = \min_{x_i}\left[\sum_i \sum_j [f(\delta_{ij})-r_{ij}]^2/S\right]^{\frac{1}{2}} ,$$

where $r_{ij} = \|x_i-x_j\|$ and S is a scale factor.

The minimization of F is with respect to the best monotonic function f(.).

The computational procedure itself starts minimization in the highest dimensional space possible, D=N-1, where N is the total number of objects. Either a regular D-simplex or some random configuration is used. Elaborate procedures have been developed to prevent premature termination in a local minimum. Starting in the highest dimensional space appears to help. Once a minimum in F is determined, one reduces the embedding dimension and repeats the process. The ID is generally determined from a F(D) vs D plot. Frequently, a well-defined "elbow" in the plot indicates the ID. Alternatively, comparison with Monte Carlo simultations is made.

MDS is a versatile and powerful approach and numerous variants have been developed, including non-metric [21] and 3-way [22] MDS. A modern presentation of the theory, with history and classification, is in [23]. Recent applications can be found in [24]. A detailed comparison of two prominent algorithms is carried out in [25]. A simple introduction to MDS concepts is in [26], while an earlier, but more technical assessment of problems and prospects is given by SHEPARD [27]. Two entire volumes have been devoted to pre-1972 theory and application of MDS [28].

A variant of MDS is SAMMON's nonlinear mapping [29]. This starts with N points in D-space and maps them into 2-space by minimizing a mapping error E:

$$E = \left[1/\sum_{i<j}^{N} (r_{ij}^*)\right] \sum_{i<j}^{N} \left[(r_{ij}^*-r_{ij})^2/r_{ij}^*\right] ,$$

where $r_{ij}^* = \|x_i-x_j\|$ is in D-dimensional Euclidean space while r_{ij} is in 2-space. HOWARTH [30] tested the method in a geological context and suggested, on the basis of Elefante's results for simplexes in 2-19 dimensions, that E could be used to obtain the ID of a given data set.

The dynamic approaches described above (with the exception of MDS) do not determine directly the ID. This is done after the original data were suitably unfolded. The static methods we now outline are specifically designed to obtain the ID.

5. Static Approaches

A. TR Algorithm

TRUNK [31] has proposed a statistical method of estimating ID with a sequence of
hypothesis tests. Consider the null hypothesis test: d=k at a point x. Then the
vector v from x to its (k+1)st nearest neighbor (NN) among the given points is projected onto the space E_k spanned by the k vectors from x to its first k NN's. If the
angle between v and this space is small, then the (k+1)st NN is assumed to lie in E_k
and the local ID about x is k or less. If the angle is large then the local ID is
greater than k. The overall ID is estimated by computing the angles for all points
and averaging them. It this average angle is less than some threshold for dimension
k then the null hypothesis is accepted, otherwise, k is incremented and the process
repeated. Although fast and requires few parameters (noise variance and thresholds),
the method seems to be limited to $d < 4$, because thresholds are available only for the
above d's; the probability of making the correct decision decreases as d increases.
Another limitation is that decisions are based on angles between <u>linear</u> subspaces.

B. Nearest-Neighbor Algorithms

Nearest-neighbor (NN) methods, standard for probability density estimation (PDE) in
statistics [32,33], have been rediscovered by physicists and mathematicians, who were
interested in characterizing dynamical system trajectories by their (fractal) dimensionality [3-10]. However, the probabilistic underpinning of NN methods isn't explicitly emphasized and this leads to ambiguities when the ID is determined from the
slope of the ubiauitous log-log plots which are generally S-shaped. On the other
hand, a NN-based method was developed for pattern analysis [11] that uses explicitly
probabilistic concepts in its derivation. In fact, it assumes that the D-dimensional
points are governed locally by a Poisson spatial process with unknown density p(.).
The most important feature in this derivation is that it demonstrates explicitly that
the dependence of k-NN distances on k <u>does not</u> follow pure power-law behavior.

Consider a set of points $\{x_i\}$, i=1,2,...,N, embedded in D-dimensional Euclidean
space. The x_i are sampled from an <u>unknown</u> distribution with probability density
p(x). The simplest estimator of p(x), $\hat{p}(x)$ is [34]

$$\hat{p}(x) = (k/N)/V , \tag{1}$$

where k is the number of NN's of x within a hypersphere of radius R_k about x. The
volume is $V = V_d R_k^d$, where $V_d = \pi^{d/2}/\Gamma(d/2+1)$ is the volume of a unit d-dimensional
hypersphere. Substitute into (1) and take logarithms to get

$$\log R_k = (1/d)\{\log k - \log N - \log[V_d \hat{p}(x)]\} . \tag{2}$$

Eq. (2) cannot be used directly to estimate d because $\hat{p}(x)$ is not independent of k.

Let $r_k(x)$ be the distance from x to its kth NN. If p(x) is continuous and nonzero at x, then the density function for $r_k(x)$ for sufficiently large N and small r>0
is [11]

$$f_{k,x}(r) = (d/r\Gamma(k))C^k \exp(-C) ,$$

where $C = Np(x)r^d V_d$. The expected value of $r_k(x)$ is $E[r_k(x)] = (k/C)^{1/d}/G_{k,d}$,
with

$$G_{k,d} = k^{1/d}\Gamma(k)/\Gamma(k+1/d) .$$

If $\bar{r}_k$ is the sample-averaged distance to the kth NN over the set of points,

$$\overline{r}_k = (1/N) \sum_{i=1}^{N} r_k(x_i) \; ,$$

then the expected value of this average is

$$E(\overline{r}_k) = (1/N) \sum_{i=1}^{N} E[r_k(x_i)] = k^{1/d} S_N / G_{k,d} \tag{3}$$

and SN is independent of k. Taking logarithms of (3) gives

$$\log G_{k,d} + \log E(\overline{r}_k) = (1/d) \log k + \log S_N \; . \tag{4}$$

The term $\log G_{k,d}$ is in the range $0 < \log G_{k,d} < 0.12$ for all k and d, the maximum occurring for $k=1$ and $d=2.17$ [11]. As an estimator for $E(\overline{r}_k)$ use $\overline{r}_k$ in (4). Using this observed value gives an estimator d of d:

$$\log G_{k,\hat{d}} + \log \overline{r}_k = (1/\hat{d}) \log k + \log S_N \; . \tag{5}$$

Plot $\log \overline{r}_k$ vs $\log k$ for $k=1,2,\ldots, k_{max}$ and least-square fit a straight line through the points. Its slope will be $1/\hat{d}$. One has to solve for $\hat{d}$ iteratively; the initial estimator $\hat{d}_o$ is obtained by setting $\log G_{k,\hat{d}}$ to zero [11]. Typically 4-5 iterations are sufficient for $\hat{d}$ to converge to a limit within 0.003. Note that $\hat{d}_o$ is frequently quite inaccurate.

<u>6. Discussion</u>

The above k-NN algorithm was tested against many of the other ID-estimators in [35]. The general conclusion was that it worked better than most of the others (MDS was not tested), at least on the sample problems the authors had considered. However, these tended to be low-dimensional (ID$<$4), with rather limited number of patterns.

The defects of the k-NN algorithm as formulated in [11] became obvious when I tried to apply it to the characterization and classification of trajectories of both integrable and non-integrable Hamiltonian systems of high dimensionality [36]. It consistently <u>underestimated</u> the known ID. This led me to consider in depth the nature of the k-NN algorithm.

The most relevant characterization of the k-NN method is that it is a <u>nonparametric</u> estimator of multivariate probability densities. This means in particular that the exact nature of the unknown p(x) is not critical; the penalty one pays for this freedom is that the efficiency of nonparametric methods is much less than optimal.

Since practical considerations limit the total number of points N, it is essential to determine the optimum rate of convergence of the estimate $\hat{d}(N,k)$ to d with N and k. This was ignored in [11], even though it is known. In fact, in order that $\hat{p}(x)$ of (1) converge to the true p(x) it is necessary and sufficient that [37]

$$\lim_{N \to \infty} (k/N) = 0 \quad \text{and} \quad \lim_{N \to \infty} k(N) = \infty \; .$$

Empirical estimates indicate that $k \propto \sqrt{N}$ is a reasonable choice. If the underlying distribution is multivariate Gaussian, then [32,33] the fastest rate at which the local mean square error of the estimated density decreases is $\sim N^{-\beta}$ if $k \sim N^{\beta}$, $\beta = 4/(4+d)$. Furthermore, the <u>variance</u> of p is $O(1/k)$ and its <u>bias</u> is $O[(k/N)^{2/d} + 1/k]$.

The constant of proportionality reduces from 0.75 for $d=4$ to 0.42 for $d=32$ [38]. For fixed sample size N, as d increases k_{opt} decreases; this may be the consequence of the emptiness of high-dimensional space.

We tested the k-NN algorithm embodied in (5) by applying it to various distributions of <u>known</u> dimensionality [39]. In particular, uniformly distributed random points were generated from the interior of the unit d-sphere as well from its surface, and from the unit d-torus, using the fast algorithms of [11], Typically, for each dimension $1 < d < 20$, ten samples of N=250, 500, 1000 and 2000 points were drawn and their <u>apparent</u> ID determined via (5). The averages and standard deviations were computed for several k's. The number of NN's chosen was proportional to $\sqrt{N}$, (i.e. $k = [\alpha \sqrt{N}]$, [ ] denoting the integer part), with 6 different α's, $\alpha = m/4$, $m=2(1)7$.

The results confirm the theoretical expectations [32,33,38]. The three most relevant findings are that

1) $\hat{d}(k,N)$ are only <u>weakly</u> dependent on the nature of the distribution,
2) $\hat{d}(k,N)$ increases <u>sublinearly</u> with increasing d,
3) $\hat{d}(k,N) > \hat{d}(k',N)$ <u>if k < k'</u>.

A crude error analysis suggests that (1/d) could be expanded in inverse powers of $\hat{d}$. In fact, fitting $1/d^2$ to a polynomial in $1/\hat{d}$:

$$1/d^2 = \sum_{i=0}^{3} a_i / \hat{d}^i$$

and determining the best $a_i(k,N)$ by least squares fitting gives a fitting error of less than 0.5% over the various distributions and (k,N)-pairs tried. For fixed N, low-degree polynomials in k give excellent fits to $a_i(k,N)$; for fixed k, equally good fits with such polynomials in 1/N can be obtained. Thus, reasonable interpolation in the k-1/N plane is possible [39].

The <u>calibrated</u> algorithm gave very respectable results for the standard chaotic attractors in the literature [39], even with as few as 500 points. It also gave the true ID within 10% for d=20 (20 coupled harmonic oscillators embedded in the 40-dimensional phase space) and N=2000 [39].

Thus, it seems that for high-dimensional sets <u>calibration is essential</u> to recover the true ID (e.g. for d=20, $\hat{d}(k,2000) \approx 13$, and the bias would decrease only as $N^{-1/12}$). The penalty incurred (the distribution dependence of the result) appears to be small [39]. The conclusion is particularly relevant for experimental time series for which <u>embedding</u> into higher- and higher-dimensional spaces is advocated to recover the ID [6,40].

It should be emphasized again that most of the currently used d-estimators in the dynamical chaos-strange attractor field are variants of k-NN multivariate density estimators. They differ from each other only in their preference of how to solve (2). None of them <u>iterate</u> on $\hat{d}$, as one does in solving (5) [11,39]. The latter's explicit formulation provides alternate estimators [39]. Thus assume that

$$(1/N) \sum_i p(x_i)^{-1/d}$$

is very weakly dependent on N, a reasonable assumption for reasonably large N. Write $r(N,k,d)$ for $E(r_k)$. Then define

$$T(N,N') \equiv r(N,k,d)/r(N',k,d) = (N'/N)^{1/d} , \qquad N,N' \text{ large} ,$$

$$U(k,k') \equiv r(N,k,d)/r(N,k',d) = \Gamma(k')\Gamma(k+1/d)/\Gamma(k)\Gamma(k'+1/d) .$$

In particular, $U(k,k+1) = k/(k+1/d)$. Thus log $T(N,N')$ vs log(N'/N) is a straight line with slope 1/d and $U(k,k+1)$ vs (1/k) has slope (-1/d) for $kd >> 1$. These, either separately, or together with (5) could be used to give a more consistent estimation of d.

7. Prospects

The most fruitful extension for improved d-estimation is to regard it as a byproduct
of probability density estimation, either _directly_ or via a finite number of its
moments. The latter can be most readily implemented for the k-NN estimator as derived through Eqs. (1)-(5). The critical step is to replace the expected value of
$r_k(x)$ by the expected value of $r_k^\gamma(x)$, $E(r_k^\gamma)$:

$$E(r_k^\gamma) = \int_0^\infty r^\gamma f_{k,x}(r)dr = (k/C)^{\gamma/d} G_{k,d,\gamma}$$

with

$$G_{k,d,\gamma} = k^{\gamma/d} \Gamma(k)/\Gamma(k+\gamma/d) .$$

Sample-averaging r_k^γ:

$$\overline{r_k^\gamma} = (1/N) \sum_{i=1}^{N} r_k^\gamma(x_i)$$

leads to the expected value $E(\overline{r_k^\gamma})$

$$E(\overline{r_k^\gamma}) = (1/N) \sum_i E[r_k^\gamma(x_i)] = k^{\gamma/d} S_N(\gamma)/G_{k,d,\gamma} ,$$

where

$$S_N(\gamma) = (1/N) \sum_i \left[N p(x_i)V_d \right]^{-\gamma/d} .$$

Taking logarithms and substituting $\overline{r_k^\gamma}$ for $E(\overline{r_k^\gamma})$ gives

$$\log G_{k,\hat{d},\gamma} + \log r_k^\gamma = (\gamma/\hat{d}) \log k + \log S_N(\gamma) ,$$

completely equivalent to (5). $0 < G_{k,d,\gamma} < 0.12\gamma$ for $\gamma > 0$ and the same fitting
procedure could be applied. Thus either one could obtain a $\hat{d}$ which is consistent for
a range of moments by fitting _simultaneously_, or a set $\hat{d}(\gamma)$ can be calculated by
individual fitting for each γ. The latter approach has been advocated very recently
[40], and purports to define a nonuniformity estimate for the dynamical system
(distribution).

Rather than using NN-type estimators, one can use explicitly the relation for the
γ-th moment of the distribution of distances r in a d-sphere of radius R [17]

$$\overline{r^\gamma} = H(d)(2R)^\gamma \Gamma[(d+\gamma+1)/2]/\Gamma(d+\gamma/2+1)(d+\gamma) , \tag{6}$$

with

$$H(d) = d\Gamma(d+1)/\Gamma[(d+1)/2] .$$

The best $d(\gamma)$ is obtained by estimating the moments from the data and finding a
non-linear least-sqaures fit to (6) for each γ. By choosing different R's (different
fractions of the data) both _local_ and _global_ information about the distribution can
be gleaned. If $d(\gamma)$ is truly dependent on γ, two possible conclusions can be
reached:
    a) The underlying distribution is hyperspherical but _nonuniform_
    b) the underlying distribution is uniform but _not hyperspherical_.
(Note that this important distinction is not made in [40]). Since one can vary R, as

well as move the center of the hypersphere of fixed radius to sample the coverage of
the data set, an idea of the nature of distribution can be built up. In addition,
<u>simultaneous</u> fitting of all moments to one overall d gives additional information.
This particular approach can be extended to other domains much more readily than the
NN approach (for cylinders see [41,42]).

All the above is tantamount to estimating, via a finite number of generalized
moments (γ does not have to be integer) the underlying probability density. A direct
tackling of the problem is possible in principle, however, in practice, two serious
obstacles bar this route. One is the "curse of dimensionality": high-dimensional
space is essentially empty. The other is that noisy and information-poor variables
quickly derail methods that are based on interpoint distances.

An exciting new class of methods has great promise. These are the <u>projection
pursuit</u> (PP) techniques, first successfully implemented by FRIEDMAN and TUKEY [43].
The original formulation has been extended since to PP-regression [44], PP-classi-
fication [45], and PP-probability density estimation [46]. A masterly review, which
puts PP on a general theoretical basis and shows how other, commonly used multivariate techniques are special cases of PP, is given in HUBER [47].

The general idea of PP is to find "interesting" low-dimensional projections of a
high-dimensional point set by numerically maximizing a certain objective function or
<u>projection index</u> (PI). By choosing different projection indices different charac-
terizations of the high-dimensional set can be achieved. An elegant and useful
classification of PI's is given in [47].

The most important feature of PP is that it can bypass the curse of dimensionality. This is achieved by extending univariate density estimation to higher
dimensions such that it involves only univariate estimation.

The PP density estimation method (PPDE) constructs estimates of the unknown p(x)
recursively: [46]

$$p(x) \sim p_M(x) = p_0(x) \prod_{m=1}^{M} f_m(\theta_m \cdot x) \tag{7}$$

where p_M is the estimate after M iterations of the procedure; p_0 is a given initial
multivariate density function; θ_m is a unit vector specifying a direction in the
D-dimensional embedding space, so that

$$\theta_m \cdot x = \sum_{i=1}^{D} \theta_{mi} x_i \; ;$$

$f_m(.)$ is a <u>univariate</u> function.

PPDE chooses the directions θ_m and constructs the corresponding functions
$f_m(\theta_m \cdot x)$.

From (7) one gets the recursion relation

$$p_M(x) = p_{M-1}(x) f_M(\theta_M \cdot x) \; .$$

The relative goodness of fit is measured by the cross-entropy term of the
Kullback-Leibler distance $W = \int \log f_M(\theta_M \cdot x) p(x) dx$. W is maximum at the same location as

$$w(\theta_M, f_M) = \int \log f_M(\theta_M \cdot x) p(x) dx \; . \tag{8}$$

Eq. (8) is maximized by

$$f_M(\theta_M \cdot x) = p^{\theta_M}(\theta_m \cdot x)/p^{\theta_M}_{M-1}(\theta_m \cdot x)$$

if $p(x)$ and the direction θ_M is known. Here p^{θ_M} and $p^{\theta_M}_{M-1}$ represent the data and current model <u>marginal</u> densities along the 1-dimensional subspace spanned by θ_M. Using this f_M for a given θ_M, the direction θ_M that maximizes (8) is found.

In practice $p(x)$ is not known. In its place we have a sample of N observations $\{x_i\}$. The cross entropy is then estimated by the log-likelihood

$$\hat{W} = (1/N) \sum_{i=1}^{N} \log p_M(x_i) \, .$$

Similarly, $w(\theta_M, f_M)$ is estimated by

$$\hat{w}(\theta_M, f_M) = (1/N) \sum_{i=1}^{N} \log f_M(\theta_M \cdot x_i) \, .$$

Details of the implementation are to be found in [46]. This includes the elimination of redundant variables, termination criteria and examples. The examples demonstrate the superiority of PPDE over k-NN methods, with or without noise. It would be particularly interesting to test PPDE as a means of finding an upper bound to ID in high-dimensional samples, or for experimental situations where embedding in higher dimensional spaces is required to reconstruct the phase portrait. Such studies are being initiated.

8. Conclusions

It appears that a useful cross-fertilization among the different ID-estimator procedures and concepts is possible. Various versions of nonlinear mapping of the original data set, together with direct (via PPDE) or indirect (via moments) estimation of the underlying probability density function and the related ID-function seem to offer the greatest promise.

9. Acknowledgement

I thank Dr. G. Mayer-Kress and the Los Alamos National Laboratory, who invited me to present a lecture at the CNLS Workshop on Dimensions and Entropies in Chaotic Systems; this spurred me to compile and present this cross-disciplinary review.

References

1. J. Balatoni and A. Rényi, Publ. Math. Inst. Hung. Acad. Sci. 1, 9 (1956).
2. E. Ott, W.D. Withers and J.A. Yorke, J. Stat. Phys. 36, 687 (1984.
3. P. Grassberger and I. Procaccia, Phys. Rev. Letts. 50, 346 (1983).
4. J. Guckenheimer and G. Buzyna, Phys. Rev. Letts. 51 1438 (1983).
5. R. Badii and A. Politi., Phys. Rev. Letts. 51, 1661 (1984).
6. P. Grassberger and I. Procaccia, Physica 9D, 189 (1983).
7. H.G.E. Hentschel and I. Procaccia, Physica 8D, 435 (1983).
8. Y. Termonia and Z. Alexandrowicz, Phys. Rev. Letts. 51, 1265 (1983).
9. P. Grassberger and I. Procaccia, Physica 13D, 34 (1984).
10. P. Grassberger, Phys. Letts. 97A, 224, 227 (1983).
11. K.W. Pettis, Th. A. Bailey, A.K. Jain, R.D. Dubes, IEEE. Trans. Pattern Anal., Machine Intell. PAMI-1, 25 (1979).
12. S. Watanabe, Trans. Fourth Prague Conf. Inform. Theory, Statist. Decision Functions and Random Processes, (1965), pp. 635.
13. G.H. Ball, AFIPS Fall Joint Computer Conf. 533 (1965).
14. K. Fukunaga and D. R. Olsen, IEEE Trans. Comput. C-20, 176 (1971).
15. R.S. Bennett, IEEE Trans. Inf. Theory, IT15, 517 (1969).

16. C. Chen and H.C. Andrews, IEEE Trans. Comput. $\underline{C-23}$, 178 (1974).
17. R.D. Lord, Ann. Math. Stat. $\underline{25}$, 794 (1954); M.G. Kendall and P.A.P. Moran, Geometrical Probability, Griffin's Statistical Monographs and Courses #10, Ch. Griffin and Co. Ltd., London (1963), pp.53.
18. D.H. Schwartzmann, Ph.D. Thesis, Dept. of Biomed, Eng. U. of California, Los Angeles (1972); D.H. Schwartzmann and J.J. Vidal, IEEE Trans. Comput. $\underline{C-24}$, 1175 (1975).
19. J.C. Gower and G.J.S. Ross, Appl. Statistics $\underline{18}$, 54 (1969).
20. F.J. Rohlf, Comp. J. (Algorithms Suppl.) $\underline{16}$, 93 (1970).
21. J.B. Kruskal, Psychometrika, $\underline{29}$, 115 (1964).
22. J.D. Carroll and J.J. Chang, Psychometrika $\underline{15}$, 283 (1970).
23. J. De Leeuw and W. Heiser, in "Handbook of Statistics", Vol. 2, pp. 285 (1982), North-Holland Publ. Co., P.R. Krishnaiah, L.N. Kanal (eds.).
24. M. Wish and J.D. Carrol. Ref. 23, pp. 317.
25. J.C. Lingoes and E.E. Roskam, Psychometrika Monograph Suppl. $\underline{38}$, 1 (1973).
26. J.B. Kruskal and M. Wish, Multidimensional Scaling (1978) Sage Publications, Beverly Hills/London.
27. R.N. Shepard, Psychometrika, $\underline{39}$, 373 (1974).
28. A.K. Romney, R.N. Shepard and S.B. Nerlove, Multidimensional Scaling, Vol. 1, Theory, Vol. 2, Applications (1972) New York, Seminar Press.
29. J.W. Sammon, IEEE Trans. Comput. $\underline{C-18}$, 401 (1969).
30. R.J. Howarth, Math. Geology, $\underline{5}$, 39 (1973).
31. G.V. Trunk, IEEE Trans. Comput. $\underline{C-25}$, 165 (1975); Inform. Contr., $\underline{12}$, 508 (1968).
32. Y.P. Mack and M. Rosenblatt, J. Multivariate Analysis, $\underline{9}$, 1 (1979).
33. M. Rosenblatt, in Smoothing Techniques for Curve Estimation, eds. Th. Gasser and M. Rosenblatt (Lecture notes in Math. No. 757, Springer, Berlin (1979)) pp. 181.
34. R.O. Duda, P.E. Hart, Pattern Classification and Scene Analysis, new York, Wiley (1973), p. 87.
35. N. Wyse, R. Dubes, A.K. Jain, in "Pattern Recognition in Practice", E.S. Gelsema, L.N. Kanal (eds.), North Holland Publ. Co., p. 415 (1980).
36. R.L. Somorjai, (unpublished 1980-82).
37. D.O. Loftsgaarden and C.P. Queensberry, Ann. Math. Statist. $\underline{36}$, 1049 (1965).
38. K. Fukunaga and L.D. Hostetler, IEEE Trans. Inform. Theory $\underline{19}$, 320 (1973). R.L. Somorjai and M.K. Ali, Physica D. (submitted).
39. F. Takens, in: Proc. Warwick Symp. 1980, D. Rand and B.S. Young, eds., Lecture Notes in Math. 898 (Springer, Berlin, 1981).
40. R. Badii and A. Politi, J. Stat. Phys. $\underline{40}$, 725 (1985).
41. J.M. Hammersley, Proc. Roy. Soc. (A), $\underline{210}$, 98 (1951).
42. J.M. Hammersley, J. Math. Phys. $\underline{31}$, 139 (1952).
43. J.H. Friedman and J.W. Tukey, IEEE Trans. Comp. $\underline{C-23}$, 881 (1974).
44. J.H. Friedman and W. Stuetzle, J. Amer. Statist. Assoc. $\underline{76}$, 817 (1981).
45. J.H. Friedman and W. Stuetzle, (unpublished, 1980); J.H. Friedman, Dept. Statist., Stanford Univ. Report LCM006 (1984).
46. J.H. Friedman, W. Stuetzle, A. Schroeder, J. Amer. Statist. Assoc. $\underline{79}$, 599 (1984).
47. P.J. Huber, Ann. Statist. $\underline{13}$, 435 (1985).

Part VI

Analysing Spatio Temporal Chaos

Characterizing Turbulent Channel Flow

A. Brandstater[1], *H.L. Swinney*[1], *and G.T. Chapman*[2]

[1] Department of Physics and the Center for Nonlinear Dynamics,
University of Texas, Austin, TX 78712, USA
[2] NASA Ames Research Center, Moffett Field, CA 94035, USA

We discuss different methods of characterizing turbulent channel flow in terms of "the number of independent degrees of freedom". These methods all suggest that the dimension of this system is greater than 10.

1. Introduction

In recent years low-dimensional strange attractors have been found in hydrodynamic systems which are closed (bounded), such as Rayleigh-Benard convection [1] and Couette flow [2]. These systems show highly coherent structures even above the onset of chaotic behavior, and attractor dimensions between two and five have been determined. The available algorithms (for computing, for example, the correlation dimension [3]) provide reasonable estimates for dimension values up to ~ 10, and some idea of the dimension for values up to perhaps ~ 20; hence they have been adequate for the analysis of chaos in closed systems, at least near the onset of chaos.

Now there is an increasing interest in open hydrodynamic systems [4], which usually show behavior that is less coherent than closed systems. We are examining turbulence in an open system, channel flow [5], which is the flow between two parallel walls of finite separation and infinite extent in the streamwise and spanwise directions. The flow undergoes a direct transition from a two-dimensional laminar flow (which has a parabolic profile) to turbulence without any intervening sequence of instabilities. The Reynolds number (Re= VH/ ν, where V is the center-line velocity, H is the channel half-width, and ν is the kinematic viscosity) at which the parabolic flow becomes linearly unstable is 5772 [6]; however, there is large amount of hysteresis--once the flow becomes turbulent, it will remain turbulent with decreasing Re down to about 1300. In practice, it is difficult to construct a channel with perturbations so small that the laminar flow can be maintained to the linear instability point [7].

Turbulent channel flow contains horseshoe-like spatial structures, so called hairpin vortices; see Fig. 1. These vortices are generated at the channel wall and decay towards the channel center. The vorticity vector has a preferred direction of 45° with respect to the wall, and each of these structures shows a leg of positive and a leg of negative vorticity. The size as well as the distribution of these vortices is unpredictable.

Coherent structures imply a reduction in the number of degrees of freedom of the flow. Because the vortices are dense at the wall and decay towards the center, it seems plausible that fewer modes would be needed to describe the flow near the wall than in the center. For the hypothetical case of infinite resolution, however, one should be able to detect all modes at any point in the fluid, even if its amplitude were infinitesimally small; then presumably the dimension would be the same for attractors constructed from measurements at any point in the flow. However, the dimension calculated for attractors

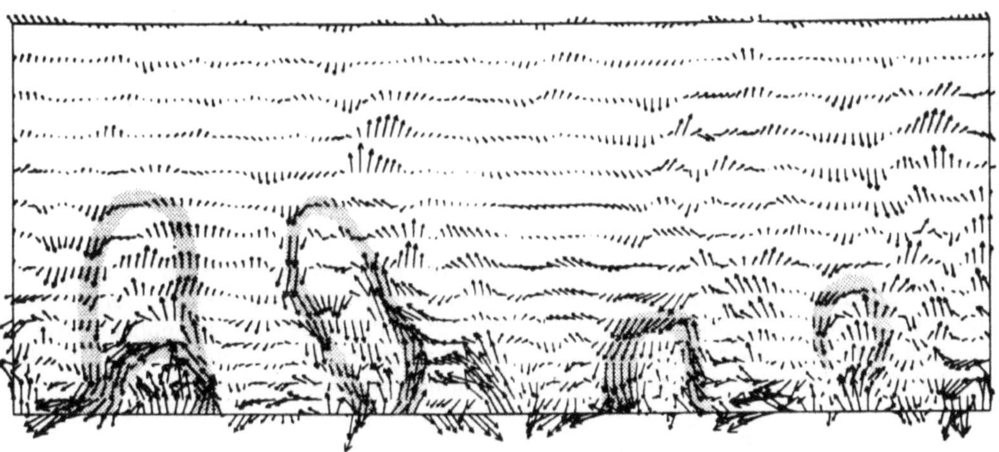

Fig.1 Projection of the vorticity vectors in a (y,z)-plane (inclined at 45°). The lower boundary is near the wall. From Moin: J. Fluid Mech. **155**, 441 (1985)

constructed from a finite number of data of finite resolution will likely increase with increasing distance from the wall.

Calculating the dimension of attractors for turbulent channel flow is much harder than for the Rayleigh-Benard or Couette-Taylor systems just beyond the onset of chaos. Unlike these closed systems, channel flow becomes highly turbulent immediately beyond the transition, at which point the attractors could have a large value of the dimension. Hence one does not expect to determine the fractal part of the dimension, but rather an approximation--is d equal to 10, 20, or greater?

The data we use for our calculations originate from numerical simulations done by MOIN and KIM[5] at NASA Ames. The calculations were made by a large eddy simulation method, which means that the large-scale motion above the grid size is calculated exactly, while at large Reynolds numbers the small-scale behavior below the grid size is modeled. An eddy viscosity model is used to simulate the action of the small scales on the large scales. But, because for Reynolds numbers as low as those used for our calculations, the motion is still confined to wavelength above the grid size, no model was necessary for the small-scale behavior; all length scales are calculated exactly.

The data from the simulations that we have analyzed thus far were obtained for Re= 2600. Time sequences are available with 1200 time steps in 848 grid points per plane parallel to the channel wall, for a total of 29 planes. The separation between time steps is rather short. For example, for the streamwise component of the velocity, the characteristic frequency, given by the variance of the velocity power spectrum, is 0.0184f_n(Nyquist frequency). The spatial separation of the grid points is sufficient: the correlation as well as the mutual information [8] between neighboring grid points is near zero.

The flow is homogeneous in the stream and spanwise directions but not in the direction normal to the wall; therefore, all the points in one plane can be considered equivalent and used for the reconstruction of the attractor. Planes of equal distance from the wall are also equivalent. We therefore have 1200×64×2 = 153,600 points per velocity component available to reconstruct the attractor. For the phase space axes we can take all 8×2 grid points in one direction (stream or spanwise) for each velocity component, appending the time sequences of the grid points in the other direction. This leads to 153,600 vectors if the phase space is less than 48 dimensional and all three velocity components are used.

2. Attractor Dimension

Using this data base we have tried to determine the attractor dimension from the time series in all grid points of a plane. We have used the k-th nearest neighbor algorithm which was introduced by PETTIS[9]. An estimate d' of the dimension d is obtained from the scaling of the average distance of k-th nearest neighbors on the attractor, $< r_k >$. If $r_k(x_i)$ is the distance between point x_i and its k-th nearest neighbor, and

$$< r_k > = \sum_i r_k(x_i),$$

then

$$< r_k > \sim G(d',k) \times k^{1/d'}$$

with

$$G(d',k) = (k^{1/d'} \Gamma(k))/\Gamma(k+ 1/d')$$

Because d' is not linearly related to $\log < r_k >$, one has to solve for d' iteratively. A first estimate d' is obtained by setting G(d',k) to zero. Because G(d',k) is very small for all k and d', the first estimate differs from the second by only about 0.1 to 0.2. Three iterations are usually enough to obtain convergence to d within the accuracy of the calculation.

The advantage of this algorithm is that the distance is the *dependent* rather than the independent variable. On different parts of the attractor the dynamic structure, noise range, and saturation range occur on different length scales. Averaging over a fixed length scale could mean adding parts of the noise region for one reference point to parts of the dynamic region for another reference point. Hence the average over several reference points, chosen randomly on the attractor, would not show a clear range of self-similar dynamic structure of the attractor. Figure 2 shows the improvement of the scaling range obtained by using the k-th nearest neighbor algorithm, compared here to the standard point-wise dimension algorithm [2].

Figure 3 shows the result of the dimension calculation for the channel flow using all three velocity components in the plane nearest to the wall. The value of d does not saturate with increasing embedding dimension, even with the embedding dimension increased to 39. These values of d are also compared in Fig. 3 with the value of the dimension deduced for random numbers. The value of d for channel flow clearly stays below the value deduced for random numbers and shows a stronger trend to saturation. The actual values of d for each phase space dimension must be higher than the calculated value, as can be seen from the values for the random number system, which are lower than the expected value d= m for m> 10. With an increasing amount of data the calculated value is expected to converge to the true value.

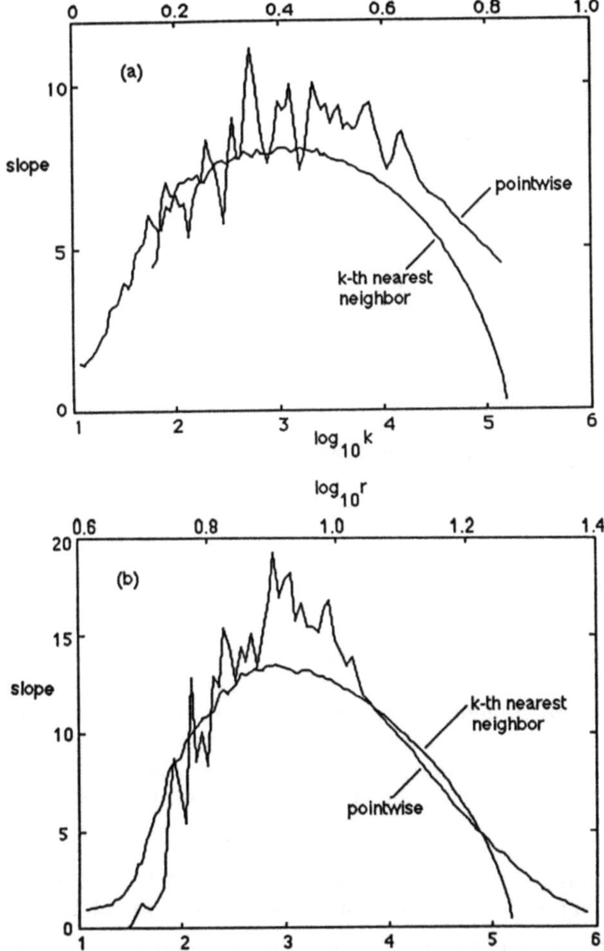

Fig.2 Local slopes $d(\log N(r))/d(\log r)$ [point-wise dimension algorithm: $N(r)$ is the number of data within a ball of radius r] and $(d(\log r_k)/d(\log k))^{-1}$ [k-th nearest neighbor algorithm] for two different values of the phase space dimension: (a) m= 9, (b) m= 21. A range of constant slope would indicate the range of self-similar dynamic structure; hence this slope gives the value of d for this phase space dimension.

3. Approximating the Dimensionality from the Trajectory Matrix

A method of approximating the dimensionality from the trajectory matrix has been proposed recently by BROOMHEAD and KING[10]. The trajectory matrix is an $n \times m$ matrix $\mathbf{X}$, composed of all n state vectors x_i along the trajectory in an m-dimensional phase space.

$$\mathbf{X} = n^{-1/2} \begin{bmatrix} x_1^T \\ x_2^T \\ \cdot \\ \cdot \\ \cdot \\ x_n^T \end{bmatrix} \qquad x_i = (v_i^1, v_i^2, v_i^3, ..., v_i^m) \, \varepsilon \, R^m$$

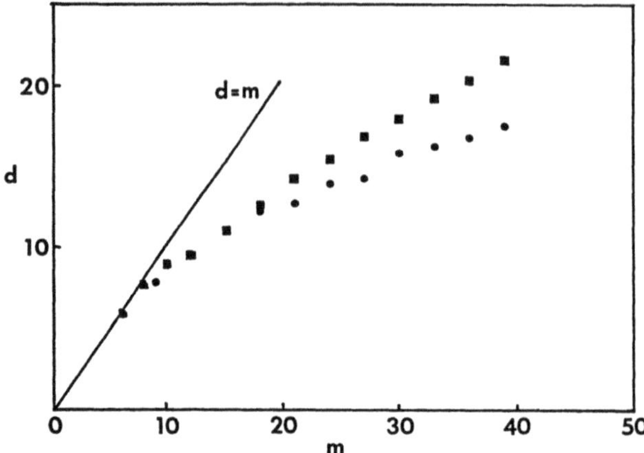

Fig.3 The value of d as a function of the phase space dimension for turbulent channel flow data [•] and random numbers [■]. In each case we chose 220 reference points randomly on the attractor (153,600 data points) without restriction concerning their position.

The question of the dimensionality of the Euclidean subspace of R^m containing the attractor is equivalent to determining the number of linearly independent vectors that can be determined from the columns of X; hence it is the rank of X. This means that one looks for the projection of the attractor onto each axis and determines whether this projection is zero or not. If it is zero, then this direction of phase space does not contribute to the dynamics of the system; hence the dimensionality of the system is lower.

Instead of determining the rank of the n×m matrix X (with n usually of the order 10^3 to 10^6), it is more convenient to determine the rank of the symmetric m×m matrix $\Theta = X^T X$, because rank X = rank Θ. It should be pointed out that the elements of Θ are time averages

$$\Theta_{ij} = \sum_{k=1}^{n} v_k^{\,i} v_k^{\,j}$$

If therefore the phase space is not reconstructed using time-delay coordinates, Θ does not contain any time information.

The rank of a matrix can be determined by a singular value decomposition [11]: that is, determine unitary matrices U and V such that

$$U^T \Theta V = \Sigma, \quad \text{where} \quad \Sigma = \text{diag}(\sigma_1, \sigma_2, ..., \sigma_m).$$

The σ_i are the singular values of Θ. If

$$\sigma_1 > \sigma_2 > ... > \sigma_{m'} > \sigma_{m'+1} = ... = \sigma_m = 0,$$

then Θ is of rank m'. In the presence of noise all singular values are shifted and are nonzero. In this case one needs an estimate of the noise level, and the rank of Θ can be determined from the eigenvalues greater than the noise level.

Because we simply look at the projection of the system on all axes, this method yields only an upper bound for the dimension of a system. For example, consider a periodic system which is represented by a limit cycle with contributions to three directions in phase space. The Broomhead and King method would then give the number three as an approximation for the dimensionality. If the trajectory matrix were created by using time delay coordinates, then each column of X would be obtained from the previous by a phase shift, given by the delay time t. With an n-periodic system, each column would then show this n-periodicity with a different phase. If t is not an integer multiple of the period of either of the system periods, then 2n modes are required to describe all columns of X. Hence the singular value decomposition for an n-periodic system (with n frequencies not necessarily irrationally related) will give at least the value 2n for the dimensionality of the system. For a triangular wave the rank of Θ is arbitrarily high, depending on the resolution, as Table 1 illustrates.

Table 1: The first 20 singular value for four different systems

singular-value #	2-periodic	3-periodic	triangular wave	channel flow
1	11.60911	15.04830	19.34676	2.358342
2	7.95255	12.42053	18.23311	1.460138
3	0.30141	1.34768	5.41345	1.392868
4	0.10744	1.13231	5.01888	1.269103
5	0.00000	0.01217	3.14670	1.227032
6	0.00000	0.00074	2.68216	1.165903
7	0.00000	0.00000	0.73765	1.147497
8	0.00000	0.00000	0.70263	1.106496
9	0.00000	0.00000	0.64177	1.098383
10	0.00000	0.00000	0.60672	1.092996
11	0.00000	0.00000	0.56587	1.090235
12	0.00000	0.00000	0.54657	1.079886
13	0.00000	0.00000	0.53320	1.069730
14	0.00000	0.00000	0.34088	1.062687
15	0.00000	0.00000	0.24959	1.055983
16	0.00000	0.00000	0.24760	1.036262
17	0.00000	0.00000	0.24440	1.028867
18	0.00000	0.00000	0.24254	1.022948
19	0.00000	0.00000	0.24093	1.010264
20	0.00000	0.00000	0.22134	1.008390

Results obtained from applying the singular value decomposition method to the channel flow data are shown in table 1. Using the data base described above, we chose a 45 dimensional phase space (only the first 20 singular values are shown in the table). *All singular values are found to be greater than zero.* However, the result for the triangular wave illustrates, the upper bound for dimension given by the singular value decomposition method can be so large that it is not useful. This could be the case for the channel flow data; the result that no limit is found on the number of singular values does not necessarily mean that the turbulent channel flow is high dimensional.

4. Proper Orthogonal Decomposition

As discussed in Section 2, the dimension algorithms fail to converge for the turbulent channel flow that we have analyzed. The algorithms failed because the system is not low dimensional. An alternative question that can be addressed is: How many modes are required to describe at least parts of the flow? We could look for a set of optimized modes that would at least extract the large-scale coherent structures. This question seems relevant to a turbulent fluid problem because large-scale coherent structures contribute most of the total energy; with increasing Reynolds number more energy is transported down to smaller scales. A method of extracting coherent structures from an ensemble of measurements of a fluid has been proposed by LUMLEY[12]. It is in principle the same method as used by Broomhead and King (a Karhunen-Loewe decomposition), only from a different point of view. The idea is to find a function $\phi^{(1)}$ under the constraint that its root mean square correlation with the members of the ensemble is maximized. For the remainder of the ensemble, which is orthogonal to $\phi^{(1)}$, find again a function $\phi^{(2)}$ under the same condition, and so on. This is a variational problem which can be reduced to an eigenvalue problem of the form

$$\int \Theta_{ij}(y,y',z,z')\phi_i^{(n)}(y',z')dy'dz' = \lambda_j^{(n)}\phi_j^{(n)}(y,z),$$

where $\Theta_{ij}(y,y',z,z')$ is a two point correlation tensor,

$$\Theta_{ij}(y,y',z,z') = \; <u_i(y,z),u_j(y',z')>$$

Here we are interested in the eigenfunction in the y-z plane with y normal to the wall and z in the spanwise direction. This eigenvalue problem has infinitely many solutions. The velocity field u can then be represented in terms of the eigenfunctions

$$u_i(y,z) = \sum_n a_n\phi_i^{(n)}(y,z) \qquad a_n = \int u_i\phi_i^{(n)}dydz \; .$$

MOIN[13] has applied this method to the channel flow data and found that the dominant eigenfunction in the y-z plane can already describe the principal shape of a hairpin vortex. Integrating over the wall region only, he found that this mode contributes 58% to the total turbulent kinetic energy, and the first five eigenfunctions contribute 95% of the turbulent kinetic energy (private communication). But Moin's interest was to select the spatial structure only. He therefore averaged Θ_{ij} over all times. The result cannot necessarily be related to the dimension of the attractor for the same reason as mentioned above in the discussion of the singular value decomposition of the trajectory matrix. But the difference between his calculations (for the y-z plane) and the singular value spectrum obtained by us (for the x-z plane) is surprisingly high. One reason might be the following: We try to construct spatial modes from an ensemble of measurements. For our data base spatial information was very sparse; neighboring grid points were independent. It should therefore be difficult to fit spatial modes to this ensemble. Moin used a different data base with grid points about eight times as dense and therefore still correlated. Therefore, it should be better possible to detect coherent structures using his data base.

5. Concluding Remarks

All approaches we have considered have failed to determine the number of degrees of freedom relevant to the dynamics of the turbulent channel flow data that have been analyzed. However, the data analyzed thus far extend over rather short times. Work is underway on data obtained for a Reynolds number of about 1400 instead of 2600; the attractor dimension should be smaller and therefore easier to determine at this lower Reynolds number. The data to be analyzed will include very long time series from laboratory measurements as well as results from numerical simulations. It should be possible to obtain a definitive value for the attractor dimension for these data.

Acknowledgements

We thank Parviz Moin for providing the data.

This research is supported by NASA-Ames University Consortium number NCA2-1R781-401.

References

1. B. Malraison, P. Atten, P. Berge, M. Dubois: J. Phys. Lett. **44**, 987 (1983)
2. A. Brandstater, J. Swift, H. Swinney, A. Wolf, D. Farmer, E. Jen,
   J. Crutchfield, Phys. Rev. Lett. **51**, 1442 (1984)
3. P. Grassberger, I. Procaccia: Phys. Rev. Lett. **50**, 346 (1983)
4. K. R. Sreenivasan: *Fundamentals of Fluid Mechanics*, Ed. S. A. Davis,
   J. L. Lumley (Springer 1985)
5. P. Moin, J. Kim: J. Fluid Mech. **118**, 341 (1982)
6. S. A. Orszag: J. Fluid Mech. **50**, 689 (1971)
7. M. Nishioka, S. Iida, Y. Ichikawa: J. Fluid Mech. **72**, 731 (1985)
8. A. Fraser, H. L. Swinney: Phys. Rev. A, to be published
9. K. Pettis, T. Bailey, A. Lain, R. Dubes: IEEE Transaction on Pattern
   Analysis and Machine Intelligence **PAMI-1**, 25 (1979)
10. D. S. Broomhead, G. P. King, submitted to Physica D
11. J. J. Dongarra, C. B. Moler, J. R. Bunch, G. W. Stewart: *LINPACK Users Guide*
    (SIAM, Philadelphia 1979)
12. J. Lumley: *Transition and Turbulence*, Ed. R. Meyer (Academic Press 1981)
13. P. Moin: American Institute of Aeronautics and Astronautics AIAA
    paper # 84-0174 (1984)

Characterization of Chaotic Instabilities
in an Electron-Hole Plasma in Germanium

G.A. Held and C.D. Jeffries

Department of Physics and Lawrence Berkeley Laboratory,
University of California, Berkeley, CA 94720, USA

Helical instabilities in an electron-hole plasma in Ge in parallel dc electric and magnetic fields are known to exhibit chaotic behavior. By fabricating probe contacts along the length of a Ge crystal we study the spatial structure of these instabilities, finding two types: (i) spatially coherent and temporally chaotic helical density waves characterized by strange attractors of measured fractal dimension $d \sim 3$, and (ii) beyond the onset of spatial incoherence, instabilities of indeterminately large fractal dimension $d \geqslant 8$. In the first instance, calculations of the fractal dimension provide an effective means of characterizing the observed chaotic instabilities. However, in the second instance, these calculations do not provide a means of determining whether the observed plasma turbulence is of stochastic or of deterministic (i.e., chaotic) origin.

1. Introduction

It is by now well established that the onset of turbulence in a wide range of physical systems can be characterized by low-dimensional chaotic dynamics.[1] That is, the evolution of these systems corresponds to motion in phase space along trajectories confined to a strange (fractal) attractor.[2] Experimentally, it is often difficult to distinguish between deterministic chaos and stochastic noise -- both are characterized by broad spectral peaks. To establish that experimentally observed behavior is indeed chaotic, it is necessary to examine the structure of the attractor itself. This requires methods of data reduction designed specifically to identify and characterize low-dimensional chaotic attractors. These include the construction of phase portraits, Poincaré sections, return maps, and bifurcation diagrams. In those cases where the chaotic behavior is characterized by an attractor of dimension greater than approximately 2.5, even these methods of analysis cannot distinguish between chaos and stochastic noise; the fractal structure becomes too dense to be discerned through visual inspection of a two-dimensional projection of a Poincaré section. In such instances, one must calculate quantitative measures of chaos Such as fractal dimensions,[3] Lyapunov exponents,[4] and metric entropy[5] of the attractor. In this paper we present the results of our efforts to calculate fractal dimensions as a means of identifying and characterizing chaos in helical instabilities of an electron-hole (e-h) plasma in germanium (Ge).

Spontaneous current oscillations in an e-h plasma in a dc electric field E_0 and a parallel dc magnetic field B_0 are known to be the result of an unstable, travelling, screw-shaped helical density wave.[6,7] Held, Jeffries, and Haller[8] have found that when this instability is strongly excited by an increasing electric field, it will undergo both period-doubling and quasiperiodic transitions to low-dimensional chaos. Experimentally, we vary the applied dc fields and record the dynamical variables I(t), the total current passing through the sample, and V(t), the voltage across it. By forming probe contacts along the length of our crystals, we are also able to monitor the local variations in plasma density.

We have found two distinct types of behavior: (i) an essentially spatially coherent and temporally chaotic plasma density wave characterized by an attractor of fractal dimension $d \sim 3$, and (ii) a spatially incoherent wave with an immeasurably large fractal dimension $d > 8$. Further, as the applied electric field E_0 is increased, we observe a transition between these two states -- characterized by a partial loss of spatial order and a jump in the fractal dimension. While the increase in fractal dimension from $d \sim 3$ to $d > 8$ is somewhat abrupt ($\Delta E_0/E_0 \sim 0.05$), the breakup of spatial order occurs gradually. It is physically reasonable that the onset of spatial incoherence (which increases the number of available degrees of freedom) would result in an increased fractal dimension. However, we cannot firmly establish that the onset of spatial disorder is *coincident* with the observed jump in fractal dimension; the possibility that these two events occur at comparable fields and yet are not directly related cannot be completely excluded.

The methods by which we determined that case (i) corresponds to a temporally chaotic, spatially coherent density wave are described in detail elsewhere.[9] We present here a discussion of the methods which we have used to determine the fractal dimension of the attractors associated with such instabilities. Following that, we discuss the difficulties which we have encountered in attempting to characterize spatially incoherent instabilities in the context of chaotic dynamics.

2. Experimental Procedures

Our experiments are performed on a $1 \times 1 \times 1$ mm^3 sample cut from a large single crystal of n-type Ge with a net donor concentration $N_D \sim 3.7 \times 10^{12}$ cm^{-3}.[8] A lithium-diffused n$^+$ contact (electron injecting) and a boron-implanted p$^+$ contact (hole injecting) were formed on opposite 1×1 mm^2 ends. Phosphor-implanted n$^+$ contacts were formed on two opposite $1 \times 10 \, mm^2$ faces. Using photolithography, we etched onto these two faces a pattern of eight pairs of contacts 0.5 mm wide and spaced by 1 mm along the length of the sample. The voltage $V_i(t)$ across a pair of these contacts is a measure of the local variation in plasma density.[7] The sample was lapped, etched, and then stored in dry air for 72 hours to allow the surfaces to passivate.

When taking data, the sample is cooled to 77 K in liquid N_2 and connected in series with a $100 - \Omega$ resistance and a variable dc voltage, which both generates the e-h plasma via double injection and creates the dc electric field E_0. The applied voltage V_0, the applied magnetic field B_0, and the angle between the two fields θ comprise our control parameters; typically $\theta = 0 \pm 3°$. In practice, we fix B_0 and θ and sweep V_0, while recording the dynamical variables $I(t)$, $V(t)$, and $V_i(t)$, which characterize the plasma behavior.

3. Low-Dimensional Attractors -- Transitions to "Weak" Turbulence

In different regions of parameter space (V_0, B_0, θ) different types of transitions to turbulence are observed. For our system we make the operational definition that a transition to "weak" turbulence is one in which the transition from periodicity to chaos is followed by a transition back to periodicity as V_0 is increased further. All such transitions that we have observed occur over a small range (i.e., ~ 1 V) of V_0, and in all such chaotic states there exists at least one fundamental peak which stands out clearly above the broad-band "noise" level of the power spectrum.

For several different values of B_0 we have observed quasiperiodic transitions to weak turbulence: as V_0 is increased, the onset of a quasiperiodic state (simultaneous oscillations at two incommensurate frequencies) is followed by a transition to chaos. The power spectra for one such sequence, taken at $B_0 = 11.15$ kGauss, is shown starting in figure 1(a) with $V_0 = 2.865$ volts: $I(t)$ is spontaneously oscillating at a fundamental frequence $f_1 = 63.4$ kHz. At $V_0 = 2.907$ volts, the system becomes quasiperiodic: a second spectral component appears at $f_2 = 14$ kHz, incommensurate with f_1 [figure 1(b)]. At $V_0 = 2.942$ volts, the system is still quasiperiodic; however, the two modes are interacting and the nonlinear mixing gives spectral peaks at the combination frequencies $f = mf_1 + nf_2$, with m,n integers [figure 1(c)].

As V_0 in increased further, we observe a series of frequency lockings[10], i.e., $(f_1/f_2) =$ rational number, until the onset of chaos is reached, indicated by a slight broadening of the spectral peaks [figure 1(d)]. As V_0 is increased further, the e-h plasma exhibits increasingly turbulent behavior [figures 1(e) and (f)]. This is followed by a return to quasiperiodicity at $V_0 = 3.125$ volts and, subsequently, simple periodicity at $V_0 = 3.442$ volts.

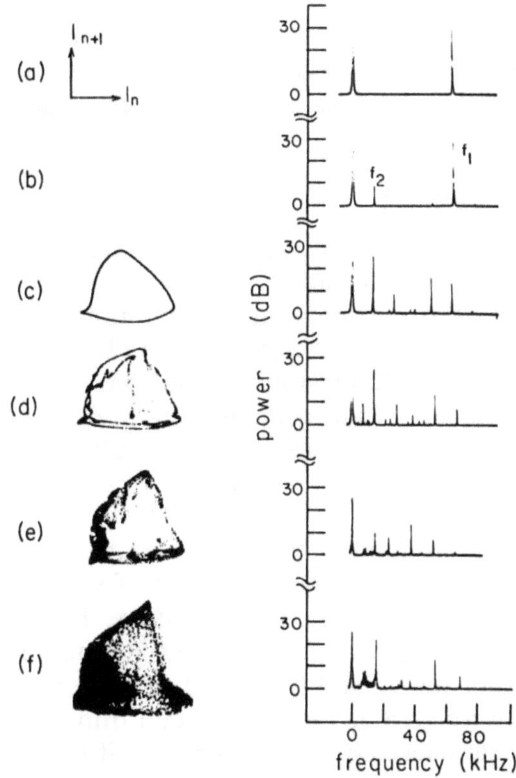

Fig. 1. Return maps, I_n vs. I_{n+1} (where $\{I_n\}$ is the set of local current maxima), and power spectra of the plasma current $I(t)$ at $B_0 = 11.15$ kGauss with increasing V_0: (a) 2.865 volts, periodic at $f_1 = 63.4$ kHz. (b) 2.907 volts, quasiperiodic with second frequency $f_2 = 14$ kHz. (c) 2.942 volts, quasiperiodic with combination frequency components. (d) 3.016 volts, onset of chaos. (e) 3.033 volts, chaotic. (f) 3.058 volts, more chaotic; the fractal dimension of this attractor, $d = 2.7$, is measured in figure 3.

Figure 1 also shows a sequence of return maps, topologically equivalent to Poincaré sections.[11] Periodic motion corresponds to a closed 1-dimensional orbit in phase space; the Poincaré section in this case is simply a point [figure 1(a)]. Similarly, when the system is quasiperiodic, corresponding to motion on a 2-dimensional torus, the Poincaré section is approximately a circle [figures 1(b) and (c)]. However, as the system becomes chaotic, we find that the Poincaré section begins to wrinkle and to occupy an extended region. This does not *necessarily* imply that the behavior is stochastic, but rather that the dimension of the strange attractor (which is one greater than the dimension of the Poincaré section) is too large to be determined by visual inspection of the Poincaré section. For these attractors the fractal dimension must be calculated quantitatively.

The fractal dimension is a measure of the number of "active" degrees of freedom needed to characterize the evolution of a system. If this evolution is described by trajectories in a G-dimensional phase space, then the fractal dimension d_F is defined as follows:[3]

$$d_F = \lim_{\delta \to 0} \frac{\log M(\delta)}{\log (1/\delta)} \tag{1}$$

where the phase space has been partitioned into cubes of volume δ^G and $M(\delta)$ is the number of these cubes visited by the attractor.[12] This measure is known variously as the capacity, Hausdorff dimension, and fractal dimension. Other, alternative, dimensions which characterize strange attractors have also been devised. These include the information dimension d_I,[3] and the correlation dimension d_C.[13] It has been proven[14] that generally $d_F > d_I > d_C$. However, in most cases where these dimensions have been calculated, all three have yielded almost identical results.[13,15,16,17]

Equation (1) assumes an attractor contained within a G-dimensional phase space. The coordinates of the phase space may be any set of variables which, when taken together, uniquely identify the state of the system. For our experiments, these variables could be the plasma density and momentum measured at many different points within the crystal (provided of course that the number of independent probes G were greater than the fractal dimension d). Experimentally, this method of characterizing the system is difficult to realize. It is not always feasible to have an arbitrary number of probes for a given system and, further, it is not known how many probes will be required. One cannot know this until the fractal dimension d_F has already been determined.

Fortunately, there is a method of reconstructing phase space from a single dynamical variable using a technique based on the embedding theorem.[1,13,15,18] If $\{V_1(t), V_2(t), \ldots, V_G(t)\}$ is a phase space constructed from G independent variables, then the reconstructed phase space $\{V_1(t), V_1(t+\tau), \ldots, V_1(t+(D-1)\tau)\}$ is conjectured to be topologically equivalent to the original phase space, for almost all τ, provided $D > 2G+1$.[18] Attractors in both the original and reconstructed phase spaces will be characterized by the same Lyapunov exponents and fractal dimensions. In our experiments, we use a reconstructed phase space derived from the measured current $I(t)$. The coordinates of our phase space are thus $\{I(t), I(t+\tau), \ldots, I(t+(D-1)\tau)\}$, where, typically, $5\,\mu s < \tau < 15\,\mu s$; we find that the calculated fractal dimensions are independent of τ. In practice, one calculates the fractal dimension d for increasing embedding dimension D until d converges with respect to D.

Calculations of fractal dimensions using the box-counting algorithm of Eq. (1) tend to be computationally inefficient.[19] Large regions of phase space are visited only rarely. Thus large numbers of data points and, consequently, large amounts of computer time are often required. Calculations on systems with d≈3 can require more than a million data points. However, it is possible to calculate the "pointwise" fractal dimension[20] (which is conjectured[3] to be equal to the information dimension) using the following, more efficient algorithm.[21] A D-dimensional phase space is reconstructed from a single dynamical variable. Next one computes the number of points on an attractor, $N(\epsilon)$, which are contained within a D-dimensional hypershpere of radius ϵ centered on a randomly selected point on the attractor. One expects scaling of the form:

$$N(\epsilon) \propto \epsilon^d \tag{2}$$

where d is the fractal dimension of the attractor. Thus a plot of $\log N(\epsilon)$ vs. $\log\epsilon$ is expected to have slope d (for sufficiently small ϵ). This procedure is carried out for consecutive values of $D = 2, 3, 4, \ldots$, until the slope has converged. This is done to insure that the embedding dimension chosen is sufficiently large (important if the dimension of the phase space is not known) and to discriminate against high dimensional stochastic noise, not of known deterministic origin.

A comparison of equations (1) and (2) illustrates the difference between the fractal and pointwise dimensions. The calculation of the fractal dimension involves determining the fraction of phase space occupied by the entire attractor. On the other hand, the pointwise dimension is defined as the scaling of $N(\epsilon)$ with ϵ, for $N(\epsilon)$ centered around a *single* point on the attractor. The conjecture that the pointwise dimension is equal to the information dimension (which, like the fractal dimension, is measured globally over the attractor[3]) implies that the scaling laws which govern the fractal structure are constant throughout the attractor. It is therefore sufficient to determine the scaling exponent at a single point on the attractor. We note that the pointwise dimension is conjectured to be equal to the information dimension, not the fractal dimension, but, as mentioned earlier, the two are found to be experimentally indistinguishable.

We have computed the pointwise dimension d for our plasma instabilities at various points along the quasiperiodic transition to chaos described above. For each of eleven values of V_0 between 2.865 volts and 3.125 volts ($B_0 = 11.15$ kGauss) we recorded N (≈ 98000) successive values of the current at 5 μs intervals [i.e., $I_n = I(t+n\tau)$, $n = 1, \ldots, 98000$; $\tau = 5\,\mu$s]. From each data set $\{I_1, \ldots, I_N\}$ we constructed $N - D + 1$ vectors $G_n \equiv (I_n, I_{n+1}, \ldots I_{n-D+1})$ in a D-dimensional phase space. In principle, one should be able to calculate the fractal dimension with Eq. (2) using data centered around a single point on the attractor G_n; that is, calculations of $N(\epsilon)$ centered around different vectors G_i should all yield the same value of d. Experimentally this is not actually observed, as discussed below.

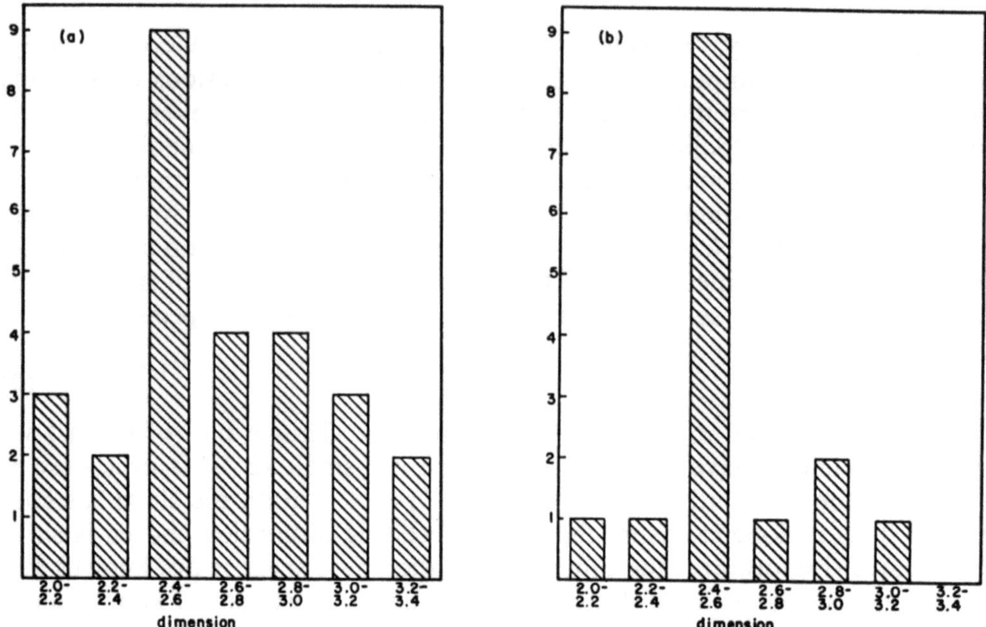

Fig. 2. Histogram of fractal dimension calculations for $V_0 = 3.058$ volts and $B_0 = 11.15$ kGauss [same operating conditions as in figure 1(f)]. (a) The fractal dimension is calculated 27 times by observing the scaling of $N(\epsilon)$ [equation (2)] around 27 randomly chosen points in reconstructed phase space. The vertical axis refers to the number of these calculations for which the fractal dimension d is found to be in each of the ranges specified on the horizontal axis. The distribution is centered around $d = 2.4 - 2.6$. (b) The same distribution as (a), except that those calculations yielding unphysical results (see text) have been removed. The distribution is still centered at $d = 2.4 - 2.6$, but it has narrowed appreciably.

For $V_0 = 3.058$ volts we constructed plots of $\log N(\epsilon)$ vs. $\log \epsilon$ for $N(\epsilon)$ centered on 27 randomly selected vectors G_i. The slopes of these 27 plots comprise 27 measurements of the fractal dimension d. A histogram of these values of d is shown in figure 2(a); the result is a distribution centered around $d = 2.4 - 2.6$. However, a careful examination of the 27 plots of $\log N(\epsilon)$ vs. $\log \epsilon$ indicates that several of these plots yield unreliable values of d, for reasons discussed below. Upon elimination of these suspect points, the width of the histogram narrows appreciably, as shown in figure 2(b). For an experimental system, there are at least three conditions under which one will not expect scaling of the form of Eq. (2) for $N(\epsilon)$ centered around certain random points on the attractor.

First, the random point may be situated in a region of the attractor which is visited only rarely. Thus, even with a large number of data points there are not enough nearby data points to resolve the fractal structure and thus to observe the scaling of Eq. (2). In such a case, the plot of logN(ϵ) vs. logϵ will have a gradually increasing slope for small ϵ, in contrast with the break to a steeper, non-convergent slope for small ϵ that is expected for chaotic systems in the presence of thermal noise.[22] This break is expected because the dynamics of all physical systems are characterized by thermal (stochastic) processes at energies below $\sim$kT; these processes are characterized by fractal dimensions on the order of the number of particles in the system.[23] We eliminate all plots which do not show the physically expected break to steeper slope for small ϵ.

A second difficulty arises when N(ϵ) is centered in a region of the attractor where the length scales over which the structure is fractal are comparable to or less than those corresponding to thermal fluctuations ($\sim$kT). In these cases the fractal structure may be "washed out" by thermal noise, resulting in a plot of logN(ϵ) vs. logϵ which does not have a well defined (convergent) slope. We discard these plots as well.

Finally, if a hypershpere N(ϵ) is centered on the attractor in a region of high lacunarity,[24] the resulting plot of logN(ϵ) vs. logϵ will not have a well defined slope.

By rejecting those plots of logN(ϵ) vs. logϵ which do not exhibit physically reasonable characteristics (i.e., a well defined slope and a break to steeper slope for small ϵ), we obtain a much sharper distribution of values for the fractal dimension d, as seen in figure 2(b). However, when we plot logN(ϵ) vs. logϵ, where N(ϵ) is the average over many hyperspheres, we find that this average slope is unchanged ($\pm$5%) by the rejection of the unphysical plots. This was found to be true for several cases. Thus, in most instances we simply plot logN(ϵ) vs. logϵ for N(ϵ) averaged over many randomly chosen hyperspheres. (This same procedure has also been utilized in studies of free surface modes of a vertically forced fluid layer[25] and Couette-Taylor flows.[15]) Figure 3(a) shows our results for $V_0 = 3.058$ volts with the embedding dimension D = 2,4,6, and 8; for D $\geqslant$ 6 the slope (and thus the fractal dimension) has converged to 2.7. The fractal dimension for all the states shown in figure 1, as well as several states not shown, are plotted in figure 3(b).

Within the chaotic regime, the fractal dimension of the attractor varies between 2 and 3. This demonstrates that the observed plasma turbulence shown in figures 1(d)-(f) may be described with only a few degrees of freedom; the behavior of the system remains largely deterministic. If the observed turbulence were due to thermal or stochastic processes, then a measurement of the fractal dimension d would not have converged for small embedding dimension D. The dimension of the attractor d could then have been on the order of the number of conduction electrons and holes in the crystal[23] ($\approx 10^{10}$).

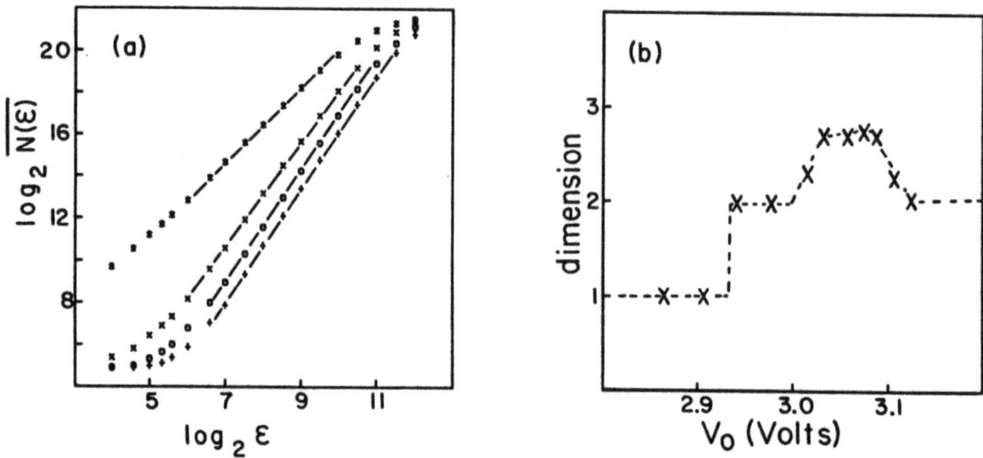

Fig. 3. (a) Plots of $\log\overline{N(\epsilon)}$ vs. $\log\epsilon$ used to determine the fractal dimension d of the chaotic attractor at $V_0 = 3.058$ volts, $B_0 = 11.15$ kGauss, using method discussed in text and Eq. (2) [averaged over 25 randomly chosen points in reconstructed phase space]. Embedding dimension $D = 2, 4, 6$ and 8 correspond, respectively, to symbols *, x, o and +; for $D \geqslant 6$, the slope converges to $d = 2.7$. (b) Dependence of measured dimension d on applied voltage V_0. $B_0 = 11.15$ kGauss. Values $d = 1$ and $d = 2$ correspond to periodic and quasiperiodic orbits, respectively. All calculations were checked for convergence with respect to embedding dimension D and number of data points sampled N. All values of d represent an average over 25 randomly selected points in reconstructed phase space.

4. Transitions to "Strong" Turbulence

With sufficiently large applied electric and magnetic fields, we find that we can drive the plasma into a turbulent state from which it will not become periodic again as V_0 is increased further. Instead, all of the frequency peaks in the power spectrum merge into a single, broad, noiselike band. We classify this as a transition to "strong" turbulence. Such a transition is shown in figure 4. At $V_0 = 10.4$ volts, I(t) is simply periodic at $f_0 = 321$ kHz, with higher harmonics present as well [figure 4(a)]. At $V_0 = 11.6$ volts, I(t) is quasiperiodic and at $V_0 = 12.1$ volts (not shown), the onset of broadband "noise" can be observed. At $V_0 = 13.8$ volts [figure 4(b)], only a few of the peaks can be seen above the noise, and when $V_0 = 21.8$ volts [figure 4(c)], only a very broad peak remains.

We find that this transition to strong turbulence is characterized by a partial loss of spatial coherence. In the right hand column of figure 4, we plot the voltage traces across two pairs of probe contacts which are separated by

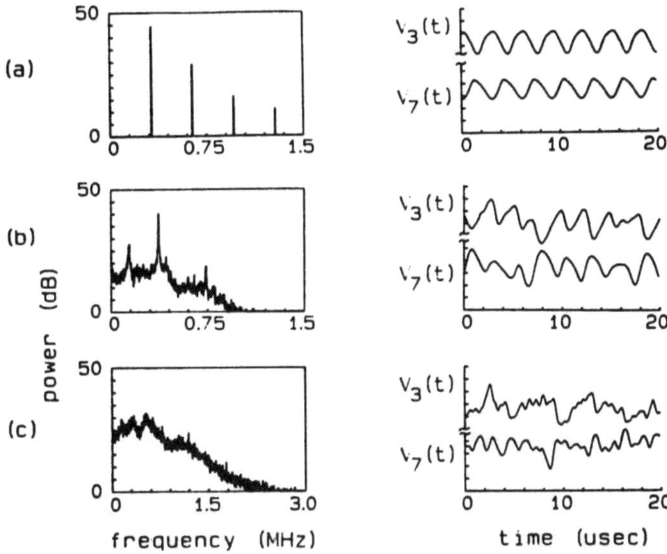

Fig. 4. Left, measured power spectra of I(t); right, measured voltages for two pairs of probe contacts separated by r = 4 mm: $V_3(t)$ and $V_7(t)$ correspond to probe pairs located 3 and 7 mm away from the p^+ contact, respectively. $B_0 = 11.15$ kGauss. (a) $V_0 = 10.4$ volts, periodic at $f_0 = 321$ kHz. At $V_0 = 12.1$ volts (not shown) temporal chaos has set in, with measured fractal dimension $d \approx 2.5$, figure 5 (a). (b) $V_0 = 13.8$ volts, power spectrum has broad base and peaks; comparison of $V_3(t)$ and $V_7(t)$ shows beginning of spatial incoherence; measured fractal dimension $d > 8$. (c) $V_0 = 21.8$ volts, power spectra very broad, more marked loss of spatial coherence, measured fractal dimension $d > 8$.

r = 4 mm, for $V_0 = 10.4$, 13.8, and 21.8 volts. In the periodic case, the wave is spatially coherent with a wavelength of approximately 8 mm (i.e., a 4 mm separation corresponds to a 180° phase shift). At $V_0 = 13.8$ volts we are just beyond the onset of the break-up of spatial order -- the basic oscillatory pattern and the 180° phase shift are approximately maintained between the two traces, but changes in the shapes and spacings of the peaks can also be observed. For $V_0 = 21.8$ volts, the wavelike structure of the traces, as well as the readily observable spatial correlation, is no longer present.

We would like to determine whether this breakup of spatial order can be characterized by chaotic dynamics: Do the spatially uncorrelated states still correspond to motion in phase space along a low-dimensional strange attractor? We have as yet been unable to answer this question definitively. Just prior to the breakup of spatial coherence, $V_0 = 12.1$ volts, the total current I(t) of the system is characterized by a low-dimensional attractor; measurements of the fractal dimension yield $d = 2.5$ [figure 5(a)]. However, just after the onset of spatial disordering, $V_0 = 12.9$ volts, the fractal dimension has

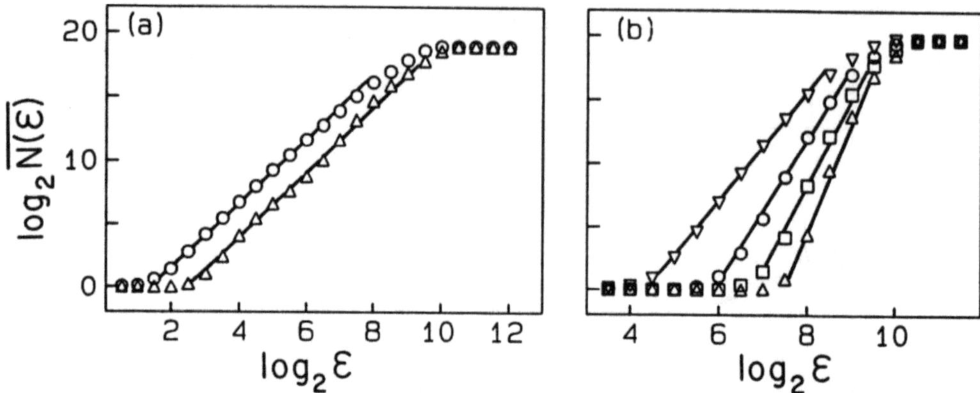

Fig. 5. Plots of $\log\overline{N(\epsilon)}$ vs. $\log\epsilon$ used to determine fractal dimension d at $B_0 = 11.15$ kGauss. (a) $V_0 = 12.1$ volts, $N = 490000$ data points; the symbols o and $\triangle$ refer to embedding dimensions D of 4 and 8, respectively. Slopes have converged to 2.5 with respect to both D and N. (b) $V_0 = 12.9$ volts, $N = 884000$; ∇, o, $\square$, and $\triangle$ refer to $D = 6$, 10, 14 and 18, respectively. Slopes have not converged with respect to either D or N. For $D = 18$ slope is 8.7.

increased to the point where we cannot calculate its value -- we can only set a lower limit: $d \geqslant 8$. This is shown in figure 5(b) where the slope has not converged with respect to either embedding dimension D or number of data points N. Figure 5(b) was taken with $N = 884000$ and required 50 hours of CPU time on a Sun microcomputer. For $V_0 = 21.8$ volts, $N = 884000$ points and embedding dimension $D = 18$, the slope is 14 and has definitely not converged.

For our fractal dimension plots of figure 5 we note that the curves become horizontal (saturate) for (i) $\epsilon > \epsilon_1$, a hypershpere large enough to include all points on the attractor and for (ii) $\epsilon < \epsilon_2$, a hypersphere so small that only the single point at its center is within it. This behavior is to be expected for all fractal dimension plots, provided ϵ is varied sufficiently; it is important to do this to ensure that all experimental data are examined.

Calculations based on time series taken across different pairs of probe contacts $V_i(t)$ yield the same fractal dimensions d as those based on total current $I(t)$, for both spatially coherent and incoherent states. Further, we find that for fixed values of our applied fields, the power spectrum measured across a pair of probe contacts $|V_i(\omega)|^2$ is essentially identical to the power spectrum of the total current $|I(\omega)|^2$. This suggests that the spatial incoherence may be due to the dispersive nature of the e-h plasma.

This difficulty in calculating large fractal dimensions is a problem incurred with very chaotic systems. The number of data points required for convergence increases exponentially with the fractal dimension of the system.[19,26] At present, although we know that our system experiences a large jump in dimensionality at the onset of spatial incoherence, we have not yet determined whether this onset is characterized by chaotic dynamics of an attractor of fractal dimension may orders of magnitude smaller than the number of degrees of freedom of the particles in the system ($\sim 10^{10}$). Other approaches for quantitatively characterizing very chaotic states (say, $d > 10$) will need to be developed before this intriguing question can be answered.

We wish to thank E. E. Haller and the members of his laboratory for the Ge samples and assistance in the sample preparation. This work was supported by the Director, Office of Energy Research, Office of Basic Energy Science, Materials Science Division of the U. S. Department of Energy under Contract No. DE-AC03-76SF00098.

References

1. for example, H. L. Swinney: Physica (Utrecht) **7D**, 3 (1983); see also *The Physics of Chaos and Related Problems,* edited by S. Lundqvist, Phys. Scr. **T9** (1985).

2. D. Ruelle and F. Takens: Comm. Math. Phys. **20**, 167 (1971); E. Ott: Rev. Mod. Phys. **53**, 655 (1981).

3. for example, J. D. Farmer, E. Ott, J. A. Yorke: Physica (Utrecht) **7D**, 153 (1983).

4. A. Wolf, J. B. Swift, H. L. Swinney and J. A. Vastano: Physica **16D**, 285 (1985).

5. J. P. Crutchfield and N. H. Packard: Int. J. Theor. Phys. **21**, 433 (1982); Physica **7D**, 201 (1983); P. Grassberger and I. Procaccia: Phys. Rev. A **28**, 2591 (1983).

6. I. L. Ivanov and S. M. Ryvkin: Zh. Tekh. Fiz. **28**, 774 (1958) [Sov. Phys. Tech. Phys. **3**, 722 (1958)].

7. C. E. Hurwitz and A. L. McWhorter: Phys. Rev. **134**, A1033 (1964).

8. G. A. Held, C. Jeffries and E. E. Haller: Phys. Rev. Lett. **52**, 1037 (1984); *Proceedings of the Seventeenth International Conference on the Physics of Semiconductors, San Francisco, 1984,* edited by D. J. Chadi and W. A. Harrison (Springer-Verlag, New York, 1985), p. 1289.

9. G. A. Held and C. Jeffries: Phys. Rev. Lett. **55**, 887 (1985).

10. for example, J. P. Gollub and S. V. Benson in *Pattern Formation and Pattern Recognition* edited by H. Haken (Springer-Verlag, Berlin, 1979), p. 74.

11. N. H. Packard, J. P. Crutchfield, J. D. Farmer and R. S. Shaw: Phys. Rev. Lett. **45**, 712 (1980).

12. Unconventional notation is used in Eq. (1) to avoid confusion with the notation of Eq. (2).

13. P. Grassberger and I. Procaccia: Phys. Rev. Lett. **50**, 346 (1983).

14. H. G. E. Hentschel and I. Procaccia: Physica **8D**, 435 (1983).

15. A. Brandstäter *et al.*: Phys. Rev. Lett. **51**, 1442 (1983).

16. H. L. Swinney and J. P. Gollub: to appear in Physica **D**.

17. This has been observed in calculations of fractal, information and correlation dimensions for a system consisting of a driven p-n junction in series with an inductor and a resistor. G. A. Held and C. Jeffries, unpublished.

18. F. Takens: "Detecting Strange Attractors in Turbulence", in *Lecture Notes in Mathematics 898*, edited by D. A. Rand and L. S. Young (Springer-Verlag, Berlin, 1981), p. 366.

19. H. S. Greenside, A. Wolf, J. Swift and T. Pignaturo: Phys. Rev. **A 25**, 3453 (1982).

20. The pointwise dimension is defined in reference 3. Similar definitions of dimension have been given in references 11, 20, 21, and 31 therein.

21. This is the method used by A. Brandstäter *et al*, in reference 15. See also reference 13.

22. A. Ben-Mizrachi, I. Procaccia and P. Grassberger: Phys. Rev. **A 29**, 975 (1984).

23. A fractal dimension of this magnitude ($\sim 10^{10}$) is experimentally unattainable for two reasons. First, an inordinate number of data points would be required, as discussed in Section 4. Second, one would need to measure signals with a very large bandwidth $\Delta f \sim 1/\tau$, where τ is the shortest fluctuation time; typically $\Delta f \sim 10^9 - 10^{14}$ Hz for stochastic noise in conducting media. Thus, fractal dimensions of this magnitude are operationally meaningless.

24. B. B. Mandelbrot: *The Fractal Geometry of Nature* (W. H. Freeman and Company, New York, 1983), p.310.

25. S. Ciliberto and J. P. Gollub: J. Fluid Mech. **158**, 381 (1985).

26. H. Froehling, J. P. Crutchfield, J. D. Farmer, N. H. Packard, and R. Shaw: Physica **3D**, 605 (1981).

Instabilities, Turbulence,
and the Physics of Fixed Points

M. Duong-van

University of California, Lawrence Livermore National Laboratory,
Livermore, CA 94550, USA

1. Abstract

By solving the recursion relation of a reaction–diffusion equation on a lattice, we find two
distinct routes to turbulence, both of which reproduce commonly observed phenomena: the
Feigenbaum route, with period–doubling frequencies; and a much more general route with
noncommensurate frequencies and frequency entrainment, and locking. Intermittency and
large–scale aperiodic spatial patterns, also observed in physical systems, are reproduced in
this new route. The fractal dimension has been estimated to be about 2.6 in the oscillatory
instability and about 6.0 in the turbulent regime.

Experimental evidence supporting Feigenbaum's route to turbulence [1,2] has become
richer since 1978. In this route, nonlinear systems manifest chaos via period– doubling
bifurcations. For example, Rayleigh–Bénard systems with low [3,5] and intermediate [6]
Prandtl numbers exhibit this route. In detailed experiments on low– aspect–ratio
Rayleigh–Bénard cells, GIGLIO et al. [6] saw four period doublings and obtained values of δ
that agree with FEIGENBAUM's [1] universal δ = 4.6692 within their experimental error.

When the aspect ratio is large, however, very different behavior is found [7–11]. As the
stress parameter (the Rayleigh number, in the case of Rayleigh–Bénard systems) is
increased, cascades of instabilities are observed, each step of which adds new complications
to the convective behavior [3,8,10]. Unstable patterns are formed and temporal chaos [8]
sets in, with alternating random bursts and quietness: this is called intermittency [3,8].
Noncommensurate frequencies arise in the Fourier spectrum of the chaotic variable, and
entrainment and locking occur as the stress parameter is varied [3,7,8].

In this paper we show that both these routes to turbulence, with all the properties just
described, can be simply simulated with a quadratic map at each site of a spatial lattice and
with a coupling between nearest–neighbor sites. This new route leads to a fractal
dimensions of 2.6 at the oscillatory instability regime and 6.0 at the turbulent regime.

Let u represent the chaotic variable: it may be a velocity component or a temperature
fluctuation of the system being studied. We build a lattice of sites with a quadratic map
$u \rightarrow Q(\lambda,u)$ at each site, and allow interaction between nearest neighbor sites through a
coupling parameter g. We assign a random value of u to each lattice site, and let the lattice
evolve in time steps $t_n = n\tau$, n = 1, 2, 3 ..., when τ is the time between two intersections
with the Poincaré plane. We find the same behavior for all quadratic $Q(\lambda,u)$: for example,

$$Q(\lambda,u) = \lambda u(1-u), \tag{1a}$$

$$Q(\lambda,u) = \lambda \sin(\pi u) \tag{1b}$$

give the same behavior. For simplicity, we use the logistic map, Eq. (1a), in this study. In
one dimension, we use the prescription [14,15]

$$u_{n+1}(m) = \lambda u_n(m)[1 - u_n(m)] + \tfrac{g}{2}[u_n(m+1) + u_n(m-1) - 2u_n(m)], \tag{2}$$

where the index m spans the N lattice points m = 1, 2, ...,N. Similarly, in two dimensions we
have

$$u_{n+1}(j,k) = \lambda u_n(j,k)[1-u_n(j,k)]$$

$$+ \frac{g}{4}[u_n(j+1,k) + u_n(j-1,k) - 2u_n(j,k)]$$

$$+ \frac{g}{4}[u_n(j,k+1) + u_n(j,k-1) - 2u_n(j,k)], \tag{3}$$

where the indices j,k span the lattice in the x and y directions, respectively.

With Eq. (2) [the same results are found for Eq. (3)], two routes to turbulence are observed, in which the Feigenbaum route is seen as a special case.

For small g (e.g., $g \simeq 0.001$), when λ approaches the accumulation point λ_∞, the Fourier spectrum of the time sequence $u_n(m)$ for a particular m shows period–doubling bifurcations [1]. (With $\lambda_\infty \simeq 3.569$ and $g \simeq 0$, for example, we obtain a period–doubling Fourier spectrum that agrees well with that obtained by GIGLIO et al.[6]) As g increases, the peaks in the Fourier spectrum become wider, as observed by MAUER and LIBCHABER [3]. In our study, this width increase is a consequence of the dissipative term controlled by g. As λ increases to 4 (for <u>any</u> value of g) the spectrum becomes flat and chaotic.

Only in special cases (such as in Rayleigh–Bénard systems with low aspect ratio) does the turbulence observed in nature follow the Feigenbaum route. More generally (as in Rayleigh–Bénard systems with large aspect ratios), the instabilities and turbulence show richer behaviors [7,11]. One observes noncommensurate frequencies in the Fourier spectrum and the phenomena of frequency entrainment and locking; complex quasi–periodic, aperiodic, and intermittent time histories of the values of chaotic variables at individual points in the system; and similar time variations in the spatial patterns formed in certain systems [3]. By iterating Eq. (2) or (in the case of the spatial patterns) Eq. (3) with values of g away from zero, we can reproduce all these phenomena, provided we restrict λ to the Feigenbaum simple fixed points region $1 < \lambda < 3$.

We built a periodic one–dimensional lattice with N = 2000, and recorded the time evolution and the corresponding Fourier transform of $u_n[13]$ for times up to $n = 2^{12}$ and for a variety of values of λ and g (m is arbitrarily chosen equal to 13). We found that for every value of g, there is a λ_{max} at which the u eventually blows up (diverges) with time n.

For illustration, we choose g = 0.915 and vary λ from $\lambda_{max} = 1.621$ to $\lambda_{min} = 1.0$. Figure 1(a) ($\lambda = 1.62$) shows the Fourier spectrum and the time history of u_n. This broad spectrum, with its intermittent bursts and quietness, appears to correspond to observations described in Ref. 3. In Fig. 1(b) ($\lambda = 1.52$) the frequency peaks become narrower, and the amplitude variations become smaller.

In Fig. 2(a) ($\lambda = 1.449$), the time history shows that the system attempts to settle to the fixed point ($u^* = 1 - 1/\lambda$) after some transient time. The competition between the approach to the fixed point (due to λ) and the diffusion away from the fixed point (due to g) gives rise to the instability observed. In Fig. 2(a) ($\lambda = 1.449$) the Fourier spectrum exhibits noncommensurate frequencies, as observed by MAUER and LIBCHABER [3].

As λ is decreased further, the frequencies are entrained (Fig. 2(b) $\lambda = 1.49$) and locked (Fig. 3(a) $\lambda = 1.48$).

We generate visible patterns with the two–dimensional Eq. (3) by use of a new graphical technique [17], scaling the u(j,k) to a 0–to–256 linear gray scale. Simulations of this sort correspond quite closely to the patterns seen by DUBOIS and BERGE [8] in their experiments with silicon oil.

In the simulation of these patterns, the u(j,k) are assigned initial (n = 1) random values between 0 and 1, resulting at most in small–scale random patterns at that time. As time increases, these patterns disappear into a highly uniform sea (when u_n approaches the fixed point); eventually, however, large–scale structures grow, evolve, and temporarily or

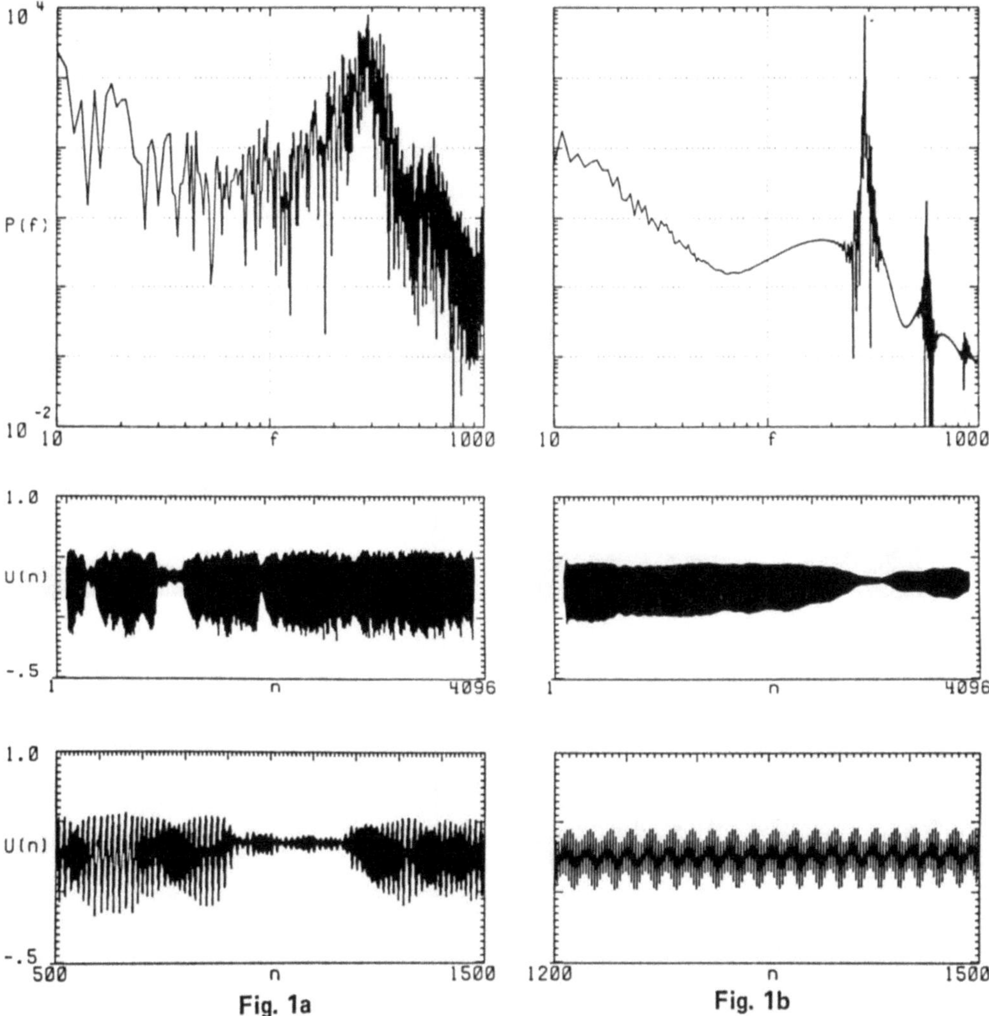

Figure 1.   (Top to bottom: Fourier spectrum of u_n; and enlargement of indicated portion of u_n.) (a) g = 0.915, λ = 1.62; (b) λ = 1.52.

permanently stabilize. Fig. 4 shows the pattern developed in a 50x50 lattice for λ = 1.5, g = .905.

We have chosen to study only the simple fixed points region $1 < \lambda < 3$ of the logistic map, Eq. (2a). As long as g = 0, this branch produces uninteresting behavior: the u_n approach the fixed point $u^* = 1 - 1/\lambda$. Without g, there is no instability in this region, and no patterns. When we turn g on, however, depending on the values of g and λ, we may   get rich and interesting behaviors clearly, in Figs. 1 and 2, g acts to keep the u_n from their tendency toward the fixed point. <u>Thus instabilities appear to result from a competition between tendencies towards the fixed points and away from it</u>, and the time history intermittency phenomenon (Fig. 1a) is, in fact, a consequence of this competition.

Influenced by the recent measurements of the fractal dimension in periodically excited air jet by BONETTI et al. [17] and in an electron–hole plasma in Ge by HELD and JEFFRIES [18], we calculate the fractal dimension of our system in the oscillatory instability regime (Figs. 2b, 3a) and turbulent regime (Fig. 1a) and we found $D_F \simeq 2.6$ and 6.0 respectively.

173

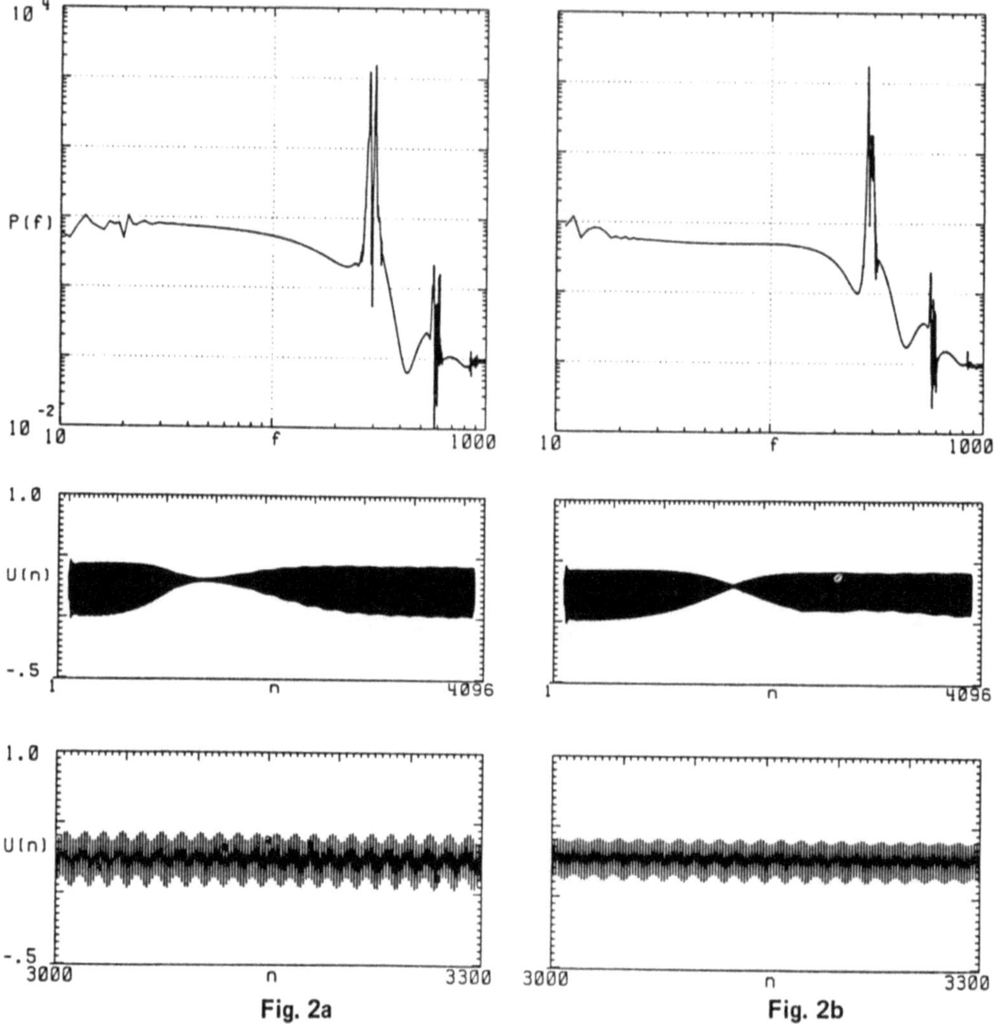

Fig. 2a　　　　　　Fig. 2b

Figure 2.　　(Same plots as in Fig. 1.) (a) g = 0.915, λ = 1.499; (b) λ = 1.490.

From the set of data $u_n(m)$ where n = 3000 and the lattice site m varies from 1 to 500, we constructed $m_{max} - D + 1$ vectors $G^m = (u^m, u^{m+1},..., u^{m+D-1})$ in a D–dimensional phase space; D is referred as the imbedding dimension of the reconstructed phase space G. Note that the coordinates in our reconstructed phase space correspond to different lattice sites and not to time delays. Next, we compute the number of points on the attractor, N(R), which are contained in a D–dimensional hypersphere of radius R. If one considers the fractal dimension a critical index of a critical phenomenon, one expects at the critical point, N(R) to scale as R^{D_F} where D_F is the fractal dimension of the attractor. This procedure is carried out for consecutive values D = 2,3,4 ..., to insure the D to be sufficiently large. We found that for g = .915, 1.47 < λ < 1.49 (around the phase lock regime) the fractal dimension is roughly 2.6. In the turbulent regime, g = .915, λ = 1.62, D_F ≈ 6.0. The D_F for the first case stabilizes to 2.6 at D ≈ 20 while for the latter case, D_F is hard to calculate even at D = 30. This difficulty was also experienced by BONETTI et al. [17] and ATTEN et al. [19]. The most interesting phenomenon observed here is the existence of a parametric transition of the D_F from 0.5 at λ < 1.47 to D_F ≈ 2.6 at 1.47 < λ < 1.49.

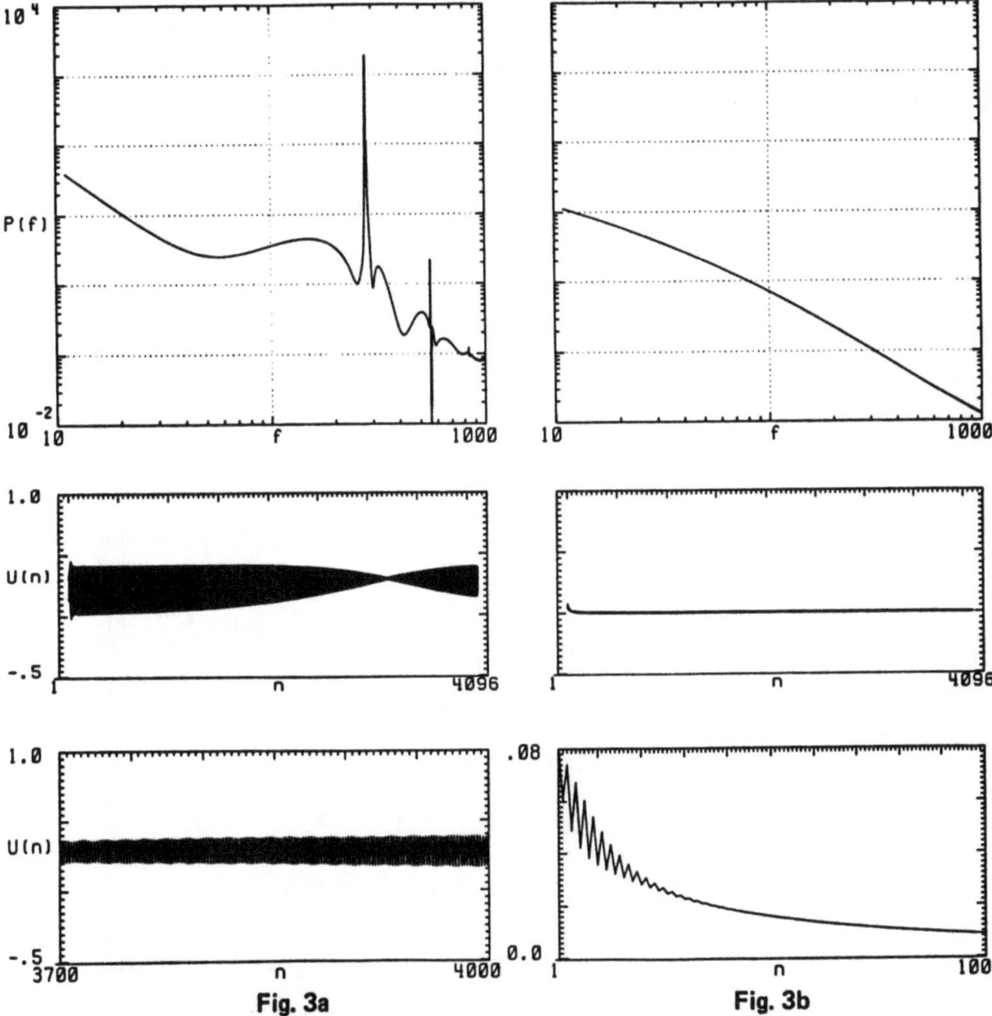

Figure 3. (Same plots as in Fig. 1.) (a) $g = 0.915$, $\lambda = 1.48$; (b) $\lambda = 1.00$.

Figure 5 shows D_F versus λ at a given g. The universal value of $D_F \approx 2.6$ observed experimentally is a consequence of the quadratic property of the logistic map. We have repeated the calculation with a 2–dimensional lattice, and varied the time delay and also the lattice size to 2000. Within the errors, the fractal dimension does not change significantly. In this paper, the errors are simply estimated by fitting to the slopes of the Log N(R) versus Log R curves. We observe a change of the width of the plateau at $D_F \approx 2.6$ with the size of the lattice. For 300 sites, the plateau width spans from $1.46 < \lambda < 1.51$ and for 2000 sites, the plateau width narrows down to $1.48 < \lambda < 1.49$. From the trend, one might speculate that in the continuum limit, the transition from $D_F \approx 0$ to a large fractal dimension (in the turbulent regime) might be continuous. Further investigations of this phenomenon and the detailed errors calculations are presently carried out.

In conclusion, we note that the variety of phenomena experimentally observed in the approach to turbulence has brought forth a variety of explanations: for example, intermittency can be explained [18] by the Lorenz model; the lock–in phenomenon can be explained by the Flaherty and Hoppensteadt model [19]. The discovery of the intermittency

Figure 4.   Patterns generated by a 50 x 50 lattice with $\lambda = 1.50$, $g = .905$ and $n = 90$.

in the Logistic map by HAKEN and MAYER–KRESS [20] is a good indication that the quadratic map on the lattice should manifest intermittency. In our model, by putting the quadratic map on the lattice with nearest neighbors coupling, we economically recover all these phenomena. Furthermore, we have found that the common fractal dimension of 2.6 is a consequence of the quadratic map.

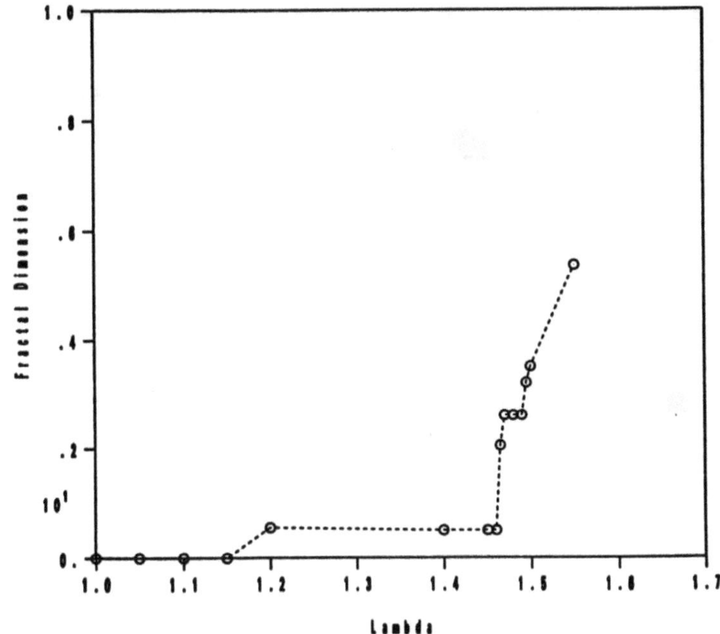

Figure 5. Fractal dimension variation as a function of the parameter λ at a fixed $g = .915$. Detailed features of the parametric transition at $\lambda = 1.47$ should be complemented by refering to Figs. 1a–3b.

This work was performed under the auspices of the U.S. Department of Energy by the Lawrence Livermore National Laboratory under contract number W–7405–ENG–48.

The author thanks P. R. Keller for the graphics used in this paper and P.W. Murphy for editorial assistance. He thanks his colleagues at Lawrence Livermore and Los Alamos National Laboratories for numerous profitable discussions. Valuable suggestions and comments by P. Cvitanovic and M. Feigenbaum are appreciated. He thanks G. Mayer–Kress for valuable advise on the finalization of this work. He also thanks the Aspen Institute for Physics for their hospitality and the participants for stimulating discussions.

2. <u>References</u>

1. M. J. Feigenbaum, Phys. Lett. <u>74A</u>, 375 (1959). I thank P. Cvitanovic for providing me with information on a more comprehensive survey of recent data on this phenomenon. Also, <u>Universality in Chaos</u>, P. Cvitanovic, Ed. (Adam Hilger Ltd., Bristol, England, 1984), p. 29.
2. M. J. Feigenbaum, Comm. Math. Phys. <u>77</u>, 65 (1980).
3. J. Mauer and A. Libchaber, J. Phys. Lett. <u>40</u>, L–419 (1979); and <u>Universality in Chaos</u>, P. Cvitanovic, Ed. (Adam Hilger Ltd., Bristol, England, 1984), p. 109.
4. A. Libchaber and J. Mauer, J. Phys. Colloq. <u>41</u>, C3–51 (1980).
5. P. S. Linsay, Phys. Rev. Lett. <u>47</u>, 1349 (1981).
6. M. Giglio, S. Musazzi, and V. Perini, Phys. Rev. Lett. <u>47</u>, 243 (1981).
7. J. P. Gollub and S. A. Benson, Phys. Rev. Lett. <u>41</u>, 948 (1978).
8. M. Dubois and P. Bergé, J. Fluid, Mech. <u>85</u>, 641 (1978), and <u>Systems Far From Equilibrium</u>, L. Garrido, Ed. (Springer–Verlag, New York, 1980), p. 381.
9. G. Ahlers and R. W. Walden, Phys. Rev. Lett. <u>44</u> 445 (1980); Phys. Lett. <u>76A</u>, 53 (1980).
10. G. Ahlers and P. R. Behringer, Supp. Prog. Theor. Phys. <u>64</u>, 186 (1978).
11. J. Mauer and A. Libchaber, J. Phys. Lett. <u>41</u>, 515 (1980).
12. K. Kaneko, Prog. Theor. Phys. <u>69</u> 1427 (1983); Prog. Theor. Phys. <u>72</u>, 480 (1984).
13. I. Waller and R. Kapral, Phys. Rev. A <u>30</u>, 2047 (1984).

14. Minh Duong-van and P. R. Keller, "Pattern Formation in Spinodal and Phase Turbulent Instabilities," Lawrence Livermore National Laboratory, Livermore, CA, UCRL–92716 (1985).
15. P. Berge, M. Dubois, P. Mannville and Y. Pomeau, J. Physique–Lettres $\underline{41}$, L–341 (1980).
16. J. E. Flaherty and F. C. Hoppensteadt, Study Appl. Math., $\underline{58}$, 5 (1978).
17. M. Bonetti et al., PRL $\underline{55}$, 492 (1985).
18. G. A. Held and Carson Jeffries, PRL $\underline{55}$, 887 (1985).
19. P. Atten et al., "Determination of Attraction Dimension for Various Flows," to be published.
20. G. Mayer–Kress and H. Haken, Phys. Lett. A, $\underline{82A}$. 151 (1981).

Part VII

Experimental Results and Applications

Determination of Attractor Dimension and Entropy for Various Flows: An Experimentalist's Viewpoint

J.G. Caputo, B. Malraison, and P. Atten

Laboratoire d'Electrostatique et de Materiaux Dielectriques,
Centre National de la Recherche Scientifique, Ave. de Martyrs, 166 X,
F-38042 Grenoble Cédex, France

Introduction

In order to characterise quantitatively the behaviour of dissipative dynamical systems we have to determine the values of information dimension, metric entropy and Lyapunov exponents associated with the limit sets in phase space. For numerically integrable dynamical systems such as iterated maps and systems of ordinary differential equations, methods are available which lead to the determination of Lyapunov exponents with an accuracy generally depending only on the power of the utilized computer. We furthermore have the values of information dimension and metric entropy by applying the conjectured formulas relating their values to the Lyapunov exponents.

The problem is quite different when we try to characterise the chaotic behaviour of a definite experimental system (we are more particularly interested in some hydrodynamic flows, keeping in mind the problem of hydrodynamic turbulence). Most often the experimentalist measures one or very few variables (for example a velocity component at a given point of a flow). From an experimental time-series of data there is at the present time no practical way of determining the Lyapunov exponents. However, studies that are just starting indicate that for differential dynamical systems where the exact dimension of the phase-space is known, it if often possible to estimate the Lyapunov exponents from a reconstruction of the local Jacobian [1, 2]. This is the reason why efforts in the past few years have focused on the question of estimating attractor dimensions and entropies. The time-delay technique [3] allows one to reconstruct an attractor of equivalent topological properties [4] in R^n from a single time-series X (t) (the variable is sampled every Δt)

$$\vec{X} (t) = (X (t), X (t - \tau), \ldots X (t - (n - 1) \tau)) \qquad\qquad (1)$$

($t \in [\Delta t, N \Delta t]$, $\tau = p \Delta t$, p integer) provided that n is large enough.

Methods looking at the effective repartition of discrete points in this phase space lead to estimates of the dimension ν and the entropy K [5, 6] from the computation of the so-called integral correlation functions C (r). Many questions nevertheless are still open concerning these methods and their practical use by the experimentalists (we deliberately adopt this point of view in this paper).

First the attractor dimension ν is estimated from the asymptotic behaviour C (r) ~ r^ν for r $\to$ 0, the function C (r) itself being the limit for a number of data N $\to \infty$. What is the departure from this asymptotic value ν when working with a finite N and examining C (r) for r values not very small ? The latter question is a very serious one, because one has no a priori estimation of the relative lengths below which there is the scaling law r^ν and often sets of curves C (r) for different embedding dimensions n are difficult to interpret.

The number N of data required also is an unanswered question. We feel that this number should increase with the attractor dimension ν but there is for the moment no rule or argument fixing either the value of N or the frequency at which data have to be sampled.

Finally there is a basic question we have to face : what is the order of magnitude of the attractor dimension that can be obtained with these methods with some confidence ?

We examine some of these questions by illustrating the different limitations we have encountered in the application of this method both to numerical and experimental data. We first indicate our practical conditions before studying two low-dimensional attractors and then a few higher dimensional attractors. The distinction we have established is a bit arbitrary because of overlapping effects, but it agrees well with what we currently know.

1. Practical Conditions And Methods

There are a certain number of parameters which are more or less imposed by the experiment one is dealing with. The signal/noise ratio is fixed by the measurement conditions. For all the experiments mentioned below it was greater than 10^4 in power. Another important parameter is the sampling period ; unless the signal can be recorded on an analog device, it is fixed for a given state of the system. We will see that this can induce systematic errors. The number N of data and the sampling rate are not independant. Indeed, there are practical limits imposed by the total duration of the experiment, particularly in situations where the typical time-scale is of the order of minutes or more. For a given sampling frequency, N should be as large as possible but compatible with reasonable computing time.

For the determination of the dimension, Grassberger and Procaccia propose to calculate :

$$C (r) = \lim_{N \to \infty} C_N (r) \tag{2}$$

$$C_N (r) = \frac{1}{N} \sum_{i = 1}^{N} C_i (r) \tag{3}$$

where

$$C_i (r) = \frac{1}{N} \# \{ j, \ \| \vec{X}_i - \vec{X}_j \| \leqslant r \} \tag{4}$$

(rate of points contained in a sphere of radius r around $\vec{X}_i$). It is expected that

$$C (r) \underset{r \to 0}{\sim} r^\nu \tag{5}$$

In practice, it is possible to approximate $C_N (r)$ (and $C (r)$) by averaging in (3) on a limited number m of origins $\vec{X}_i$ instead of N. Furthermore, if the measure is ergodic on the attractor, then for almost every point $\vec{X}_i$ of the attractor : $C_i (r) \underset{\substack{r \to 0 \\ N \to \infty}}{\sim} r^\nu$ [7].

The metric entropy can be defined from the information

$$I_b = - \sum_{(i_1 \ldots i_b)} p (i_1 \ldots i_b) \log p (i_1 \ldots i_b) \tag{6}$$

associated for a given partition of the phase-space and sampling period Δt with all the sets of b elementary boxes. $p (i_1 \ldots i_b)$ is the probability of the state of the system to be in box i_1 at time t, in box i_2 at time $t + \Delta t$, ... The metric entropy is :

$$K = \sup_{B, \Delta t} \left[\lim_{b \to \infty} \frac{I_b}{b . \Delta t} \right] \tag{7}$$

where the supremum is taken on all possible partitions B and sampling periods Δt.
In view we use methods of the Grassberger-Procaccia type it is convenient to define a lower bound K_2 of K :

$$K_2 = \sup_{B, \Delta t} \left[\lim_{b \to \infty} - \frac{\mathrm{Log}}{b \, \Delta t} \left[\sum_{(i_1 \ldots i_b)} p^2 (i_1 \ldots i_b) \right] \right] \tag{8}$$

When working with the maximum norm in a phase-space of dimension n reconstructed by the time-delay technique ($\tau = \Delta t$) the relation [6] :

$$C_{n + b} (r) \underset{\substack{r \to 0 \\ b \to \infty}}{\sim} r^\nu e^{-(n + b) \Delta t K_2} \tag{9}$$

allows one to obtain K_2.

Note that increasing n or b is strictly equivalent, and that applying the Grassberger-Procaccia technique then leads to the determination of both ν and K_2.

All the data which have been analysed were converted to integers of 12 or 15 bits in the fashion of experimental data. A study done on a limit cycle with added noise showed two regimes for the curves $C (r)$. For r values way below the noise scales ε, we have $C (r) \sim r^n$ which is typical of the white noise. For r values well above ε, $C (r) \sim r$ (signature of the limit cycle). The cross-over is broad and broadens with increasing embedding dimensions. However, for attractors of dimension 2 and higher, the effect of the noise on the curves $C (r)$ will generally be

negligible because the scales in r which can be observed will be much larger than ε
[8].

All norms being equivalent in R^n we have used the maximum norm,which is the more effective for the determination of K_2 [9]. Furthermore, this and the use of integer values for the data enabled us to write a procedure to compute C_i (r) in assembly language,which is very rapid.

For typical runs, the number of data ranged from 10^4 to 3.10^5, the number of origins from 10^2 to 10^3. This gives a total number of distances computed from 10^6 to 10^9.

One trick which proved very useful in making these computations accessible even with a very small machine (PDP 1103) was to compute C_i (r) for $r \leqslant r_0$, r_0 being a distance above which the finiteness of the attractor can be felt. If C (r_0) = 0,1 then this procedure discards about 90 % of the calculation of the distances $|| \vec{X}_i - \vec{X}_j ||$ in the course of the computation.

For the estimation of the exponent in (5) we plot Log C (r) as a function of Log r for a geometrical sequence (r_j) of values of r. We then look for a region in which the plot behaves linearly and compute the slope with a least square fit. When necessary, we will also report plots of estimations of $\underline{d\ [\ Log\ C\ (\ r\ )\ ]}$. For this, we calculated for each point r_j of the sequence : $\qquad d\ [\ Log\ r\ ]$

$$\Delta_j = \frac{Log\ C\ (\ r_{j\ +\ 1}\ ) - Log\ C\ (\ r_{j\ -\ 1}\ )}{Log\ (\ r_{j\ +\ 1}\ ) - Log\ (\ r_{j\ -\ 1}\ )} \qquad\qquad (10)$$

(This symmetrical approach reduces the fluctuations from one point to the other).

The errors we will quote in the following will be the mean-square deviation from the average slope < Δ_j > for a given r region. They are just an indication and a more thorough - yet unnecessary for the study of the gross effects in the curves C (r) - approach would be to estimate them with the use of a new algorithm [10].

2. <u>Two Low-Dimensional Attractors : A Rayleigh-Benard Convection Regime,</u> <u>The Lorenz Attractor</u>

Let us examine a set of curves C (r) for a fixed value of the time-delay $\tau = p\ \Delta t$ and for increasing values of n as in Figure 1. This plot was made for the Henon model (a = 1,4 b = 0,3 p = 1, Δt = 1) and is extracted from [11]. When incrementing n - assuming we have reached the asymptotic regime (9) - of one unit, the values of C (r) decrease by the factor exp (- $K_2\ \tau$) in the scaling region (C (r) ~ r^ν).

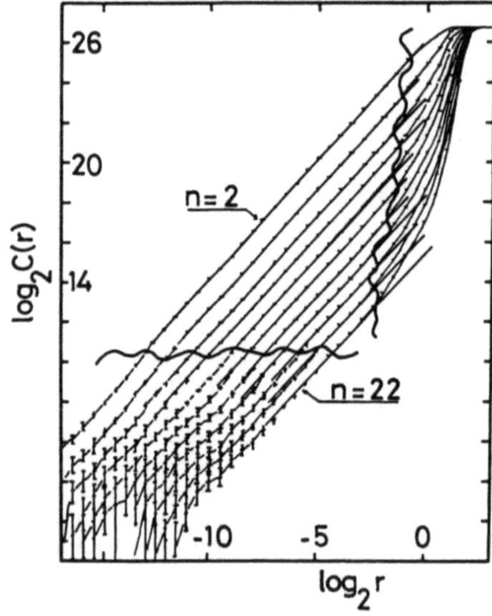

<u>Figure 1</u> : Set of curves log C (r) versus log r for the Henon map for n = 2, 4, ..., 22 using the euclidian norm.
(Δt = 1, p = 1, N = m = 15.10^3) from [11]. We have indicated the limits of the accessible scaling region.

Therefore,in this region the curves C (r) for increasing n are parallel,as shown
by the lines drawn with a triangle and shifted downwards by a constant height in the
log-log plot. Now, at least with the maximum norm, all curves C (r) tend to 1 for
r tending to the maximum extension of the attractor. Consequently,in the range of
large r values, C (r) increases steeper and steeper as n is increased and any
attempt to deduce ν from the slopes in this region is nonsense.

This effect introduces a first limitation by fixing the r range,which is interesting for the scaling behaviour below a value r_0 which for the example shown does not
seem to decrease much with n. Now, let us consider the limited number of data and
of distances that can be calculated (here N x m $\simeq N^2$ = 2,2 10^8). This fixes a range
accessible in C (r) > C_0 = q_0/Nm ($q_0 \sim$ 100 to 300 from experimental observation).
The two effects pointed out above define a quadrant in the plane (r, C (r)) in
which we will be able to extract information from the system. For an attractor of ·
given dimension ν and entropy K_2, reconstructed with a given time-delay p Δt, and
for a resolution N x m fixed, there is a limited range of values of n, one can
explore and find both ν and K_2 in.

These considerations are well illustrated first by the Lorenz system and by a convection regime in a Rayleigh-Benard experiment done by Bergé and Dubois [12] for
which we have evidenced a low-dimensional attractor [13, 8]. In the convection
experiment, the variable analysed is semi-local, it is the deviation of a light beam
crossing the cell. The signal has a characteristic time-scale T_0 ($\sim$ 10 s) and has
been sampled at about 30 pts/T_0. Figure 2-a shows the corresponding curves C (r)
for n = 10 to 37 (p = 1). We are sure to have reached the scaling range because of
the independance of the local slope versus n for a certain range in r (Fig. 2-b).
The curvatures observed for low values of C (r) in Fig 2-a are currently under
investigation and presumably arise from the influence of the experimental noise
affecting the signal. The typical noise scale for 12 bits data is about
r = 2^{12}/100 = 40.

We show in Figure 3-a the curves C (r) for the z (t) variable of the Lorenz
system (R = 28, σ = 10, b = 8/3). The typical time-scale T_0 (determined by the
peak of the power spectra) is 0.74 and the values of the parameters were
Δt = 0.02 and p = 1. We confirmed the good appearance of the scaling law in Fig. 3-a
with the examination of the local slopes (Figure 3-b) which showed a coincidence
over a certain range in r for n = 30 to 80. It is interesting to notice that for
n = 15 the local slope plot presents a certain concavity,which means a curvature for
C (r). The average exponent between r = 10^2 and 10^3 is 1.9 which is beneath the
exponents extracted for higher values of n. This effect showed that even if one
respects the Takens theoretical embedding criterion, (n > 2 ν + 1) [4] there are
practical problems that occur when working with a finite resolution. For a given
reconstruction of the attractor, we may need to explore very small scales in r to

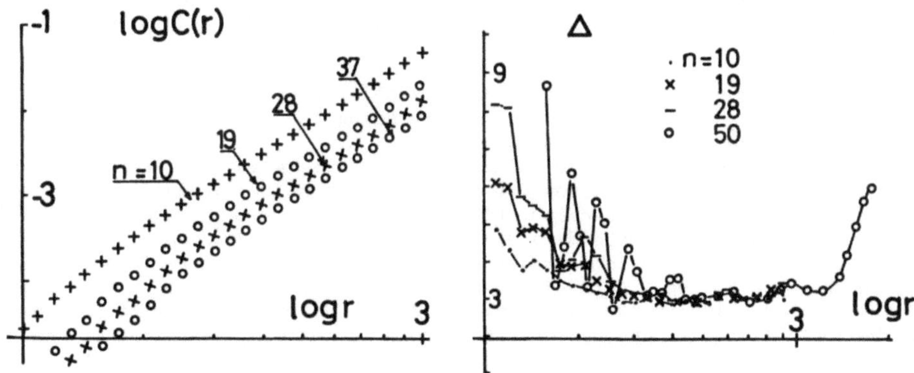

<u>Figure 2</u> : 2-a Set of curves log C (r) versus log r for a Rayleigh-Benard convec-
tion regime for n = 10 to 37.
2-b Local slopes Δ_i as a function of log r for the curves C (r) shown
in Fig. 2-a. For these computations, the conditions were the
following : Δt = T_0/30, p = 1, N = 8.10^3, m $\simeq$ 600.

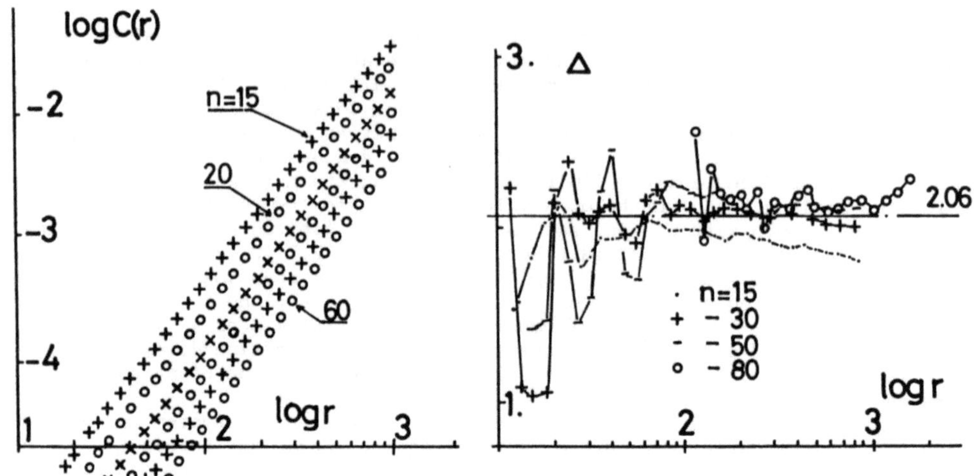

Figure 3 : 3-a Set of curves log C (r) versus log r for the Lorenz system
(r = 28, σ = 10, b = 8/3, integration time 10^{-2}) for n = 15 to 60.
3-b Local slopes Δ_i as function of log r for the curves C (r) shown in
Fig. 3-a. For these computations $\Delta t = \frac{T_0}{37}$, p = 1, N = 30 10^3, m ≃ 600.

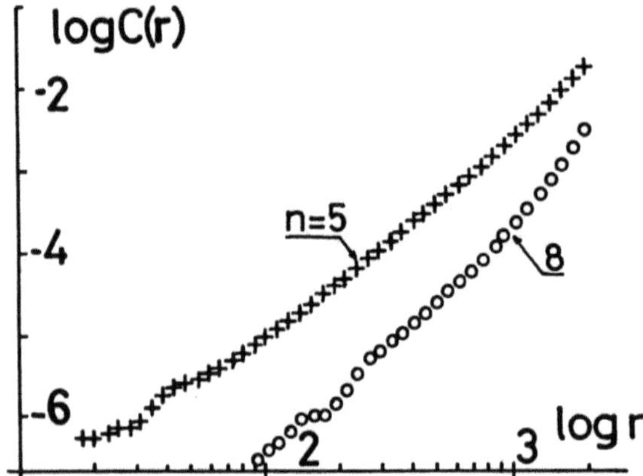

Figure 4 : Curves log C (r)
versus log r for the Lorenz
system with a large time-delay
(Δt = 0.05, p = 20) for
n = 5 and 8 (same other
parameters as in Fig. 3).

evidence the scaling zone. In this case, the time-delay p Δt = 0.02 and the dimension n are very small, and this may result in stretching the cloud of points along
the diagonal and underestimating the dimension.

When increasing n or p to very high values as for Figure 4 in the case n = 5 and
τ = 1 (p = 20, Δt = 0.05) the curves C (r) seem to conserve a scaling behaviour
in a certain region of r. But the computation of the slope reveals a value of 2.3 for
$10^2 \ll r \ll 10^3$. Increasing the embedding dimension yields a value even higher for the
averaged slope (2.6) for the same scales of r and furthermore there is a visible
curvature in the plot of log C (r) for large values of r. For n = 5, C (r) also
presents a curvature which is more difficult to detect without turning to the analysis of the local slope. It can be shown that this increase of the slope of the
curve log C (r) for a given r region ($10^2 \ll r \ll 10^3$) for the Lorenz system follows a universal law and depends only on the total length of signal
T = (n - 1) p Δt used to reconstruct the attractor (Figure 5). (T has been normalized by the typical time-scale T_0). This plot suggested that we were in the region where the plots log C (r) increase steeper and steeper with n, presenting no
scaling behaviour. A clear demonstration of this appeared when we increased the
resolution N x m for one of the computations presented in Figure 4

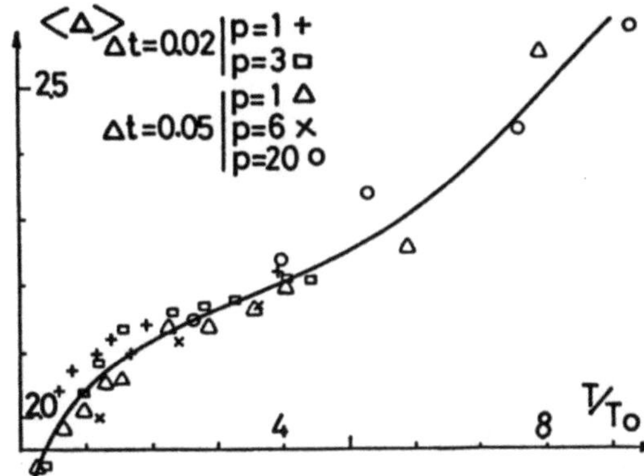

Figure 5 : Universal curve showing the behaviour of the average local slope $\langle \Delta_i \rangle$ for $10^2 \leqslant r \leqslant 10^3$ as a function of the total length of signal used to reconstruct the phase-space $T = (n - 1) p \Delta t$ normalized to the pseudo-period for different values of the sampling period Δt and delay p (same parameters as in Fig. 3).

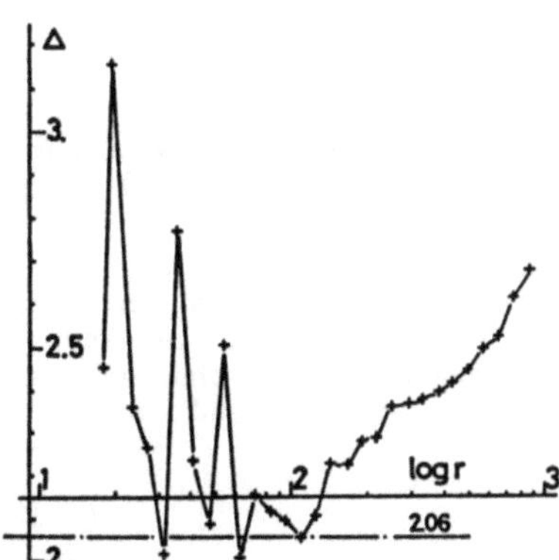

Figure 6 : Local slope Δ_i as a function of log r for the C (r) curve for the Lorenz system with n = 5 as in Fig. 4 but with $m = 25 \ 10^3$.

($N = 3.10^4$, $m = 2,5 \ 10^4$, n = 5, p = 20, $\Delta t = 0.05$ ($T \sim 5 \ T_0$)). The "local" slope for the curve C (r) displayed in Figure 6 exhibits a monotonic decrease from $r = 10^3$ to $r = 10^2$ - revealing a curvature where we had previously assumed there was a linear behaviour - followed by what seems to be a plateau (2.09 ± 0.04) for $60 \leqslant r \leqslant 120$ which for lower values of r is overcome by statistical fluctuations. This calculation showed that even with a time-delay of the order of the entropy characteristic time (1/K) one is able to obtain a scaling region and recover the correct value for ν. All that is needed is to increase the resolution N x m to attain scales in r smaller than r_0. Figures 5 and 6 suggest that for $r \leqslant 10^2$, we will observe a plateau $\langle \Delta_i \rangle = \nu$ and that a departure from this behaviour will occur for a given value of T/T_0 indicating that the limiting value r_0 of the scaling region may depend only on T. At this point, we can comment on the value of the time-delay $\tau = p \ \Delta t$, used for the reconstruction of the attractor. In theory, there is no limitation to its value [4]. In practice, there is always some noise present in the signal, which fixes the scale ε of r below which one will get no information on the system. For very large time-delays and embedding dimensions, r_0 will be very small and will eventually reach the noise scale ε. For that value of T, we will not be able to extract ν or K_2 from the data because the fractal structure will be blurred by the noise.

Despite of all the limitations mentioned above it was possible to estimate K_2 for both the Lorenz system and the Rayleigh-Benard experiment [9]. The procedure was to compute and average :

$$F_q \ (n) \ (r) = \frac{1}{q} \ \text{Log} \ [\ \frac{C_n \ (r)}{C_{n + q} \ (r)} \ ] \qquad\qquad (11)$$

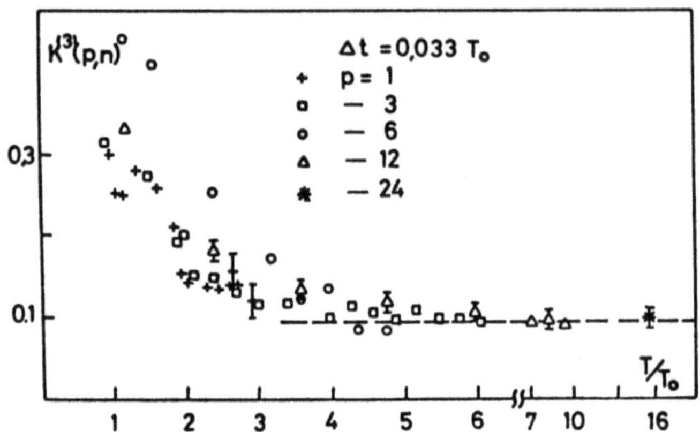

<u>Figure 7</u> : Convergence of $K^{(3)}$ (p, n) to a limit value as a function of
T = (n - 1) p Δt normalised to the intrinsic time-scale T_0, for the Rayleigh-
Bénard case.

theoretically in the scaling region (C (r) ~ r^ν), in practice in a region where
the ratio $\dfrac{C_n}{C_n + q}$ does not depend on r. This signifies that the plots C_n (r) though
being curved are shifted downwards by a constant value as n varies to n + q
(this imposes the choice of q). The averaging in r was done in order to limit statistical fluctuations. To attain the limit b $\to \infty$ in (7), we had to increase the
length (n - 1) Δt (for p = 1) of the sequences from which we build the vectors.
We did this by taking large values of p. Figure 7 shows the behaviour and convergence of :

$$K^{(3)} (p, n) = \frac{1}{p\,\Delta t} < F_3^{(n)} (r) >_r \qquad\qquad (12')$$

as a universal function of T/T_0 (T = (n - 1) p Δt ; T_0 natural time-scale ; T/T_0
appears to be the important convergence parameter) for the Rayleigh-Benard convection regime. It is clear that the limit b $\to \infty$ in (7) has been obtained and that
it does not depend on r or Δt.

3. The Case Of High-Dimensional Attractors

We have shown in the case of low-dimensional attractor what was the influence of
the different parameters n, p Δt, N. In particular, it was assumed and verified that
taking a delay p or studying a signal with a sampling period p Δt gave the same integral correlation function. In practice we obtain such an equivalence for high-dimensional systems only for high enough r. For lower values of r a "tail" of slope
about 1 is generally observed for the signal with the low sampling period (Δt).
This effect is due to the fact that the nearest neighbors of a point $\vec{X}_t$ are the
points ($\vec{X}_t \pm \Delta t, \vec{X}_t \pm 2\,\Delta t \ldots$). The reconstructed trajectory in phase-space
consists of "loops" separated by a minimum distance D_m. The distance between nearest
neighbors d_p is smaller than D_m. Between those scales, one "sees" an individual
loop and C (r) ~ r. We first evidenced this effect on a velocity measurement data
relative to a grid turbulence experiment [8] where the dynamics was known to be
very high dimensional. A 7-dimensional differential system obtained by truncating
the Navier Stokes equations with spatially periodic boundaries and a time-periodic
forcing [14] also displayed the same behaviour (Fig. 8) for a parameter value
R = 4 10^3 and for a sampling period of about $T_0/140$. Let us remark that this feature is not characteristic of high-dimensional systems. The Lorenz data would certainly exhibit such an effect for a sampling frequency high enough.

Because of the "tail", the scaling region will be reduced or masked. The log
C (r) vs. log r plot will appear curved, showing that we are not in the scaling
region. The situation is very similar to the one studied above and illustrated in
Figure 5. When we increased the sampling period to about $T_0/5$, and plotted C (r)

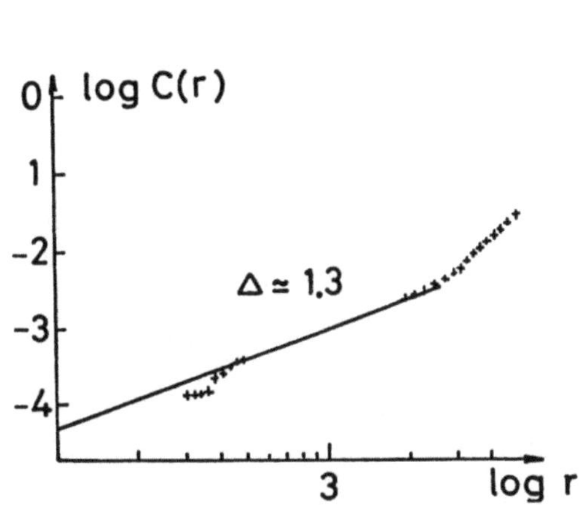

Figure 8 : Plot of log C (r)
versus log r for the 7-dimen-
sional Franceschini model
(R = 4000) exhibiting a
clear region of slope 1
($\Delta t = \dfrac{T_0}{140}$, n = 13, p = 7,
N = 5 10³ m = 100).

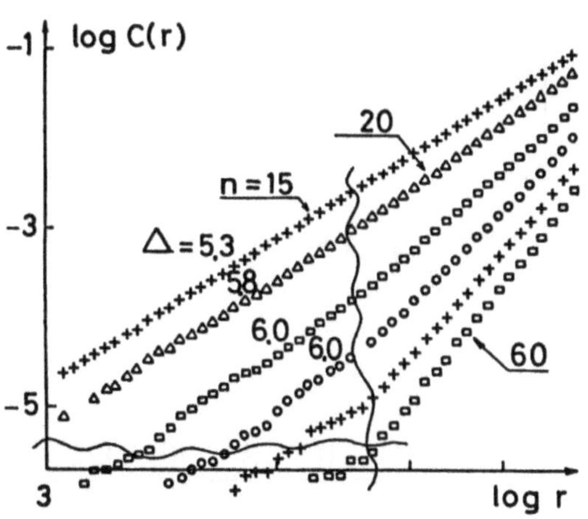

Figure 9 : Set of curves
log C (r) versus log r
for the Franceschini model
for a sampling period of
$T_0/5$ for n = 15 to 60.
The slopes Δ for the scaling region are indicated
(p = 1, N = 10^4,
m = 600).

versus r for p = 1 and n = 15 to 60 we observed the disappearance of the tail and
a region in r where the slope of C (r) determined by a least square fit appeared
to become independent of n as n $\geqslant$ 30 (Figure 9). The local slopes (defined by
(10)) exhibited very large fluctuations because of a statistical problem,due to
the fact that we had a relatively small number of data points (N = 10^4) and therefore it was meaningless to compute an error bar on the exponent from the mean-
square deviations of Δ_i. When increasing the number of data points to
N = 15.10^4 and for n = 30, 40 a clear scaling region appeared with an exponent
around 6.0 for more than half a decade in r (Figure 10). The scaling region is
clearer for this computation than for the previous one. This shows that the method
is efficient when   adequately used,because the information dimension estimated in
[15] from the Kaplan-Yorke formula is around 6.2.

   From the curves plotted in Figure 9, we can try and estimate the entropy K_2 associated to the data and compare it with the value computed from the sum of the positive Lyapunov exponents from [15] using the Pesin formula : K = Σ λ^+ $\cong$ 38,5. For
n = 40 and 50 we obtain $K^{(10)}$ (1, n) = 48 $\pm$ 5. We only have a qualitative agreement

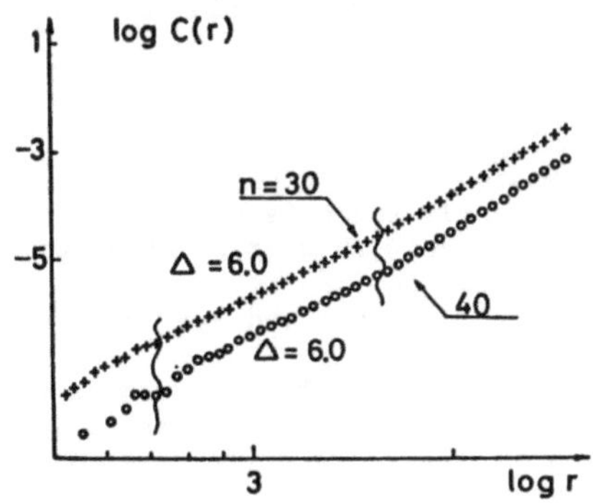

Figure 10 : Same as Fig. 9 for
n = 30, 40 except
(N = 150.10^3, m = $1,2\ 10^3$).

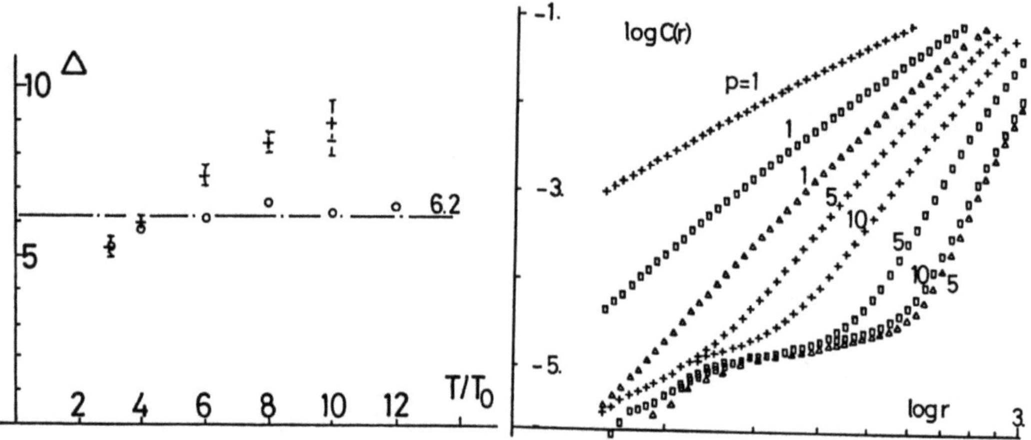

Figure 11 : Plot of the exponent Δ for the Franceschini model as a function of T/T_0 for two different sampling periods and scales of r.
$\Delta t = T_0/29$, $3.10^3 \leqslant r \leqslant 5.10^3$ (-) ; $\Delta t = T_0/5$, $3.10^3 \leqslant r \leqslant 5.10^3$ (+), $r \leqslant 2000$ (o).

Figure 12: Plot of log C(r) versus log r for the Electrohydrodynamic convection regime for a sampling period of $T_0/30$, for n = 10 (+), 20 ( ), 30 (Δ), and different delays. (N = 150 10^3, m = 150).

because of the fact that an insufficient number of periods T_0 has been explored so that the limit $b \to \infty$ in (7) has not been obtained. A huge increase in the number N of data points and origins m would then be necessary in order to attain the convergence.

The whole situation is well summarized in Figure 11, where we have reported the average maximum slope on certain r regions as a function of T/T_0 for two different sampling periods. The last point (-) suggests that there may be a saturation of the slope to a value of the order of 9. At this point we would like to warn against a wrong exploitation of the saturation of the apparent slope versus n. Here we clearly have no scaling zone, and the pseudo-saturation results from a combination of the two curvature effects due to the "edge" effect for high r values and the transition to the régime of slope about 1 for low values of r.

As mentioned above, we first evidenced the effects of the sampling frequency on an experimental signal. In particular, we studied the case of an electro-hydrodynamic convection experiment where the instability is due to an ion injection in a insulating liquid submitted to D.C. voltage which is the driving parameter. The variable analysed is the fluctuation of the total electric current crossing the cell. For a small cylindrical aspect-ratio $\Gamma \sim 1$, the transition to turbulence is typical of systems with a small number of degrees of freedom [16]. Furthermore, the power-spectra present the exponential decrease, which seems to be characteristic of deterministic chaos [17].

The regime we have analysed corresponds to a value 5 times the instability threshold value.

Figure 12 is a set of curves log C (r) versus log (r) for this data, for different values of the delay p and embedding dimension n. These plots exhibit a "tail" of slope 1 and a global curvature, and consequently there is no scaling behaviour [8]. The sampling period was $T_0/30$. In accordance with what we observed for the Franceschini model, we increased the sampling period to $T_0/3$ by taking one data point out of ten. The curves log C (r) versus log r for p = 1 and n = 50 to 150, of Fig. 13 show that the tails have disappeared.

Notice that the distances between points in phase-space have been raised, so that the scales in r we explore are higher than before. For n = 100 to 150, there is a region for C (r) > 10^2/N x m where the behaviour of the plots C (r) is roughly linear and the slopes as indicated in Fig. 13 are around 30.

188

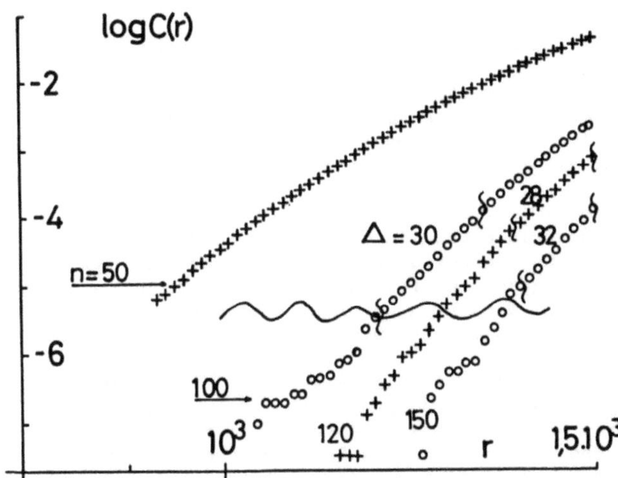

It would be senseless to conclude anything from this set of curves because the
average slope Δ for a fixed r interval, increases slightly with n, and this is an
indication that we are not in a scaling region. A larger number of data N, would be
needed in order to attain the scaling region (assuming that it exists). However,
when analysing such high-dimensional systems it is important to always keep in mind
as a reference the result of the dimension analysis for white noise [8]. For the
latter data, because of edge effects, the curves C (r) are not straight lines and
present a monotonic curvature. For n = 100, and in the same range
$10^{-5} \leqslant$ C (r) $\leqslant 10^{-3}$, the average slope for the C (r) curve for white-noise
p = 1 N = 150.10^3, m = 30) is $\Delta \simeq 40$ which is not much higher than the value shown
in Fig. 13. To characterise a high-dimensional attractor, it is then necessary to
imbed the signal in a phase-space of dimension n for which the average exponent for
white-noise with the same values of N and m is well above the dimension of the
attractor.

Also, it would be necessary in order to have a good confidence in the results for
high-dimensional experimental data to compare these with the ones for high-dimensional "numerical" attractors.

<u>Conclusion</u>

We have studied the method of Grassberger and Procaccia on a certain number of
examples and shown the influence of the different parameters. In particular, systematic errors may occur if the curves log C (r) versus log r are not analysed care-
fully enough because of a possible curvature in C (r). However, low-dimensional
attractors for which the entropy is not very large ($\sim 10^{-1} T_0^{-1}$) are relatively
easy to characterise. For the general case, it would be interesting to have more
quantitative rules. For example, we could try lo look for a sort of Shannon rule for
the choice of the sampling frequency (Δt)$^{-1}$ in order not to have a tail of slope
1. One observation from the study on the Lorenz model and the Franceschini model is
that there was convergence for the slope in the plots C (r) for large r values for
$T \sim \nu T_0$. One could then wonder whether the best way to proceed would not be to
sample the data at a period $\Delta t \sim T_0$ and to imbed it in a phase-space of dimension n
not very much higher than $\nu + 1$.

Another important problem is to know which is the maximum value r_0 below which
there is the scaling zone (C (r) $\sim r^\nu$). We can build examples where r_0 is as
small as one wishes : $r_0 \simeq 10^{-2} r_{max}$ (maximum scale in r) in the case of a torus
T_2 where the frequencies f_1, f_2 have respective amplitudes 1 and 10^{-2}. To characterize this torus T_2 it will be necessary to explore scales in r below 1 % of the
maximum scale. If one assumes that for physical reasons the ratio r_0/r_{max} remains
relatively constant, then with a certain resolution N x m there will be a maximum
dimension ν_{max} for the attractor which can be characterized, and above which no
scaling zone will be apparent in the curves C (r). If $r_0/r_{max} \simeq 1/10$, then for

$N \times m = 10^7$ then $\nu_{max} \simeq 7$. However, we cannot exclude more favorable cases where r_o is closer to r_{max}, for which the latter limit would be greater.

With the number of data N, there are also practical problems involved. To store a given sample size N it is necessary to have an experiment which is stable over a period of time about $N \times \Delta t$ and this is not always possible (imagine the example of the Rayleigh-Benard convection regime examined above, where $\Delta t \simeq T_o \sim 10$ s and $N = 10^5$).

It is premature to propose a straightforward methodology in applying the embedding technique and the Grassberger-Procaccia method in order to determine an attractor's dimension from experimental data. One cannot do without a rather long systematic study for each case. However, we suggest the following interactive procedure :

. Determine the typical frequencies of the data by a spectral analysis.

. Identify the fundamental frequency or a typical "return" time T_o, and sample the signal to about a few points per T_o.

. For the first trial choose a certain number m of origins ($\geqslant 10^2$ for example)- this fixes the resolution $N \times m$ - and test the independence of C (r) towards m.

. Compute a set of curves C (r) for p = 1 and increasing values of n, and examine the slopes of the plots to see if the scaling zone has been reached. If not, then try to take a larger number of data N and origins m and repeat the procedure until adequate convergence has been obtained ...

References

[1] J.A. Vastano, Private Communication, and this Conference.

[2] M. Sano and Y. Sawada, Phys. Rev. Lett., 55, p. 1082 (1985).

[3] H. Froehling, J.P. Crutchfield, D. Farmer, N.H. Packard and R. Shaw, Physica, 3 D, p. 605 (1981).

[4] F. Takens in "Dynamical Systems and Turbulence", Lecture Notes in Math., 898, Springer, Berlin (1981).

[5] P. Grassberger and I. Procaccia, Phys. Rev. Lett., 50, p. 346 (1983).

[6] P. Grassberger and I. Procaccia, Phys. Rev. A, 28, N° 4, p. 2591, October 1983.

[7] D. Eckman , J.P. Ruelle , Rev. Mod. Phys., 57, July 1985.

[8] P. Atten, J.G. Caputo, B. Malraison and Y. Gagne, Journal de Mécanique Théorique et Appliquée, Special Issue (1984), p. 133 - 156.

[9] J.G. Caputo and P. Atten, to be published.

[10] F. Takens "On the numerical determination of the dimension of an attractor", pre-print.

[11] P. Grassberger and I. Procaccia, Physica, 13 D, p. 34 - 54 (1984).

[12] M. Dubois, P. Bergé and V. Croquette, J. Physique Lett., 43, p. L-295 - L-298 (1982).

[13] B. Malraison, P. Atten, P. Bergé and M. Dubois, J. Physique Lett., 44, p. L-897 - L-902 (1983).

[14] V. Franceschini, Physica, 6 D, p. 285 (1983).

[15] R.K. Tavakol and A.S. Tworkowski, Physics Letters, 102 A, p. 273 (1984).

[16] B. Malraison and P. Atten in "Symmetries and broken symmetries", Pub. N. Boccara (IDSET, Paris), p. 439 (1981).

[17] B. Malraison and P. Atten, Phys. Rev. Lett., 49, p. 273 (1982).

Transition from Quasiperiodicity into Chaos in the Periodically Driven Conductivity of BSN Crystals

S. Martin and W. Martienssen

Physikalisches Institut der Universität Frankfurt,
Robert-Mayer-Strasse 2-4, D-6000 Frankfurt/Main, F. R. G.

The instabilities in the electrical conductivity of barium sodium niobate (BSN) crystals are studied in the presence of ac and dc fields. Transitions from quasiperiodicity into chaos via phase locking are observed. Phase portraits, Poincaré sections and return maps constructed from measured voltage signals illustrate the emergence of a strange attractor from a torus. The dimension and entropy of the attractor are determined as a function of the control parameter.

1. Introduction

In the last few years there has been growing interest in the study of nonlinear oscillatory and chaotic instabilities of the electrical conduction [1-6]. Recently, we reported on experiments concerning self- generated voltage oscillations and chaotic behaviour of the conductivity in barium-sodium-niobate (BSN) single crystals [7]. We present here new experimental results obtained by applying ac and dc fields to BSN crystals.

We observe transitions from quasiperiodicity into chaos via phase locking. Construction of phase portraits and Poincaré sections illustrate the emergence of a strange attractor from a torus. Return maps obtained from the Poincaré sections show the increase of the nonlinearity responsible for driving the system into chaos. Finally, the chaotic state is analyzed numerically, using the Grassberger-Procaccia method [8], in order to determine the dimension and entropy of the strange attractor as a function of the control parameter.

2. Experimental Procedure

We study BSN single crystals which are grown and prepared in our laboratory. The samples are placed in a heating oven and subjected to an annealing process at $800°C$ in a humid oxygen atmosphere. During this annealing process the crystal reacts with the oxygen atmosphere, which leads to a change in its specific conductivity. This is necessary for the development of the electrical instabilities which are observed in the temperature range of $300°C$ to $600°C$.

The experiments with BSN crystals are performed at fixed oxygen partial pressure and sample temperature. The current density is sinusoidally varied with time about an offset value and the voltage across the crystal is measured. The voltage signals are sampled by a Nicolet 4094 digital oscilloscope and stored for further analysis or are directly processed by a Wavetek 5830 digital signal analyzer.

In all experiments discussed here the oxygen partial pressure is fixed at about 100 mbar and a temperature is chosen at which self-generated voltage oscillations appear with only a constant current density passing through the sample. The ac-current density is then applied and transitions into chaos induced by varying either the amplitude or the frequency of the ac-field. Furthermore, chaos is also observed by varying the dc-offset of the current density at a fixed amplitude and frequency of the ac-field. In all three cases we observe transitions into chaos via the Ruelle-Takens-Newhouse route.

3. Quasiperiodic and Phase-Locked States

Measurements of the transition of the electrical conductivity of BSN into chaos show that the quasiperiodic state is often interrupted by a phase-locked state prior to the onset of chaos. In Fig. 1 the ratio of the frequencies f_0/f_1 ("winding number" Ω) obtained from Fourier spectra is plotted as a function of the current density amplitude. f_0 is the intrinsic frequency of the sample and f_1 the external frequency of the applied ac-field.

As the driving amplitude is increased the winding number decreases gradually, which is an indication of quasiperiodicity. At about $0.12 mA/cm^2$ the winding number attains a value of $\Omega = 1.555$ and does not change up to an amplitude of $0.14 mA/cm^2$. This corresponds to a phase-locked state in which the winding number is rational, here it is $\Omega = 14/9$. In all our experiments we observe that the phase-locked state is followed by a quasiperiodic state, in which the Fourier

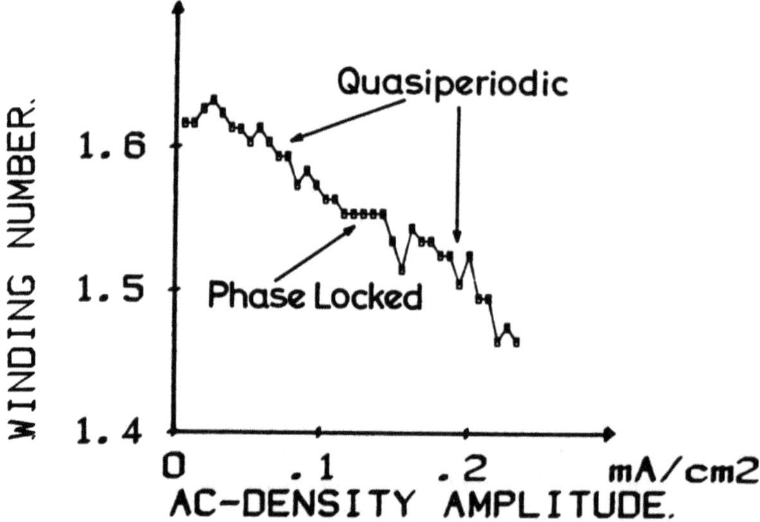

Fig. 1. Winding number $\Omega = f_0/f_1$ vs. current density amplitude of the ac-field. f_0 = intrinsic oscillation frequency, f_1 = frequency of the driving ac-field. The quasiperiodic state is interrupted by a phase locked state ($\Omega = 14/9$). Above $0.22 mA/cm^2$ the quasiperiodicity gives way to chaos, where f_0 develops into a broad band. Experimental parameters: $T = 470°C$; $f_1 = 220 mHz$; $j_0 = 0.65 mA/cm^2$ (dc-offset).

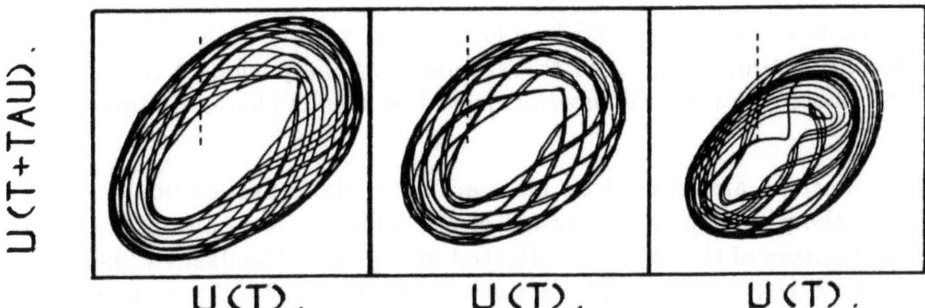

Fig. 2. Phase portraits constructed from measured voltage signals at three different values of the current density offset: a) $j_0 = 0.26mA/cm^2$; b) $j_0 = 0.39mA/cm^2$; c) $j_0 = 0.65mA/cm^2$. In (a) the system is in a quasiperiodic, in (b) near a phase-locked and in (c) prior to a chaotic state. The sequence illustrates the destruction of the torus via a phase-locked state. The dashed lines indicate the position of planes perpendicular to the $U(t)$ direction used to construct the Poincaré sections of Fig. 3. Experimental parameters: $T = 520°C; \Omega = 1.62; j_1 = 0.065mA/cm^2$ (ac-amplitude); Number of orbits $= 20; \tau = 1sec$.

spectra still contain two incommensurate frequencies. However, broadband noise superimposed at low frequencies increases gradually until, above an amplitude of $0.22mA/cm^2$, chaos becomes fully developed.

4. Phase Portraits, Poincaré Sections and Return Maps

In order to study the development of chaos from quasiperiodicity we construct phase portraits and Poincaré sections. In Fig. 2 phase portraits of measured voltage signals are shown for three different values of the dc- offset of the current density and in Fig. 3 the corresponding Poincaré sections. These measurements are made with the winding number held constant near to the reciprocal of the golden mean by adjusting the frequency of the driving field.

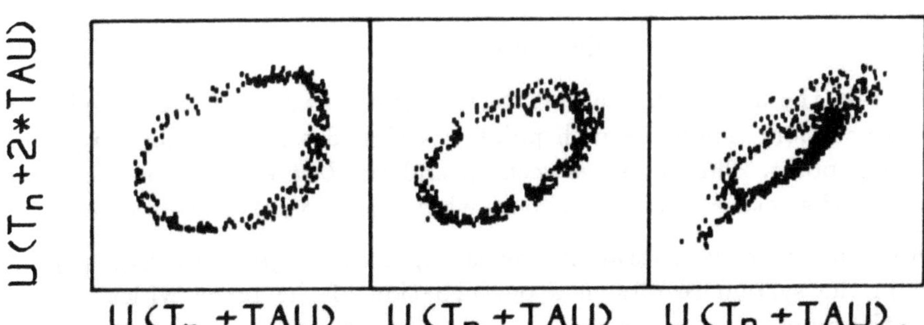

Fig. 3. Poincaré sections obtained from intersecting positively oriented orbits with planes indicated in Fig. 2. In (a) and (b) closed loops illustrate the cross section of the torus, in (c) the border of the loop becomes diffuse as the torus' smooth surface is destroyed. The experimental parameters are the same as in Fig. 2 except the number of orbits is here about 400.

The phase portraits are constructed by plotting the voltage signal $U(t+\tau)$ vs. $U(t)$, where the delay time τ is chosen to be about 1/4 the period of the intrinsic oscillation. The phase portraits all show projected images of the attractors onto a plane.

The Poincaré sections are formed by intersecting a plane perpendicular to the $U(t)$ direction in the phase space composed of the $U(t), U(t+\tau)$ and $U(t+2*\tau)$ directions. The position of the plane is indicated in Fig. 2 by the dashed lines. The positively directed orbits in the 3D phase space intersect the 2D plane spanned by the $U(t+\tau)$ and $U(t+2*\tau)$ directions.

In Fig. 2a we see the movement of the trajectory on the torus as quasiperiodic, i.e. the trajectory never closes but covers the whole torus. The corresponding Poincaré section in Fig. 3a is a closed loop which is just the cross section of the torus.

In Fig. 2b we see how the trajectory covers the torus near a phase-locked state. After the same number of orbits as in Fig. 2a the torus is not uniformly covered. The Poincaré section in Fig. 3b is again a closed loop representing the cross section of the torus. Note that the Poincaré sections are obtained from a larger number of orbits than the phase portraits, so that the nearly phase-locked state of Fig. 3b doesn't differ from Fig. 3a significantly.

In Fig. 2c a phase portrait is shown of a state at the onset of chaos. We see here how the torus is deformed and loses its simple structure. The corresponding Poincaré section in Fig. 3c illustrates the decrease in the smoothness of the attractor's surface: the points of the section are distributed diffusely around the closed loop.

For the motion of the trajectory on a torus a 1D return map can be constructed using polar coordinates (1):

$$\Theta_{n+1} = [\Theta_n + 2\pi\Omega] \cdot mod 2\pi$$

$$\Omega = \text{winding number.} \tag{1}$$

In order to obtain the angles Ω_n we first determine the center of the closed loop of the Poincaré section. Then the n-th point is assigned a radius vector r_n from the center to the point. The angle between r_n and the $U(t+\tau)$ axis is chosen to be Ω_n. In Fig. 4 the return maps are shown where Ω_{n+1} vs. Ω_n are plotted.

For an unperturbed torus a straight line is expected as given by Eq. (1). A nonlinearity in the system results in a perturbation of the torus which can be described by the map:

$$\Theta_{n+1} = [\Theta_n + 2\pi\Omega + f(\Theta_n)] mod 2\pi \tag{2}$$

where $f(\Theta_n)$ represents the nonlinearity. The magnitude of $f(\Theta_n)$ is an indication of the strength of the nonlinearity in the system.

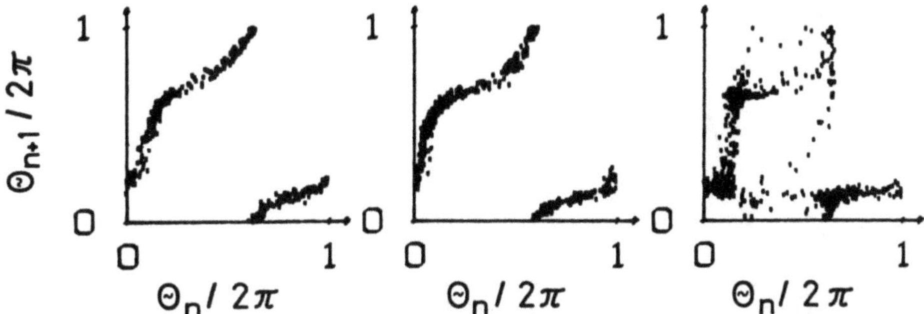

Fig. 4. Return maps constructed from the Poincaré sections of Fig. 3. In (a) the map shows a slight nonlinearity, in (b) the nonlinearity becomes more pronounced and in (c) the map even loses its invertibility which is an indication of the onset of chaos. The stray of data in (c) is due to the diffuse structure of the attractor (Fig. 3c) which results in an error for Θ_n. The experimental parameters are the same as in Fig. 3.

The return map of the quasiperiodic state in Fig. 4a contains a nonlinearity superimposed on the linear increase of Θ_{n+1} with Θ_n. In Fig. 4b the amplitude of the nonlinearity is larger corresponding to an increase of $f(\Theta_n)$ in Eq. (2). In Fig. 4c we not only see a further increase in the nonlinearity of the map, but the map has also become noninvertible for $\Theta_n \to 0$, which is an indication of the emergence of a strange attractor (9).

By plotting $(\Theta_{n+1} - \Theta_n)$ vs. Θ_n we observe that the return maps of Fig. 4 are periodic: $f(\Theta_n + 2\pi) = f(\Theta_n)$. This type of map, generally referred to as the circle map, has been extensively studied by P. Bak et. al. (10) and proposed for describing the dynamical behaviour of CDW-systems. We note however, that $f(\Theta_n)$ is in our case not sinusoidal but piecewise linear having a nonsymmetric sawtooth form.

5. Chaotic State

In order to analyze the chaotic state numerically we use the method introduced by Grassberger and Procaccia (8) and compute the correlation sum:

$$C(r, m) = \lim_{N \to \infty} \frac{1}{N^2} \sum_{\substack{i,j \\ i \neq j}} H(r - \alpha \cdot |\vec{X}_i(m) - \vec{X}_j(m)|)$$

where H = Heavyside function
     r = Hypersphere diameter
     α = Normalization factor
     N = Number of points
     $\vec{X}_i(m)$ = Point in phase space of dimension m
     $\vec{X}_i(m) = (U(t_i), U(t_i + \tau), ..., U(t_i + (m-1) * \tau))$
     τ = Delay time

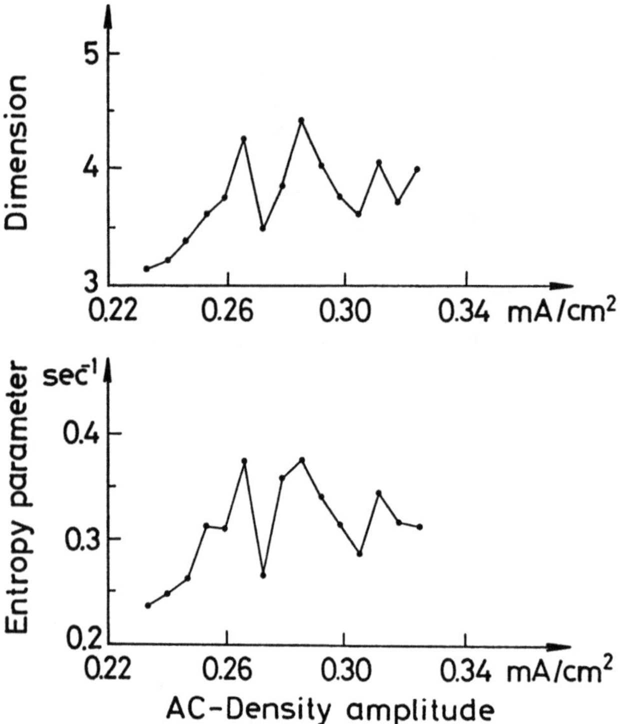

Fig. 5. The dimension and entropy of the attractor determined with the Grass-berger-Procaccia analysis from measured voltage signals. As the control para-meter current density amplitude is varied, both the dimension and the entropy show similar behaviour. The correlation sum is computed from sets of $N = 2000$ points and the delay time is chosen equal to 1/4 the period of the driving field. The dimensions and entropies are limit values for large embedding dimensions $m = 7, 8, 9, 10$. The experimental parameters are the same as in Fig. 1.

From the correlation sum we obtain the correlation exponent ν_m and the entropy parameter $K_{2,m}$ for a chosen embedding dimension m. By increasing m we deter-mine the limit values:

$$\nu_m \xrightarrow[m \to \infty]{} D_2$$

$$K_{2,m} \xrightarrow[m \to \infty]{} K_2 \leq K$$

where D_2 is the correlation dimension of the attractor and K_2 a lower bound to the Kolmogorov entropy K.

In Fig. 5 the dimension and entropy of the attractor, computed from measured time signals, are plotted vs. the current density amplitude of the ac-field. Note that both the dimension and the entropy vary in a similar manner as a function of the control parameter. This suggests that the changes in the static properties of a strange attractor are closely associated with the changes in the dynamic behaviour

of the nonlinear system. The dimension initially increases monotonically and then
fluctuates between 3.4 and 4.4. Likewise, the entropy parameter shows an initial
increase and then fluctuates between 0.25 sec-1 and 0.37 sec-1.

The dimension and entropy values obtained here are lower than those determined
from experiments with only dc-fields applied to the BSN crystals [7]. From this
we conclude that without the presence of ac-fields chaos can develop only by a
complex interaction of a number of nonlinear oscillation modes. The application
of an ac-field, however, induces a nonlinear interaction between an oscillation mode
and the external field, resulting in the emergence of a lower dimensional attractor.

6. Conclusion

The experimental results presented here show that BSN represents an interesting system for studying electrical instabilities in the solid state. Especially, the
analysis of the transition into chaos and of the strange attractor suggest that the
chaotic behaviour observed by applying ac- and dc-fields to the BSN crystals can
be described by a 1-dimensional circle map. This yields preliminary evidence for
the possibility of applying simple models to the complex phenomena observed in
this nonlinear dynamical system. Further experiments will have to be conducted in
order to confirm other universal properties of the transition from quasiperiodicity
into chaos.

We would like to thank Marian Martinez for the preparation of this manuscript.
This work is supported by the Deutsche Forschungsgemeinschaft via the "Darm-
städter/Frankfurter Sonderforschungsbereich Festkörperspektroskopie".

1. For a review see: H. G. Schuster, "Deterministic Chaos," VCH publishers,
    Weinheim (1984).

2. R. M. Fleming and C. C. Grimes, Phys. Rev. Lett. 42, 1423 (1979).

3. S. W. Teitsworth, R. M. Westervelt, and E. E. Haller, Phys. Rev. Lett. 51,
    825 (1983).

4. G. A. Held, C. Jeffries, and E. E. Haller, Phys. Rev. Lett. 52, 1037 (1984)

5. S. E. Brown, G. Mozurkewich, and G. Grüner, Phys. Rev. Lett. 52, 2277
    (1984).

6. R. P. Hall, M. Sherwin, and A. Zettl, Phys. Rev. 29B, 7076 (1984).

7. S. Martin, H. Leber and W. Martienssen, Phys. Rev. Lett. 53, 303 (1984).

8. P. Grassberger and I. Procaccia, Phys. Rev. 28A, 2591 (1983).

9. P. Bak, T. Bohr, M. H. Jensen and P. V. Christiansen, Solid State Comm.
    51, 231 (1984).

10. P. Bak, T. Bohr, and M. H. Jensen, Proc. of the 59th Nobel symposium, "The
    physics of chaos and related phenomena," Gräftaavallen, Sweden (1984),
    Physica Scripta.

Dimension and Entropy for
Quasiperiodic and Chaotic Convection

H. Haucke, R.E. Ecke, and J.C. Wheatley

Los Alamos National Laboratory, Los Alamos, NM 87545, USA

Abstract

High quality experimental data have been taken on a convection cell containing a dilute ^{3}He-^{4}He solution. We discuss some problems with the determination of dimension and entropy for experimental data, and compare the results to detailed Poincaré sections. At the chaotic transition, we show the behavior of dimension and entropy as a function of Rayleigh number.

Rayleigh-Bénard convection has been perhaps the most intensively studied non-linear experimental system. Dimension and entropy calculations on convection data have been carried out by MALRAISON, ATTEN, BERGE and DUBOIS [1] as well as GIGLIO et. al. [2]. Work done on dimensionality of attractors for other hydrodynamic systems includes that of BRANDSTÄTER, et. al. [3] for Couette-Taylor flow and GUCKENHEIMER and BUZYNA [4] for geostrophic flow. Theoretical and practical aspects of dimension and entropy measurement are discussed in a recent review article by ECKMANN and RUELLE [5]. We report here on systematic observations of convection in a ^{3}He-superfluid ^{4}He solution, in which we have achieved extremely low levels of noise and drift. Measurements were made over the transition region for the breakdown of a two-torus into a chaotic attractor. Calculations of dimension and entropy as a function of an external stress parameter are presented.

The fluid used in this work is a dilute solution of 1.46% ^{3}He in superfluid ^{4}He. One can treat this solution as a one-component fluid and derive a Rayleigh number R, analagous to the Rayleigh number used as the stress parameter for convection in ordinary fluids, and proportional to the top-bottom temperature difference. A second dimensionless parameter, the Prandtl number, is required to fully characterize the system in the one-component Boussinesq approximation. Theoretical [6] and experimental [7,8] justification for the one-component treatment of these solutions, and definitions of the appropriate Prandtl and Rayleigh numbers may be found in previous publications [7].

In this work we are using a rectangular geometry with height d = 0.80 cm, length 2.0 d, and width 1.4 d. The sidewalls are thermally insulating relative to the fluid, whereas the top and bottom boundaries are thermally conducting. In operation, the bottom boundary is maintained at a fixed temperature by feedback control. A fixed heat current is applied to the top boundary. Details of cell construction and operation can be found elsewhere [7].

The time series discussed below are all derived from a single differential thermocouple sensor. One end of the thermocouple is attached to the copper plate forming the upper boundary of the cell; the other to a small, cylindrical, thermally isolated copper plug inserted into the center of the upper plate. The sensor is sensitive to local temperature gradients but not to fluctuations in the mean cell temperature. This is one reason for our very low noise level.

With a small heat flow through the cell, the fluid is motionless and conducts heat diffusively. Increasing the heat flow increases the temperature difference between the top and bottom boundaries,and thus the Rayleigh number. When the critical Rayleigh

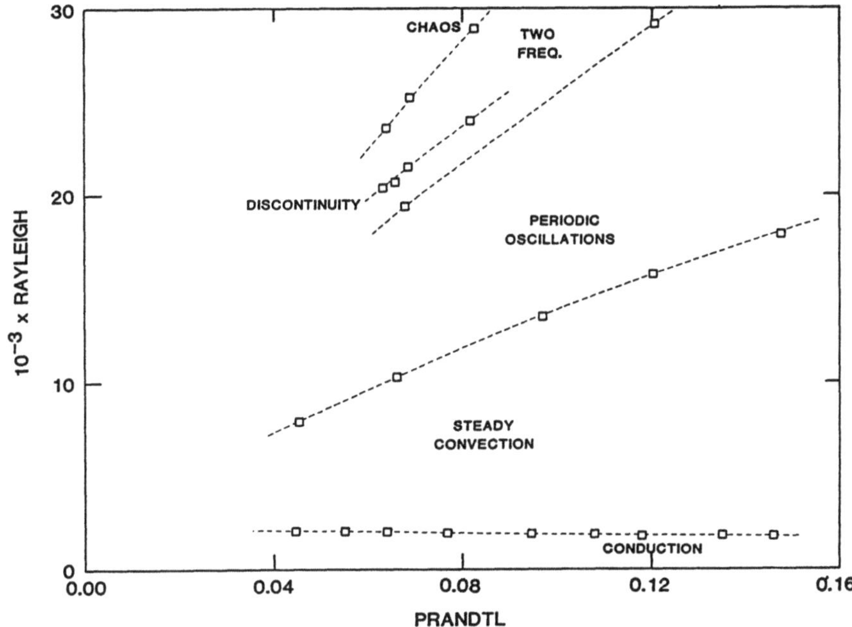

Fig. 1. Phase diagram of transitions on the Rayleigh-Prandtl number plane

number R_c is reached the fluid begins to convect. Further increasing the Rayleigh number excites first one oscillatory instability and then a second, as shown in Fig. 1. In the two-frequency region we observe mode-locking. As the Rayleigh number is increased through this region, a large discontinuous change in both oscillatory frequencies is observed,and the shape of the attractor changes drastically. We will concentrate henceforth on the region above this discontinuity and below the chaotic onset.

We define a winding number $W = f_2/f_1$, where f_1 ~0.7 Hz and f_2 ~0.1 Hz are the fundamental frequencies of the first and second instabilities seen with increasing R. Figure 2 shows a plot of W versus normalized Rayleigh number. Many flat steps are visible. At the upper right the system has become chaotic. While we find that the chaotic transition is "continuous" or gradual and thus hard to identify precisely, on the scale of Figs. 1 or 2 this is not a problem. For each point in Fig. 2 the frequencies f_1 and f_2 are determined by acquiring a time series, generating the power spectrum, and fitting peak frequencies. A more detailed discussion of the mode-locking will appear elsewhere [9] but a brief overview is given here. As an example of simple quasiperiodic data, we show in Fig. 3 the attractor for a state well below the chaotic onset. This projection is produced using the usual "delay coordinate" method [10]. The delay time used in constructing the phase space we denote by τ; τ = 5.81 seconds. for Figs. 3-5. Despite its rather contorted appearance, this is a nice two-torus. By making a suitable cut (illustrated by the shaded plane in Fig. 3) we produce the simple Poincaré section of Fig. 4. Using the parametrization shown yields the return map, Fig. 5.

As R/R_c is increased the system becomes increasingly complex and a 1-D mapping no longer is adequate to describe the system. Hysteresis and Hopf bifurcations from locked states are observed [9], and indicate a 2-D mapping is necessary to characterize the state. Continuing upwards in R/R_c, we observe chaotic states having broadband spectral noise and Poincaré sections with no apparent structure. We have used several dimension algorithms, described below, to characterize the states near the chaotic onset.

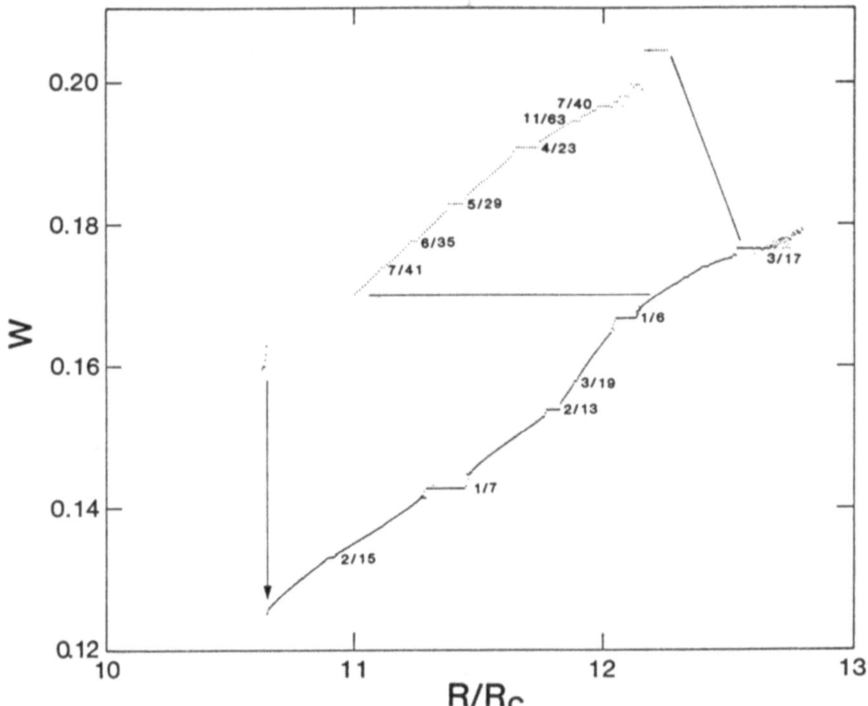

Fig. 2. A plot of winding number as a function of normalized Rayleigh number showing a number of mode-locking steps, $R_C = 2017$. The arrow shows the discontinuous jump in W corresponding to the "discontinuity" in Fig. 1.

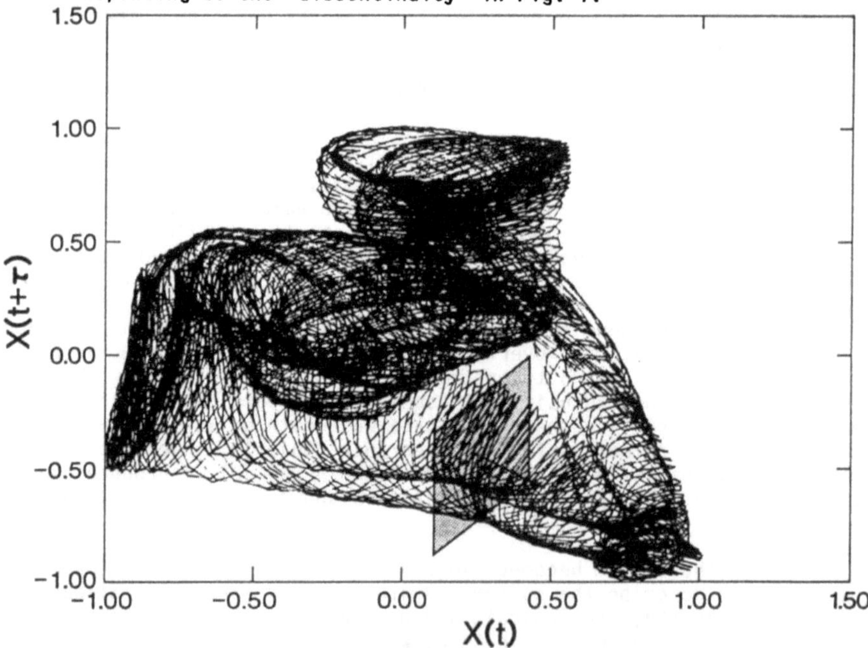

Fig. 3. Delay coordinates are used to generate a phase space. This shows a two-dimensional projection of a quasiperiodic attractor and the cutting surface used to produce Fig. 4

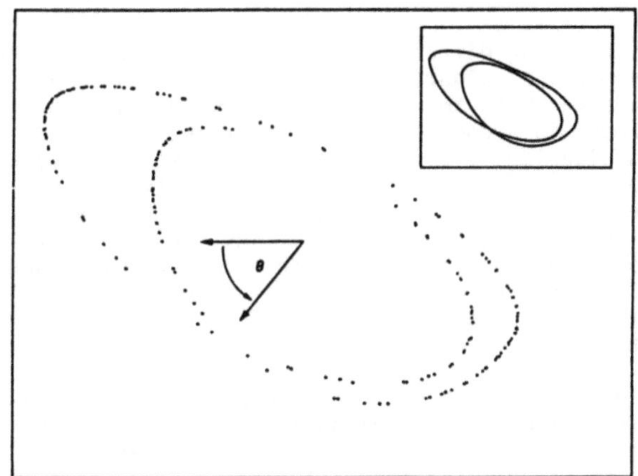

Fig. 4. This Poincaré section illustrates the simple behavior present well below the chaotic onset

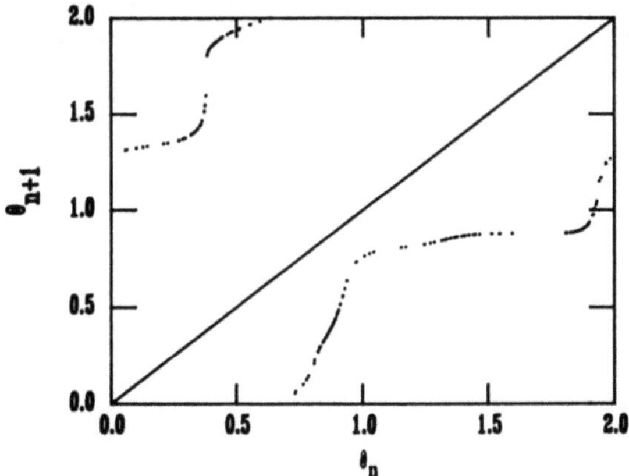

Fig 5. A return map corresponding to the data in Fig. 4

     In all our dimension work we rely on the standard "delay coordinate" method [10] for generating a phase space. In all of what follows we have used 33,000 point time series 38.5 minutes in length. We have applied three different dimension algorithms to our data. In all of these, we first randomly select a subset of points on the attractor as our reference points (enough for good statistics but not excessive computation time). Our most extensively used method is due to GRASSBERGER et. al. [11]. In this case, for each reference point the distance to all other points is calculated. These separations are then accumulated for all the reference points and the number of separations greater than or equal to ϵ, called $C(\epsilon)$, is calculated. A second algorithm follows TERMONIA [12]. Given some number of points n, R(n) is the radius of the smallest ball around a reference point that contains n points. R(n) is determined for each reference point and then averaged. It is the averaging process that distinguishes this algorithm from Grassberger's. Finally, one can calculate the number of points N inside a ball of radius ϵ circumscribed about a reference point and average $\ln(N(\epsilon))$ over the chosen set of reference points. This is known as the "pointwise" dimension. Additionally, three definitions of distance may be used: the max norm, defined as the largest of all the coordinate differences; the mod norm, defined as the sum of the absolute values of each coordinate difference; and, finally, the standard Euclidian distance, defined as the square-root of the sum of the squares of the coordinate differences.

We explored the variation of dimension using different methods of calculating point separations, and using the different dimension algorithms as described above. We found no systematic variation of dimension with the different norm definitions. However, a systematic variation was found in a comparison of the three different algorithms. The Grassberger-Proccacia algorithm yielded dimension values which were usually 10 to 15% below the values obtained using the other two algorithms. The parameters used in the algorithms were a delay of τ = 1.05 seconds, five hundred reference points and embedding dimensions ranging from 3 to 10. The convergence with increasing embedding dimension was approximately $\pm$ 5% for all three algorithms. None of the algorithms proved superior in the sense of more apparent scaling regions or less scatter in the extracted dimension. Therefore, we recommend the Grassberger-Proccacia algorithm with the max norm for ease and speed of computation.

As noted above we used a randomly selected subset of reference points to calculate the dimension. We found that there was appreciable variation in the dimension when this subset was too small,and that the variance in the average dimension decreased as the square root of N_{ref}, the number of reference points. In Fig. 6, the variation of dimension with the choice of a particular set of reference points and the number in that set is shown. In both cases τ = 7 seconds and max norm distances were used in computation of the Grassberger-Proccacia algorithm for an embedding dimension of five. For each histogram 25 different sets of reference points were used. The left pair of histograms are for a simple quasiperiodic state which yields an average dimension of 2.25. The variance in the average dimension is 0.15 and 0.05 for N_{ref} = 10 and 100, respectively. For a chaotic state (the right pair of histograms), the variance is 0.23 and 0.07 for N_{ref} = 10 and 100. Although the absolute variance is greater for the chaotic data, the fractional effect is about the same, 7% and 2% for N_{ref} = 10 and 100 respectively.

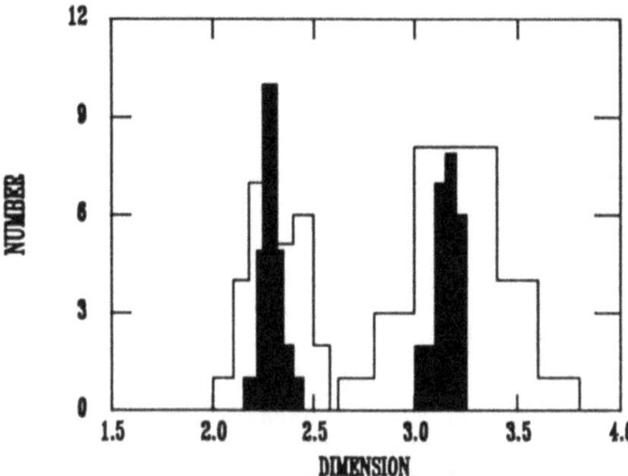

Fig. 6. Variation of distribution of dimensions with size of reference subset for simple quasiperiodic data (left) and chaotic data (right). N_{ref} = 100 (shaded), N_{ref} = 10 (outlined). Twenty-five different subsets were used

The variation of the dimension with the value of the delay used in the phase space reconstruction was also investigated. Again we used the Grassberger-Proccacia algorithm with max norm distance, an embedding dimension of five, and 200 reference points, and fit over a fixed region of $C(\epsilon)$. As seen in Fig. 7, there was only slight variation for the simple quasiperiodic state but a substantial increase in the dimension of the chaotic state for τ > 7 seconds. In addition there was a dip at τ > 1 second in both data sets,corresponding to a delay time commensurate with the shorter period of ~1.2 seconds. Peaks in the dimension for the chaotic data at τ = 7.5 seconds and 15 seconds may be due to commensuration with the long period of ~8 sec. We would in general expect that picking a value of the delay close to a oscillation period would reduce rather than increase the dimension. However, the general trend of increasing dimension after some finite delay is reached is typical for states having positive

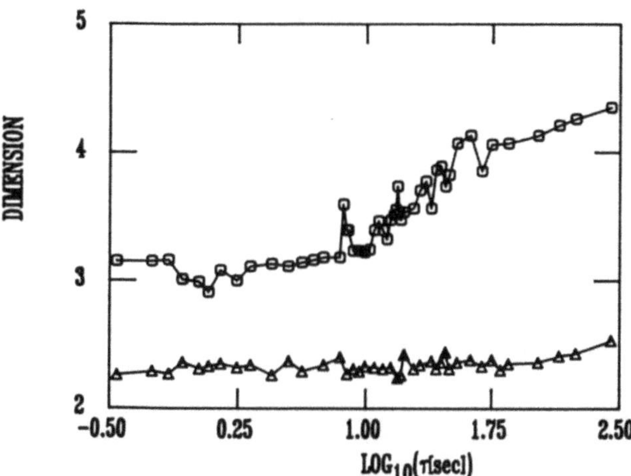

Fig. 7. Variation of fractal dimension with delay τ for an embedding dimension of five and N_{ref} = 200. Lines are guides to the eye for quasiperiodic data ($\triangle$) and chaotic data (O)

entropy. Positive entropy requires that the attractor has some finite width to its probability density along the dth coordinate even when d is much greater than the fractal dimension. For consistency, we use GRASSBERGER'S [13] measure of entropy K_2. $K_{2,d}(\epsilon)=(1/\tau)\ln(C_d(\epsilon)/C_{d+1}(\epsilon))$ and K_2 is the limiting value for large d and small ϵ. For sufficiently small ϵ this gives $\ln(C_d) - \ln(C_{d+1}) = \tau K_2$. As Grassberger has described, this produces an offset at small ϵ on the usual log $(C_d(\epsilon))$ vs log (ϵ) plot (Fig. 8). Since all the curves intersect at $\epsilon = 1$ this distorts the slopes close to the origin, and for large enough τ the scaling region is driven below the limits of resolution.

Extraction of the dimension from three different data sets is illustrated in Fig. 8 along with their associated Poincaré sections. The scaling regions chosen for the $\log_{10}(C(\epsilon))$ vs. $\log_{10}(\epsilon)$ plots are indicated. The Grassberger algorithm with N_{ref} = 500 and max norm were used. Cases a) and b) correspond to the previous quasi-periodic and chaotic data sets; c) is a particularly difficult instance. The ability of the dimension algorithms to distinguish locked, quasiperiodic and chaotic states was comparable to one's own ability based on original inspection of power spectral density plots and Poincaré sections for the three different kinds of states.

In Fig. 9 we show the behavior of dimension with increasing R/R_c calculated just as for Fig. 8. One can see an average trend of increasing dimension, as well as rapid variations produced by periodic windows. This is similar to behavior found in the supercritical circle map [14]. We intend to explore the scaling of dimension with R/R_c and the periodic window structure in the supercritical regime in some future study.

In our entropy calculations we benefited from a paper by CAPUTO and ATTEN [15]. This prompted us to try the algorithm put forth by GRASSBERGER [13]; we prefer it over TERMONIA'S method [16]. Grassberger's algorithm calculates a quantity K_2 which is a lower bound on the Kolomogorov entropy. We used the maximum norm in our calculations as suggested in refs. [9] and [13]. In Fig. 10 we show plots of $K_{2,d}(\epsilon)$ for the same quasiperiodic and chaotic data as Figs. 6-8, with N_{ref} = 3000. The chaotic data gives a conspicuous plateau region and then falls off as the amount of data becomes insufficient, whereas the quasiperiodic data decreases smoothly. A value of $\epsilon = 0.15$ was used (with the total excursion of the time series normalized to unity). Figure 11 shows the behavior of K_2 with varying R/R_c. Three types of points are used. The open and closed circles denote values obtained at d = 100, τ = 7 sec., N_{ref} = 300, and $\epsilon = 0.15$ (averaged over $90 \leq d \leq 100$). In some cases the data seemed to indicate a small positive entropy but no plateau was found, and these are the open circles. The

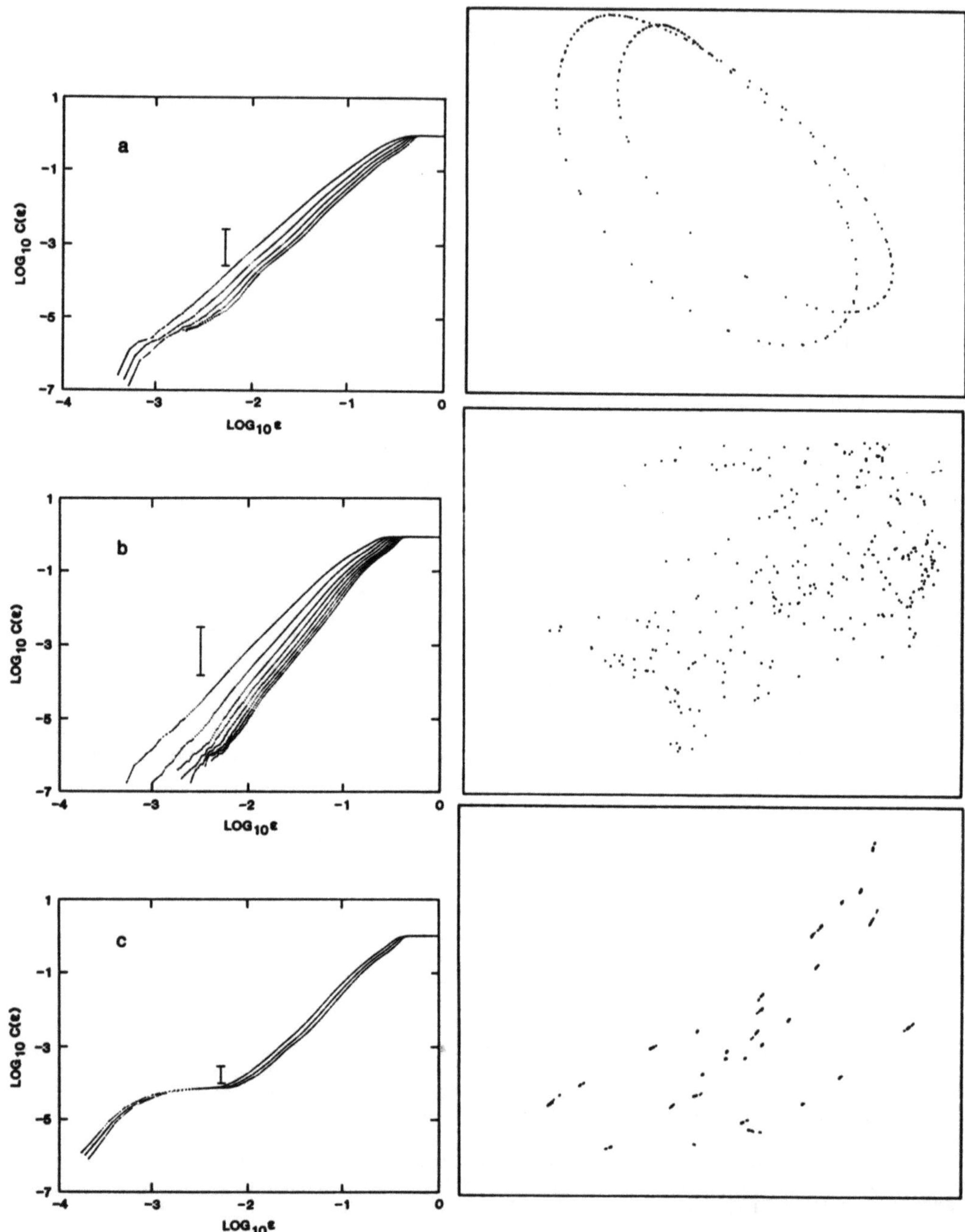

Fig. 8.   The left-hand column shows plots of $\log_{10} C(\epsilon)$ vs. $\log_{10} (\epsilon)$ for three data sets.   We evaluated the Grassberger dimension by fitting a straight line to the indicated portions of the plots.   On the right are the corresponding Poincaré sections. These sections were produced in the same manner as Fig. 4.   a) Simple quasiperiodic data, d = 3-7; b) Chaotic data, d = 3-10; c) An example of a data set where the dimension is difficult to determine, d = 5-7

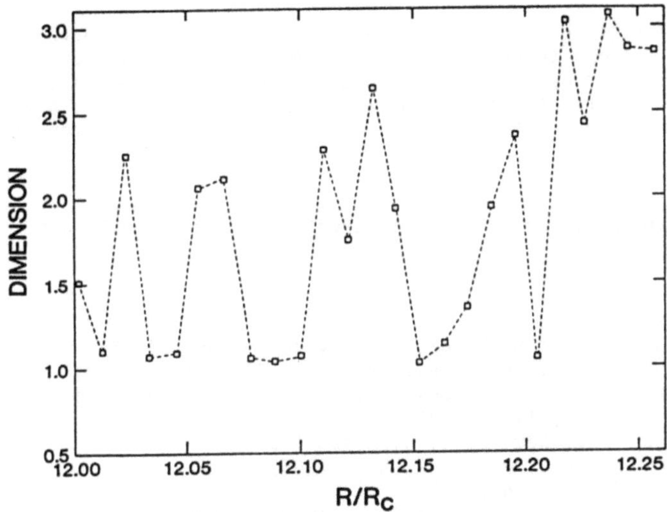

Fig. 9. The Grassberger dimension is shown as a function of normalized Rayleigh number

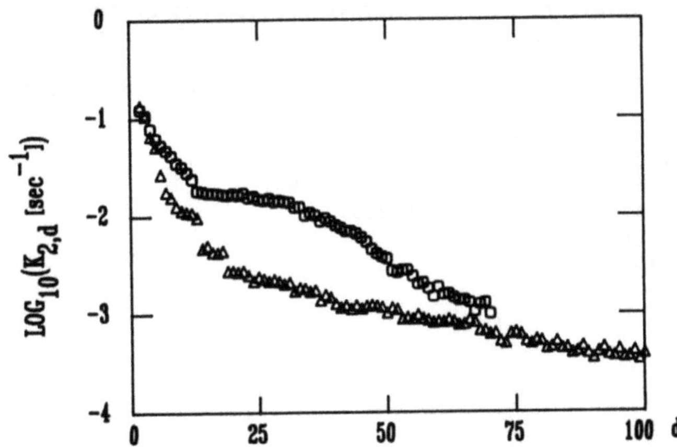

Fig. 10. $K_{2,d}(\varepsilon)$ is plotted as function of the embedding dimension d for quasi-periodic ($\triangle$) and chaotic (O) data sets. Note the plateau in the chaotic case

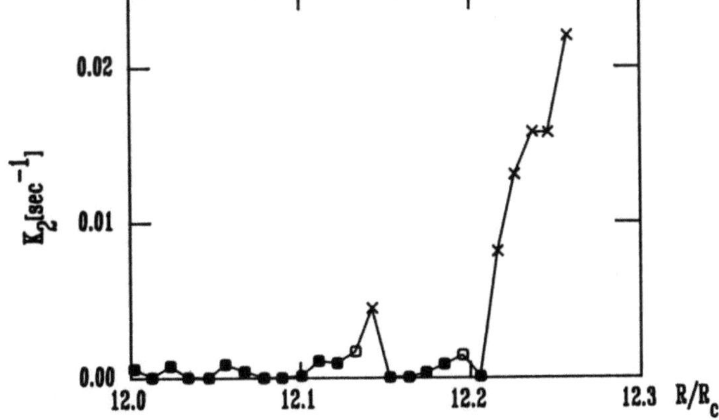

Fig. 11. K_2 versus normalized Rayleigh number

205

x points are values obtained by fitting to a plateau region. An attempt to check the scaling with τ showed that the entropy tended to increase somewhat as τ was reduced. This is probably due to the fact that as τ is decreased one examines more phase space points per unit time, which has an effect similar to decreasing ϵ. We are probably underestimating the entropy noticably for the x cases due to the relatively large value of ϵ.

We are pleased to acknowledge valuable conversations with D. Farmer and E. Jen.

References

1. B. Malraison, P. Atten, P. Berge, and M. Dubois: J. Physique L-897 (1983)
2. M. Giglio, S. Musazzi, and U. Perini: Phys. Rev. Lett. **53**, 2402 (1984)
3. A. Brandstäter, J. Swift, H. L. Swinney, A. Wolf, J. D. Farmer, E. Jen, and P. J. Crutchfield: Phys. Rev. Lett. **51**, 1442 (1983)
4. J. Guckenheimer and G. Buzyna: Phys Rev. Lett. **51**, 1438 (1983)
5. J. P. Eckmann and D. Ruelle: Rev. Mod. Phys. **57**, 617 (1985)
6. A. Fetter: Phys. Rev. B**26**, 1164 and 1174 (1982)
7. Y. Maeno, H. Haucke, R. Ecke, and J. C. Wheatley: J. Low Temp. Phys. **59**, 305 (1985)
8. Y. Maeno, H. Haucke, and J. C. Wheatley: Phys. Rev. Lett. **54**, 340 (1985)
9. R. Ecke, H. Haucke, and J. Wheatley: to be published in Proceedings of "Perspectives on Nonlinear Dynamics" workshop (1985); H. Haucke, R. Ecke, and J. Wheatley: to be published
10. N. H. Packard, J. P. Crutchfield, J. D. Farmer, and R. S. Shaw: Phys. Rev. Lett. **45**, 712 (1980)
11. P. Grassberger and I. Procaccia: Phys. Rev. Lett. **50**, 346 (1983)
12. Y. Termonia and Z. Alexandrowicz: Phys. Rev. Lett. **51**, 1265 (1983)
13. P. Grassberger and I. Procaccia: Phys. Rev. A**28**, 2591 (1983)
14. M. Jensen and I Procaccia: Phys. Rev. A**32**, 1225 (1985)
15. J. G. Caputo and P. Atten: preprint
16. Y. Termonia: Phys. Rev. A**29**, 1612 (1984)

Experimental Study of the Attractor of a Driven Rayleigh-Bénard System

J. Stavans, S. Thomae[†], and A. Libchaber

The James Franck Institute, The Enrico Fermi Institute,
The University of Chicago, Chicago, IL 60637, USA

We present a study of the geometrical and measure properties of attractors generated
by Rayleigh-Bénard convection with two oscillators present. One oscillator was induced by the flow while the second was imposed externally. The winding number was
tuned to two different irrational numbers. The evolution of the attractors with
increasing non-linear coupling between the oscillators was examined. Their dynamical
properties were extracted and compared with those of circle maps.

1. Introduction

A wide variety of physical systems exhibit self-similar behavior either in the space
or in the time domain. In the case of the time domain, attention has been focused
on quantifying the degree of stochasticity in dissipative dynamical systems which
undergo a transition to chaotic behavior.

Recently, the quasiperiodic route to chaos with fixed winding number has been
observed experimentally. The experiment explored the behavior of two oscillators
which are non-linearly coupled. Moreover it was accurate enough as to allow the
measurement of the fractal properties of both the locked regions on the critical line,
and those of the attractors of the motion in phase space.

In this paper we investigate the geometrical and measure properties of the attractor in more detail. In the following section we give a short introduction to the
phenomenology of the problem. Then we describe the essential features of the experiment. In the fourth section a short account of the theoretical concepts relevant to
the evaluation of the experimental data is given. Finally, we present the results
of the observation. A more detailed description of the way our data was analyzed
will be published elsewhere [1].

2. Phenomenology

Let us consider two oscillators which are nonlinearly coupled and assume that initially the coupling is weak. Then two possible situations may arise: the winding number,
i.e., the frequency ratio between both oscillators, can be either a rational or an
irrational number. In the former case, the signal from the experiment is periodic.
The oscillators are locked. In the latter case, the signal exhibits a beating
pattern and the motion is quasiperiodic. In contrast to a linear coupling, the interaction shifts the frequencies of the oscillators so that locked states have finite
width: when one of the frequencies changes within a finite range, the other frequency
changes as well so that the winding number does not change. Defining a parameter
space by plotting the amplitude of the interaction versus the inverse frequency of
one of the two oscillators, one can outline regions, called Arnold tongues, where
the oscillators are locked. The width of these regions increases with the amplitude
of the coupling. There is one tongue for each rational number.

[†]Institut fur Festkorperforschung, Kernforschungsanlage
Posfach 1913, D-5170 Julich

Eventually the tongues overlap defining a line, the "critical line". The tongues are ordered through the Farey construction, namely: for any two tongues with winding numbers p/q and p^1/q^1 (parents), the tongue $(p+p^1)/(q+q^1)$ is found between them. This tongue is the one with the smallest denominator of all tongues between the two parents. The intersection of all tongues with the critical line forms a Cantor-like set characterized by a fractal dimension D. Since a discussion of this set does not fall on the same lines as the rest of this paper, we bring up here the results already published in [2] for completeness. To calculate D one proceeds as follows [3]: denote by S the length of the interval between two locked band parents on the critical line. This interval includes inside it the Farey daughter corresponding to the chosen parent bands. Denote by S_i, i=1,2 the length of the intervals between the daughter band and each parent respectively. Then to a very good approximation (2%) D is obtained by solving:

$$\left\{\frac{S_1}{S}\right\}^D + \left\{\frac{S_2}{S}\right\}^D \approx 1$$

D was measured at two points on the critical line: around the golden mean $\sigma_g=(\sqrt{5}-1)/2$ and around the silver mean $\sigma_s=\sqrt{2}-1$. The result were D=0.86+-3% and D=0.85+-3% respectively. Up to experimental error, the results are the same showing that D is a global property of the critical line. The results agree up to 2% with theoretical predictions based on circle maps [4,5].

It has been experimentally found [6] that, when one is inside a tongue and the nonlinearities are increased, one reaches a chaotic regime via a period doubling cascade. The question our experiment addressed is how does one reaches chaos if one follows a quasiperiodic state with fixed winding number in analogy with the locked case. More precisely, we fix the winding number to be the golden mean and in the right part of Fig. 1 we show the typical evolution of the attractors in three dimensional phase space for increasing amplitude. We shall elaborate on this figure later in the text.

3. The Experiment

The experiment was performed in the context of Rayleigh-Bénard convection. In a Rayleigh-Bénard experiment, a horizontal layer of fluid enclosed by two horizontal plates is heated uniformly from below. The fluid's density at the bottom of the layer is then smaller than at the top. When the temperature gradient across the layer is high enough, convection sets in due to the unstable density gradient. A rescaling of the equations of motion shows that the problem is controlled by two non-dimensional numbers: the Rayleigh number R and the Prandtl number P. While R is proportional to the temperature gradient and thus contains geometrical information about the system, P is an intrinsic property of the fluid. We used mercury (P=0.025) as a working fluid. When convection sets in at $R=R_c$ it takes the form of horizontal rotating rolls whose lateral dimension is of the order of the layer's thickness d=0.7cm. Adjacent rolls rotate in opposite directions. In our case the cell had the size 0.7cm X 0.7cm X 1.4cm and two convective rolls were present.

When R is increased beyond R_c, the convective roll pattern eventually becomes unstable. For low P fluids like mercury the system undergoes a Hopf bifurcation into a time dependent periodic mode called the oscillatory instability OI. The mode is characterized by an AC vertical verticity otherwise absent in the static roll pattern. Its period is of the order of d^2/k where k is the heat diffusivity of the fluid. The oscillation is one of our two oscillators. For our cell, the frequency of the OI w_i was typically 230mHz.

The second oscillator was introduced electromagnetically using the fact that mercury is an electrical conductor. An AC electrical current sheet was passed through the mercury and the whole system was immersed in a horizontal magnetic field (H=200G) parallel to the rolls, axes. The geometry of the electrodes and field was such that the Lorentz force on the fluid produced vertical vorticity. In this way

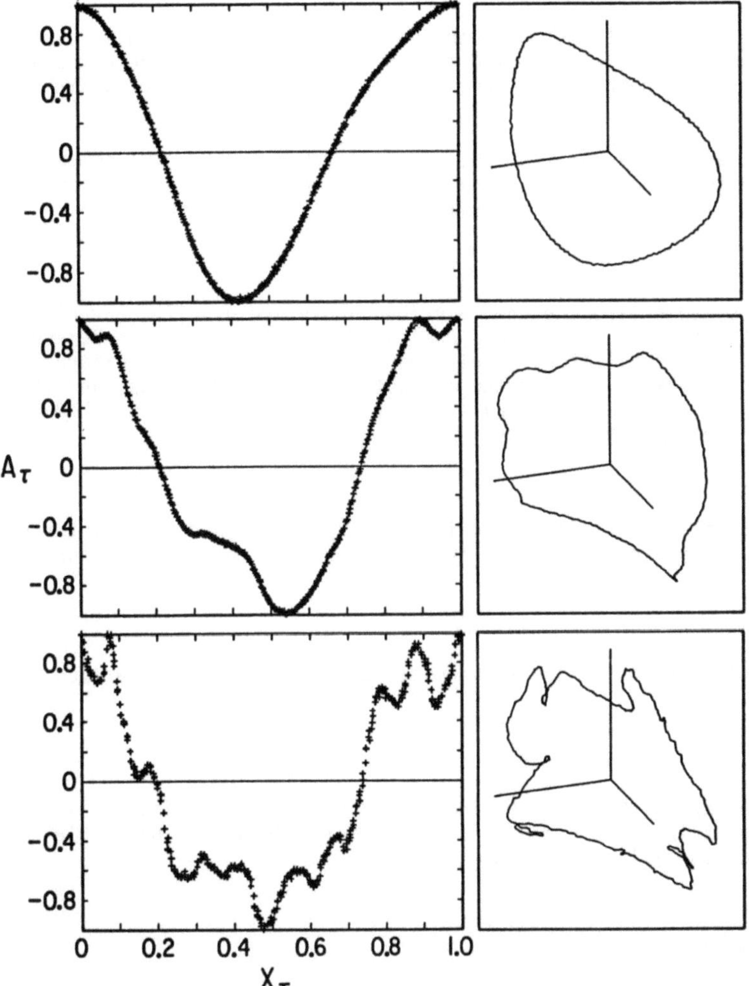

Fig. 1   Observable function and corresponding limit cycles in three dimensional phase space for quasiperiodic states with winding number $\sigma=\sigma_G\pm2.10^{-4}$.   a) 2.26mA., low amplitude, b) 7.82mA., intermediate amplitude and c) 11.81mA., nearest approximation to the critical line.

the two oscillators were dynamically coupled. The value of the currents used was typically 20mA. During the experiment R was held fixed at $R=4.09R_c$ so that the OI had a high enough amplitude. In our case, the critical temperature difference across the layer at which convection started was 3K. The temperature stability of the plates in the experiment was controlled to 10^{-5}. A change of one millidegree induced a change of 10μHz in w_i. A signal was obtained from the experiment by means of a thermal probe located in the bottom plate of our cell.

The nonlinear interaction between both oscillators was controlled by the amplitude of the injected current. The introduction or change of the external excitation's amplitude induced a change in w_i and in the amplitude of the OI. The change in the latter was not appreciable as long as the initial amplitude was high enough. Due to these changes, the external oscillator's frequency w_e was adjusted after every change in the current's amplitude in order to achieve a particular winding number $\sigma=w_i/w_e$, namely either the golden mean σ_g or the silver mean σ_s. The winding number was

measured by observing the frequencies of both oscillators in a fast Fourier spectrum
of the time signal, and by following locked states corresponding to the rational
approximants of the winding number. Winding numbers were approximated to within
2×10^{-4}.

Time series $\{A_\tau\}_{\tau=1}^{N}$ were obtained from the experiment by strobing the temperature
signals at the excitation frequency w_e. The data was then stored for subsequent
analysis. Due to the stability of the experiment N was typically 2500.

4. Theoretical Concepts

Strongly dissipative systems like the one described above are usually characterized
by low dimensional attractors. Since the dynamics of the given Rayleigh-Bénard
system is dominated by two oscillators namely the intrinsic OI and the external
periodic driving, one naturally expects the attractor to be a two-torus, and its
Poincaré section to be a circle. Dynamical systems on a circle are described by
circle maps F(s), defined by the property

$$F(s+1) = 1 + F(s). \tag{1}$$

Note that s and s+1 denote identical points in phase space. Consecutive intersections ..., s_τ, $s_{\tau+1}$, ... of the continuous trajectories with the Poincaré surface
are related by

$$s_{\tau+1} = F(s_\tau). \tag{2}$$

If F is continuous and invertible, the s-dynamics is completely characterized by
the winding number

$$\sigma = \lim_{\tau \to \infty} \frac{s_\tau - s_0}{\tau}, \tag{3}$$

i.e., all such circle maps with the same winding number can be transformed into
each other by continuous invertible coordinate transformations. In particular,
there is a transformation called conjugating function

$$s = h(x) \tag{4}$$

relating F to the rotation

$$x_{\tau+1} = R_\sigma(x_\tau), \quad R_\sigma(x) = (x+\sigma)_{\mathrm{mod}\ 1} \tag{5}$$

by

$$F_\sigma(s) = h(R_\sigma(h^{-1}(s))). \tag{6}$$

In an experiment one usually does not observe circle map variables like x or s.
One rather measures an observable A, which is a function of x or s, respectively.
Because s and s+1 are physically equivalent, the observable function A(s) must be
periodic,

$$A(s+1) = A(s). \tag{7}$$

Theoretical predictions are usually made for s rather than A [7,8,9]. Hence an
investigation of A(s) may be helpful in relating theory and experiment more closely.
Actually the phenomenology of the transition to chaos can only be fully described by
discussing the behavior of A(s) as well as that of $F_\sigma(s)$. While A(s) controls for
a given winding number the geometrical form of the limit cycle, F(s) determines via
the dynamics (2) the stationary probability density ρF. The probability to find the
system in a state between the points s and s+ds on the limit cycle is given by
$\rho F(s)$ ds. On the approach to chaos A(s) may either remain a smooth function or become "rough", thus creating the ears well known from numerical simulations.

Since the uniform rotation R_σ has a constant invariant density, ρF is related to the derivative of the inverse conjugating function by

$$\rho F = \frac{d}{ds} h^{-1}(s). \tag{8}$$

Close to the onset of chaos, numerical simulations revealed a rich structure in $h(s)$ [7]. Via (8) this must also show up in the density ρF. Actually we expect ρF to consist of interwoven singularities supported by fractals of varying dimension. Objects of this kind are concisely described by the $f(\alpha)$-analysis recently proposed by Halsey et al. [10]. Here f denotes the fractal dimension of the support of singularities of strength α.

5. Experimental Results

We started the data evaluation by plotting the measured time series $A_0, A_1, A_2, \ldots$ versus a time series $x_0, x_1, x_2, \ldots$ generated by a uniform rotation R_σ with the same winding number. Actually, this procedure rendered possible a very accurate determination of σ. Even very small deviations in $\sigma(10^{-5}$ for $N \simeq 1000)$ resulted in a considerable blur of the graph of $A(h(x))$. Essentially this procedure is nothing more than a Lissajous figure where in one axis one has our attractor and on the other axis a uniform rotation.

Then we ordered the points (x_τ, A_τ) according to increasing values of x, obtaining thus a function $\tau(n)$, with the property

$$x_\tau(n) \leq x_\tau(n+1). \tag{9}$$

The 3-tuplets $\vec{P}_\tau = (A_\tau, A_\tau+1, A_\tau+2)$ spanning the system's limit cycle in phase space [11] could thus quickly be brought into spatial order $\vec{P}_\tau(1), \vec{P}_\tau(2), \vec{P}_\tau(3), \ldots$. This is usually not a trivial problem if the limit cycle is very folded. Figure 1 shows the typical evolution of an observable function and the corresponding limit cycle for increasing excitation amplitude. As can be seen, the observable function tends to develop more and more local extrema and the limit cycle becomes increasingly wiggly. The resolution of our experiment was, however, not high enough to reveal whether this evolution finally leads to a fractal structure. For all the attractors we observed including those with different winding numbers such as σ_S, the trajectories in phase space were non-selfintersecting when the data was embedded in three dimensional space.

Adding up the Euclidean distances between points $\vec{P}_\tau(n)$ and normalizing the circumference of the cycle to unity, we constructed the curvilinear coordinate s and hence the conjugating function h. With (6) one can now compute the circle map $F_\sigma(s)$. Figure 2 shows a typical F_σ. F_σ can actually be plotted directly from the signal; the results of both procedures are the same within the experimental error. All the maps we observed were very close to R_σ, in the sense that they only exhibited small amplitude modulations on top of a uniform rotation. This is in contrast with numerical simulations where significant amplitude variations are present.

Figure 3 shows the density ρF we obtained by analyzing a time series of 2000 consecutive points A_τ. In comparison with the densities we obtained in numerical simulations ρF looks smoother. This may be due to small fluctuations of the winding number during the data acquisition. Such fluctuations would tend to smear out the fine structure of ρF. Indeed, a drift affects more severely the most dense as well as the most rarefied regions of the attractor. Nevertheless a complex structure can be observed. The information in this structure can conveniently be extracted by means of an $f(\alpha)$ analysis [10]. Such an analysis yields a very complete characterization of the attractor since it renders both its fractal and metric properties. The results of this analysis will be published elsewhere [12].

We also considered the function $h(s) - s$ and its spectrum (Figure 4), which have been discussed in [7] and in [13]. The slow decay of the peaks at the Fibonacci

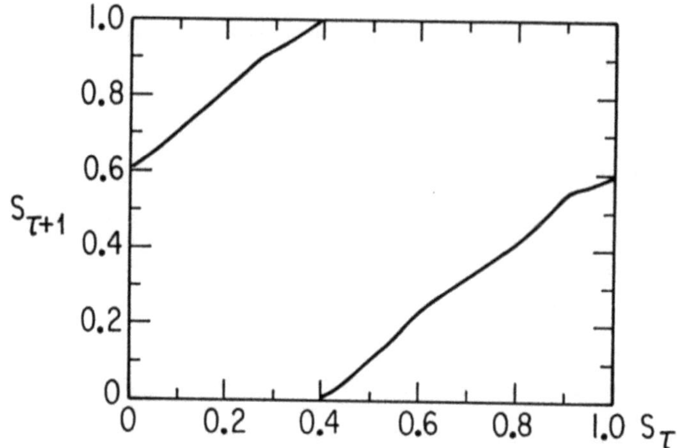

<u>Fig. 2</u> Circle map obtained from an attractor generated in a nearest approximation
to the critical line. The value of the injected current is i = 20.36mA.
The convection state is different from the one used to obtain the attractors
in Fig. 1. The winding number measured from the observable function is
0.618094. The map was obtained from 500 points out of the attractor.

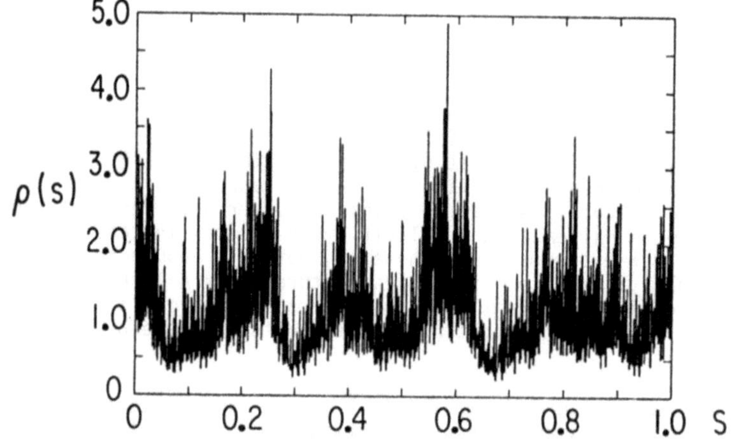

<u>Fig. 3</u> Density of points versus curvilinear coordinate along the attractor for the
attractor corresponding to Fig. 2.

frequencies (3, 5, 8, ...) indicates the proximity to the onset of chaos. Apart
from the peaks at the Fibonacci numbers one can also observe other peaks with
equivalent height. Again this can be due to the reasons mentioned above. Indeed,
we performed the same analysis for 500 points once at the beginning of the run and
once at the end. The run consisted of 2500 points. The winding number determined
was 0.618063 at the beginning whereas it shifted to 0.618088 in the end of the run.
The results for the spectra were different; the relative heights of the peaks were
changed. The spectrum shown corresponds to data around the middle of the run where
the winding number was 0.618096. For reference, the golden mean is 0.6180339 ...
The prominence of frequency peaks other than those at the Fibonacci numbers, and
the non-uniformity in height of the latter is also due to the small number of points
taken in the Fast Fourier transform algorithm (400 points). Numerical simulations
were performed with the circle map and the results agree with this conclusion. The
spectra obtained by taking 400 or 2000 points are both qualitatively and quantitatively different [14].

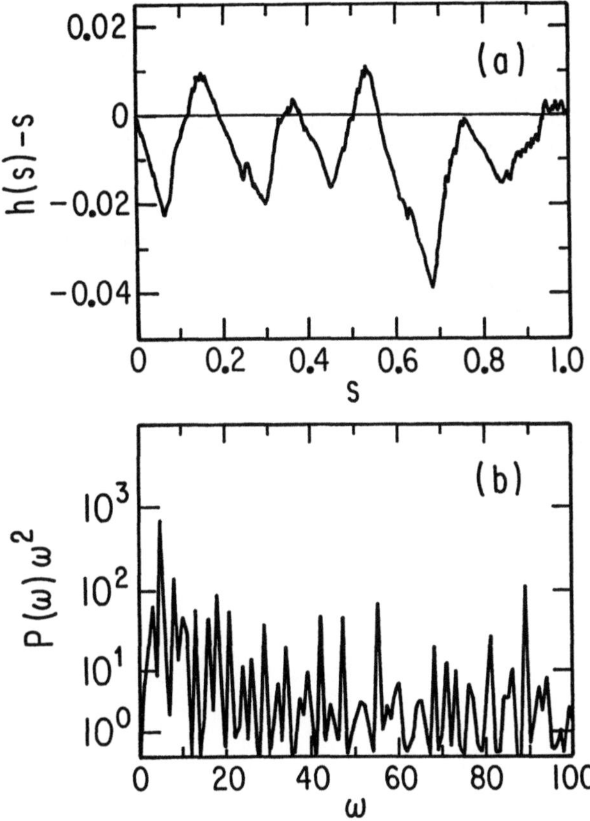

Fig. 4 a) Conjugation function transforming a uniform rotation into the attractor of Fig. 2. b) Scaled spectrum $P(\omega)\omega^2$ for the conjugation function.

6. Conclusions

This analysis we have performed on the experimentally obtained attractors seems to indicate that the real system can be faithfully described by a circle map model. As far as universality properties are concerned, the results in [2] indicate that both map and experiment belong to the same universality class. A more complete verification of the fact that both map and experiment belong to the same universality class will be published elsewhere [12].

Acknowledgements

We acknowledge very useful discussions with M. H. Jensen, T. C. Halsey, L. P. Kadanoff, F. Heslot, I. Procaccia and G. Zanetti.

This work was supported by NSF DMR83-16204 and partly by the Materials Research Lab at the University of Chicago under Grant NSF DMR 82-16892.

References

1. G. Zanetti and S. Thomae, to be published.
2. J. Stavans, F. Heslot and A. Libchaber, Phys. Rev. Lett. 55, 596 (1985).
3. See for example H. G. E. Hentschel and I. Procaccia, Physica 8D, 440 (1983).
4. M. H. Jensen, P. Bak and T. Bohr, Phys. Rev. Lett. 50, 1637 (1983); Phys. Rev. A30, 1960 (1984).

5. P. Cvitanovic, M. H. Jensen, L. P. Kadanoff and I. Procaccia, preprint.
6. A. Libchaber, C. Laroche and S. Fauve, Physica 7D, 73 (1983); J. Physique Lett. 43, L211 (1982).
7. S. J. Shenker, Physica 5D, 405 (1982).
8. M. J. Feigenbaum, L. P. Kadanoff and S. J. Shenker, Physica 5D, 370 (1982).
9. D. Rand, S. Ostlund, J. Sethna and E. Siggia, Phys. Rev. Lett. 49, 132 (1982).
10. T. C. Halsey, M. H. Jensen, L. P. Kadanoff, I. Procaccia and B. I. Shraiman, preprint.
11. J. P. Eckmann and D. Roulle, Rev. Mod. Phys. 57, 617 (1985).
12. M. H. Jensen, A. Libchaber, L. P. Kadanoff, I. Procaccia and J. Stavans, to be published.
13. G. Zanetti and G. Torchetti, Lett. Nuov. Cim., 41, 90 (1984).
14. We thank G. Zanetti for drawing our attention to this point.

Dimension Measurements from Cloud Radiance

P.H. Carter, R. Cawley, A.L. Licht, and J.A. Yorke*[†]
Naval Surface Weapons Center, White Oak, Silver Spring, MD 20903,USA

M.S. Melnik

Institute for Physical Science and Technology, University of Maryland,
College Park, MD 20742, USA

Infrared emissions from clouds exhibit chaotic behavior as a function
of angular distance at a fixed time. Preliminary results for dimensions
of the graphs of intensity <u>vs</u> angle for emissions at 3-5 μm and 8-12 μm
are reported for a sample cloud.

1. Introduction

Since the remarkable observations of LOVEJOY[1], in which cloud
perimeters, fitted over three and one-half orders of magnitude, were
determined to have dimension 1.35, there has been increased interest in
the characterization of clouds as fractals. The modern physical notion
of a fractal in its broadest sense is largely due to MANDELBROT[2], while
the basic corresponding mathematical notion of sets having non-integer
dimension is much older[3]. The simplest idea of a cloud as a fractal is
an abstraction from the model of a cloud as an irregular set of droplet
points. Sometimes also notions of scale invariance are stressed, and
these are reflective of the general irregular nature of cloud morphology
types, of cloud wispiness or fluffiness.
There are numerous physical ways in which broadly fractal notions
might find their way into cloud physics, however, and this includes the
use of ideas common to dynamical systems theory, as well as the more
visually grounded intuition described above. In this paper we report
preliminary experimental results of infrared cloud radiance measurements,
which display irregular fractal-like behavior in a straightforward, yet
novel context. In particular, we find the "fractal" (box) dimensions of
the graphs of cloud radiance <u>vs.</u> azimuthal observation angle, at fixed
time (to better than 0.1 sec.), to be 1.16 for 3.8 - 5.0 μm light, and

---

*On leave from Department of Physics, University of Illinois,
Chicago, IL 60680

[†]Permanent address: Institute for Physical Science and Technology,
and Department of Mathematics, University of Maryland, College Park,
MD 20742

1.11 for 7.3 - 11.9 μm light. We do not report correlation dimension values. We begin with a brief description of Lovejoy's work for contrast, and follow that with physical considerations pertinent to the present context of cloud radiance. We follow with a description of the experiment and of our results.

The only quantitative experimental work on fractal properties of clouds up to now appears to be LOVEJOY[1]. He used radar pictures of tropical rain areas, with 1 km x 1 km resolution and citing perimeter values 3 km $\lesssim$ P $\lesssim$ 1000 km, along with infrared satellite pictures sampled on a grid 4.8 km x 4.8 km, with 50 km $\lesssim$ P $\lesssim$ 300,000 km. Each of the two data sets lay on parallel straight lines in log-log plots of area A <u>vs.</u> perimeter. The radar data are more sensitive to structure in the perimeter owing to the higher resolution. Assuming the perimeter to have fractal dimension D, the parameter at length resolution L_1 will be greater than that measured at resolution L_2 by the factor $(L_1/L_2)^{1-D}$. When this correction, viz. $(1 \text{ km}/4.8 \text{ km})^{-0.35}$ = 1.73, is applied to the infrared satellite data the two lines become coincident, and the correlation coefficient of the common least mean square slope value D = 1.35 is 0.994. The fractal characterization of cloud and rain areas constituted in this result is a straightforward geometrical one. The characterization from the data we describe here for cloud radiance is different, however, for we treat infrared intensity <u>vs.</u> line of sight direction (angle) as a "time" series and make that the object of our analysis.

The visual appearance of a cloud is a condensation of complex processes of light scattering by assemblies of droplets (or ice) having a distribution of sizes. Cloud condensation nuclei (CCN) typically constitute a small 1 or 2 percent of the aerosol present, with diameters ranging from about 1 μm to 10 or 20 μm. Smaller droplets constitute haze, while somewhat larger ones comprise rain. Finally, maritime cloud droplet sizes generally are 2 to 3 times larger than land. (See WALLACE and HOBBS[4]). The visible region of the spectrum, 0.3 - 0.7 μm, is well supplied by sunlight, with the black body peak λ^o_{max} = ~ 0.5 μm, and absorption is weak; so the cloud we see is scattered light. On the other hand, for T = 300°K λ_{max} = 0.2014 hc/kT = 9.91 μm, which is in the middle of the 8 - 12 μm atmospheric transmission window; hence in this "longwave" band a cloud is even notilucent. Note also that at longer wavelengths we "see" an effectively reduced droplet set since, especially in the 2.6 - 6.3 μm region, absorption is strong and multiple scattering suppressed. The "midwave", 3-5 μm cloud radiance band is made up of both scattered sunlight and thermal emission in generally comparable proportions, which results from competition between the tail of the intense

solar spectrum and the relatively nearby but weaker cloud emission
peak. Finally, in all cases the received intensity, in addition to
dependence on the source strength and the scattering dependences upon
wavelength and cloud particle size distribution and density, is affected
also by aerosol scattering in the line of sight. The latter can be
significant for clouds with low angles of elevation, where the range can
be 40 to 50 km.

The measurements were taken at the Coast Guard Lighthouse on Montauk
Point, Long Island, New York, in August of 1983, under the Navy's
Background Measurement and Analysis Program (BMAP)[5]. The sensor
consisted of two bore-sighted telescopes, one to measure 3.8 - 5.0 μm
radiation and one to measure 7.3 - 11.9 μm radiation. Each telescope
contained an optical bandpass filter in front of a vertical array of
sixteen infrared detectors. The longwave HgCdTe detector produced 1/f
noise, which required data collection circuits having only quasi-dc
response, with 0.5 - 1000 Hz 3dB passband. The midwave InSb detector
electronic passband was 0 - 1000 Hz at 3dB.

The telescope boresights were 70 feet above sea level, and the center
of the field of view was elevated 5° above the sea horizon. A mirror let
each telescope scan across 2.2° of azimuth at a speed of 36.0°/s, for a
scan time of 60 ms. The intensity was sampled every 0.096 mR, which
resulted ultimately in about 300 useable data points per scan. The
sampling time was much less than the dwell time. Since the angular
resolution was 0.33 mR (both azimuth and elevation), there were 3.44
intervals of 0.096 mR per dwell.

The recorded random noise level for each detector, determined with a
room temperature hood over the detector, was 2.5 mV. The data
digitization was 0 - 10 = 12 bits (0 - 4096 decimal) so that one count
was also about 2.5 mV. Since maximum relative intensity variations
spanned about 50 counts, typically, the random noise error was 2 - 3%.
Absolute calibration of the data is not available, but that does not
matter as long as the calibration relationship is sufficiently regular,
since the graph dimension is unaffected by scale change (see below).
The digital data-acquisition system had the capacity to record only 16 of
the 32 simultaneous channels, but a switch permitted recording of 8
channels of each color. Figures 1 and 2 show 12 plots of intensity vs.
azimuth at elevation angle increments of 0.33 mr for midwave and longwave
radiance. The figures may be thought of as (slightly vertically
displaced) 6 line video scans.

The box-counting dimension of the graph of intensity I vs. angle x is
defined by

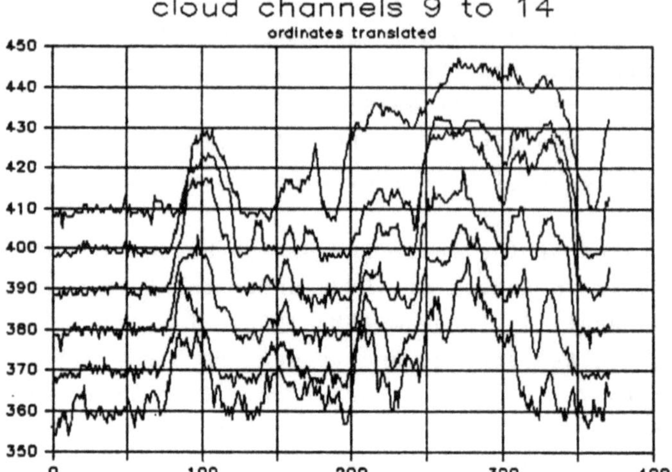

Fig. 1. Intensity _vs._ azimuth for 3-5 μm radiance (arb. units), at successive angles of elevation.

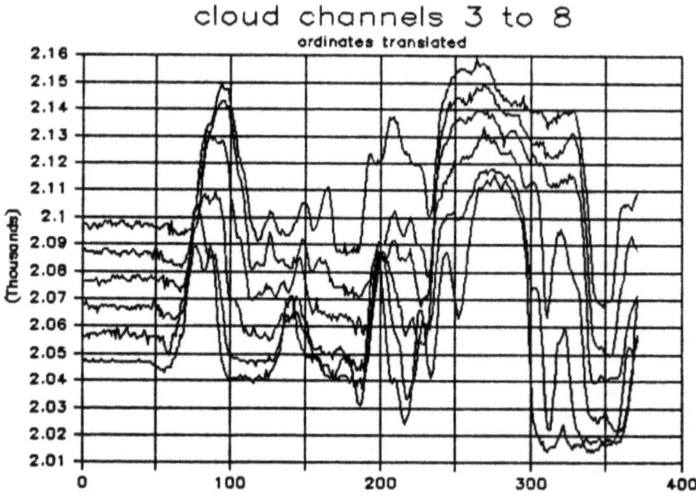

Fig. 2. Intensity _vs._ azimuth for 8-12 μm radiance (arb. units), at successive angles of elevation.

$$D = \lim_{\varepsilon \to 0} \frac{\log N(\varepsilon)}{\log \varepsilon^{-1}}, \tag{1}$$

where ε denotes the box size, defined as an angular bin, $\varepsilon \equiv x_{i+1} - x_i$, $i = 1, .., n$, and $N(\varepsilon)$ is the number of boxes necessary to cover the curve,

$$N = \sum_{i=1}^{n} \left(1 + \left[\frac{h_i}{\varepsilon}\right]\right), \tag{2}$$

218

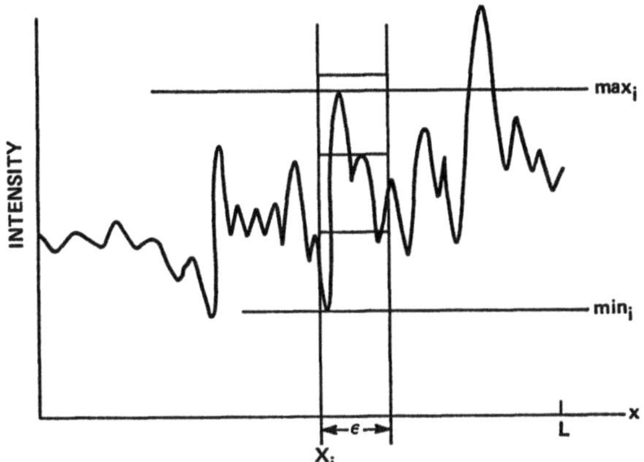

Fig. 3. Procedure for calculating box-dimension of graph.

in which the bracket denotes the integer part and h_i is the difference of
maximum and minimum intensity values in the _ith_ bin (cf. Fig. 3 -- note
the positioning of the boxes is relative to ε-local minima and maxima,
rather than that of an imposed a grid overlay[1].) We assume the limit in
(1) exists.

The dimension of a compact set is the same as the dimension of the
image of the set under any transformation that is differentiable and has
differentiable inverse. In particular, it is easy to show that D is
unchanged by linear transformation of intensity I and angle x. We give
the main elements of an explicit proof for this linear transformation
case,since the physical validity of the measurement procedure depends
critically on this fact, and in particular also since the issue sparked a
modest controversy at this conference. Since both ε and h_i are
differences, the limit (1) is trivially unaffected by addition of a
constant value to I or x. That leaves only invariance to scale
changes. It is sufficient to consider the contribution from the second
term in the summand of (2), and in fact to consider it only for the case
that

$$a(\varepsilon) \equiv \frac{1}{n} \sum_{i=1}^{n} \left[\frac{h_i}{\varepsilon}\right] \rightarrow \infty, \qquad \text{for } \varepsilon \rightarrow 0. \tag{3}$$

---

[1] Note also an artifact of the digitization of the data: the probability
that h_i/ε is an integer, when ε is an interger, is 1/ε. This effect
would bias N(ε) to low values for small ε and lead to lower estimates of
D if, in the summand of (2) the one were to be dropped whenever h_i/ε is
integral. Its retention in (2) assumes that the probability for h_i/ε to
be an integer is zero, which is correct for continuous h_i.

For, combining (1) and (2), and noting that[2] $n = 1 + [L/\varepsilon]$, where L is the length of the angular interval,

$$D = \lim_{\varepsilon \to 0} \frac{\log n}{\log \varepsilon^{-1}} + \lim_{\varepsilon \to 0} \frac{\log (1 + a(\varepsilon))}{\log \varepsilon^{-1}}$$

$$= 1 + \lim_{\varepsilon \to 0} \frac{\log a (\varepsilon)}{\log \varepsilon^{-1}}. \qquad (4)$$

Hence $D \geqslant 1$, with strict inequality only if (3) holds.

Let $L \to L' = kL$, $I \to I' = cI$, with $k > 0$, $c > 0$ since formal sign changes of intensity and angle are trivially dealt with separately. Then $h_i \to h_i' = ch_i$, $\varepsilon \to \varepsilon' = k\varepsilon$, and n is unchanged in the summation in (3). Using $x - 1 < [x] \leqslant x$ one has $a(\varepsilon) \to a'(\varepsilon')$, which is bounded between two functions of ε each of which gives the same limit in (4) as $a(\varepsilon)$ does. Specifically,

$$a(\varepsilon) \to a'(\varepsilon') = \frac{1}{n} \sum_{i=1}^{n} \left[\frac{k}{c} \cdot \frac{h_i}{\varepsilon}\right] \equiv \frac{k}{c} \bar{a}(\varepsilon) \qquad \text{and} \qquad (5)$$

$$-1 + \frac{k}{c} \cdot \frac{1}{n} \sum_{i=1}^{n} \left[\frac{h_i}{\varepsilon}\right] < a'(\varepsilon') < \frac{k}{c} \cdot \frac{1}{n} \sum_{i=1}^{n} \left(1 + \left[\frac{h_i}{\varepsilon}\right]\right), \qquad (6)$$

so that under the linear transformation

$$\frac{\log a(\varepsilon)}{\log \varepsilon^{-1}} \to \frac{\log k/c + \log \bar{a} (\varepsilon)}{\log k^{-1} + \log \varepsilon^{-1}} \qquad (7)$$

with $\bar{a}(\varepsilon)$ satisfying

$$a(\varepsilon) - \left(\frac{k}{c}\right)^{-1} < \bar{a} (\varepsilon) < a(\varepsilon) + 1. \qquad (8)$$

The inequalities (8) are now a basis for proving the claimed result, that D is unaffected.

Figures 4 and 5 show plots of $\log N(\varepsilon)$ against $\log \varepsilon$ for one of the tracings from each of Figs. 1 and 2. For the midwave scans the rms average for the 8 channels was $D = 1.16$ while for the longwave scans the corresponding 8 channel average was $D = 1.11$. This difference appears significant, but more data must be analyzed before that can be stated confidently.

---

[2] In regard to this expression for n, similar remarks to those made in Footnote 1 may be made here also. Here, however, the presence or absence of the one is inconsequential.

220

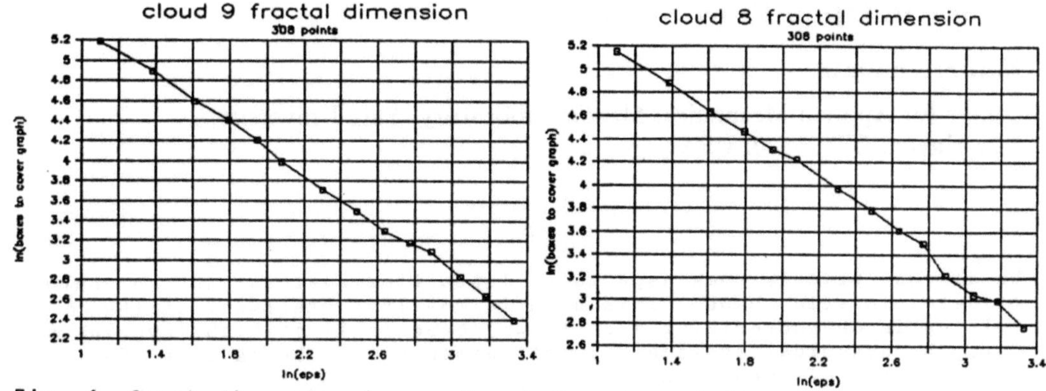

<u>Fig. 4</u> Graph dimension (neg. slope) for channel 09.

<u>Fig. 5.</u> Graph dimension (neg. slope) for channel 08.

Acknowledgments

We wish to thank B. V. Kessler for providing us with the data and for technical assistance concerning the experiment. This work was supported by the NSWC Independent Research Program, the Office of Naval Research and the Naval Air Systems Command.

References

1. S. Lovejoy, Science <u>216</u>, 185 (1982). See also "The statistical characterization of rain areas in terms of fractals" by the same author in Proc. 10th Conf. on Radar Meteorology, A.M.S., Boston, 1981.

2. B. Mandelbrot, "The fractal geometry of nature," W. H. Freeman, New York, 1977.

3. F. Hausdorff, Math. Annalen <u>79</u>, 157 (1918).

4. John W. Wallace and Peter V. Hobbs, "Atmospheric Science," Academic Press, New York, 1977.

5. M. S. Longmire, "A final technical report on calibration and use of clutter data for simulation," Western Kentucky University, Bowling Green, KY 42101.

Chaos in Open Flow Systems

K.R. Sreenivasan

Center for Applied Mechanics, Mason Laboratory, Yale University,
New Haven, CT 06520, USA

We discuss briefly some aspects of 'open flow systems' in the context
of deterministic chaos. This note is mostly a statement of the difficulties in characterizing such flows, especially at high Reynolds numbers, by dynamical systems. Brief comments will be made on the fractal geometry of turbulence.

1. Introduction

One of the most fascinating phenomena in fluid mechanics is the transition from a steady laminar state to a turbulent state. Our concern
here is a brief discussion (in the context of deterministic chaos) of
this transition process (or processes), and of aspects of the fully
turbulent state itself. We shall concentrate entirely on 'open flow
systems', or 'unconstrained' flows, e.g., wakes, jets, boundary layers,
channel and pipe flows, etc.

It is not obvious in what sense one can think of open flow systems
as genuine dynamical systems. We recall from [1] that such flows
could behave in generically different ways from the 'closed flow systems'. In all closed flow systems the boundary is fixed so that only
certain class of eigenfunctions can be selected by the system; this
does not hold for open flow systems in which the flow boundaries are
continuously changing with position. Thus, while in closed flow systems each value of the control parameter (for example, the rotation
speed of the inner cylinder in the Taylor-Couette problem) characterizes a given state of the flow globally, this is not true of open systems. Consider as an example the near field of a circular jet. For a
given set of experimental conditions, the flow can be laminar at one
location, transitional at another and turbulent at yet another (downstream) location. This usually sets up a strong coupling between different phenomena in different spatial positions in a way that is pecu-
liar to the particular flow in question. Secondly, the nature and influence of external disturbances (or the 'noise', or the 'background
or freestream turbulence') is more delicate and difficult to ascertain
in open flows: the 'noise', which is partly a remnant of complex flow
manipulation devices upstream and partly of the 'long range' pressure
perturbations, is not 'structureless' or 'white', no matter how well
controlled. Finally, it is well-known that closed flow systems can be
driven to different states by means of different start-up processes;
for example, different number of Taylor vortices can be observed in a
Taylor-Couette apparatus depending on different start-up accelerations
[2]. This type of path-sensitivity in a temporal sense does not apply
to open systems, where the overriding factor is the path-sensitivity
in a spatial sense (i.e., the 'upstream influence').

These remarks notwithstanding, it has been shown in Refs. 1 and 3
that it is worthwhile examining transition in open flow systems from
the point of view of low-dimensional chaos. The usual way of esta-
blishing this connection is via the analysis of the time history of a
single dynamical variable such as a velocity component obtained at a
fixed (Eulerian) point in the flow [4]. We should stress that this
procedure is inadequate especially for the open flow systems. Two re-

marks ought to suffice. First, since the dynamical instabilities in open flows are <u>most often</u> convective in nature, analysis of temporal Eulerian quantities does not carry with it much information on the <u>evolution</u> of the system. Deissler & Kaneko [5] have pointed out that a flow which gives every appearance of being chaotic may nonetheless have no positive Lyapunov exponents in the Eulerian frame of reference. Perhaps a more relevant method of characterizing the evolution of the flow in terms of a dynamical system would be to use the Lagrangian information obtained, say, by measuring the velocity of a fluid parti-cle as it moves about in the flow. To say the least, accurate mea-surements of this type are hard to make.

The second point to be made is that most open flow systems pos-sess strong spatial inhomogeneities in a direction normal to the flow. (Indeed, these inhomogeneities are responsible for processes that maintain the flows against viscous dissipation.) For this reason, it is <u>a priori</u> unclear to what extent the temporal information obtained at one selected point fixed in the flow can represent the global dy-namics. One might think that a simultaneous measurement (at a given time or as time sequences) of a dynamic quantity such as velocity, made at many spatial points in the flow, might solve this problem. This is not so: one does not even know how to construct a dynamical system from such empirical data.

It therefore appears worth enquiring explicitly whether, in open flow systems, attractors constructed from Eulerian point measurements, using the usual time delay techniques, are chaotic; that is, whether they are characterized by low dimensions, and possess (at least!) one positive Lyapunov exponent. This is done in section 2. In section 3, we examine the variation with the flow Reynolds number of the di-mension of the attractor, and comment briefly on the dimension at large Reynolds numbers. In section 4, brief remarks will be made on two aspects of turbulence that can be ascribed fractal dimensions.

2. Chaotic attractors for open flows: low Reynolds numbers

Chaotic attractors are characterized by at least one positive Lyapunov exponent and by relatively low dimensions that do not con-tinously increase with the embedding dimension. We have made point measurements of velocity signals in several different flows and con-structed attractors using the time delay technique; we have obtained the correlation dimension ν according to the Grassberger-Procaccia algorithm [6], and the largest Lyapunov exponent according to the al-gorithm given in Wolf et al. [7]. (Spurred by a talk that Harry Swinney gave in Kyoto in 1983, we wrote versions of a program to cal-culate the largest Lyapunov exponent, but have now switched over to the method of Ref. 7.) Since both these procedures are now well-known, we shall not describe them here.

In Table 1, we list some basic information for four flows. A crucial factor in obtaining the correlation dimension is the choice of the optimum time delay τ. We simply varied τ over a wide range, and used a τ in the range where its precise value is not critical. We show in Fig. 1 the correlation dimension as a function of τ. Clearly, too large a τ will result in the increase of ν.

Figure 2 shows the convergence with the number of iterates of the largest Lyapunov exponent for the wake, calculated using an embedding dimension of 6; other embedding dimensions yield essentially the same asymptotic value, even though the initial behaviors could be quite different. It should be remarked that the dimension and the Lyapunov exponents usually converge (for the calculations typified by Table 1) relatively fast; total signal durations of the order of $2000\tau_o$, where τ_o is the zero-crossing time scale of the auto-correlation function, was found to be usually sufficient.

Table 1: Typical data for low Reynolds number open flow systems

Flow	$Re=U_o d/\nu$	Correlation dimension, ν	Largest Lyapunov exponent, λ_1
wake behind circular cylinder[1]	67	2.6	0.65 bits/orbit
axisymmetric jet (unexcited)[2]	1000	6.3	0.95 bits/orbit
axisymmetric jet (excited)[3]	1000	3.2	——
curved pipe[4]	6625	6.0	0.40 bits/orbit

[1] d = diameter of the cylinder, U_o = upstream flow speed; data were obtained 10 diameters downstream, 1 diameter off-axis.

[2] d = diameter of the nozzle, U_o = nozzle exit velocity; data were obtained in the potential core 2 diameters downstream of nozzle exit.

[3] no Lyapunov exponent was computed because we lost the data sets immediately after computing the dimension.

[4] d = pipe diamter, U_o = section average velocity; the data correspond to the centerline of the pipe.

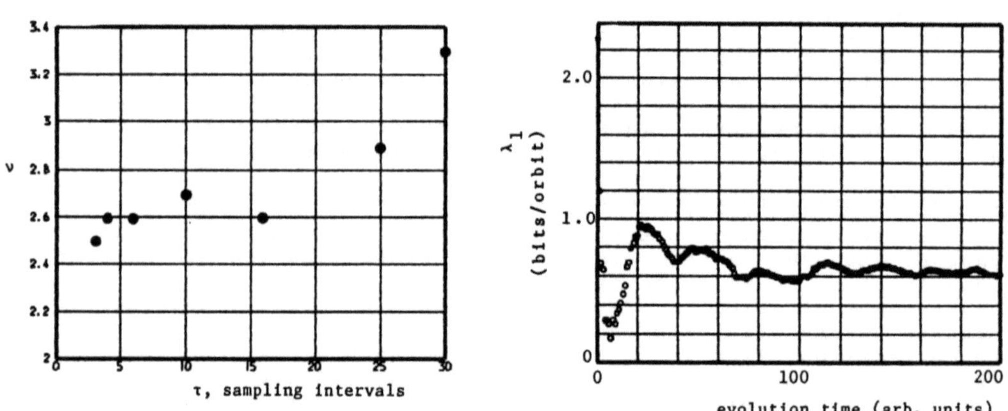

Fig. 1 The variation of the correlation dimension as a function of the time delay τ used to construct the attractor.

Fig. 2 Variation of the largest Lyapunov exponent with the evolution time.

From many such calculations, we conclude that if one constructs attractors using a single Eulerian dynamical quantity via time delay techniques, such attractors do possess (at low Reynolds numbers) characteristics of chaotic dynamics. Perhaps, Eulerian quantities do preserve some information on the dynamical evolution, in some loose sense akin to Poincaré sections!

We shall remark that these calculations do not unequivocally establish that tubulence is chaotic (in the sense of extreme sensitivity to initial conditions). Our findings could perhaps be interpreted equally well in terms of 'external noise amplification' in the system. Much more work is needed before one can determine the extent to which

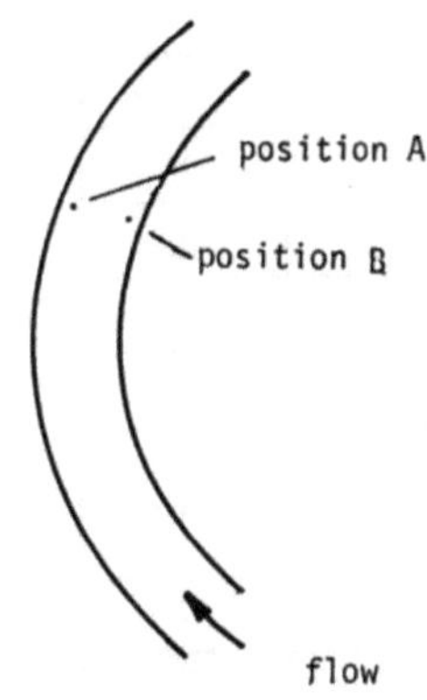

Fig. 3a Measurement stations for the curved pipe. Flow at the measurement stations is fully developed. Configuration details can be found in [3].

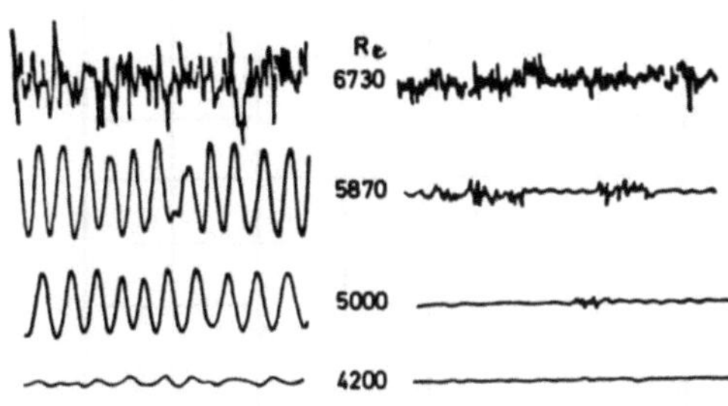

Fig. 3b Streamwise velocity fluctuations at several Reynolds numbers at position A (the right set of signals, measured 0.25 radius from the outer wall) and position B.

this last mentioned factor competes with the intrinsic sensitivity to initial conditions as the mechanism for the generation of turbulence. We should also reiterate the variation with spatial position of the characteristics of the 'Eulerian attractors'. For the curved pipe, Fig. 3 shows samples of streamwise velocity history at two spatial locations (but at the same streamwise section in the so-called fully developed region). Clearly, attractors constructed from signals at these two different locations can be expected to have different dimensions and spectra of Lyapunov exponents. For an Re of 6625, the data are as shown in Table 2. At the least, these data suggest that the interpretation of the dimension as an indicator of the dynamically significant degrees of freedom of flow needs some qualification.

Table 2: The spatial variation of the characteristics of the 'Eulerian attractors' at two different spatial positions in the same flow at the same streamwise location at the same Re. Data are for curved pipe; details as in Fig. 3.

position A		position B	
ν	λ_1, bits/orbit	ν	λ_1, bits/orbit
6.0	0.4	2.7	0.17

3. Dimension calculations at higher Reynolds numbers

If we persist with dimension calculations at higher Reynolds numbers — using the same technique, in spite of its shortcomings — they become uncertain because:

(a) The number of data points required for convergence, and the number of steps involved in dimension calculations go up;

(b) One cannot in general find a proper range of time delays over which the results are sensibly independent;

(c) There is no guarantee that the dimension calculations asymptote to constant values as the embedding dimension increases.

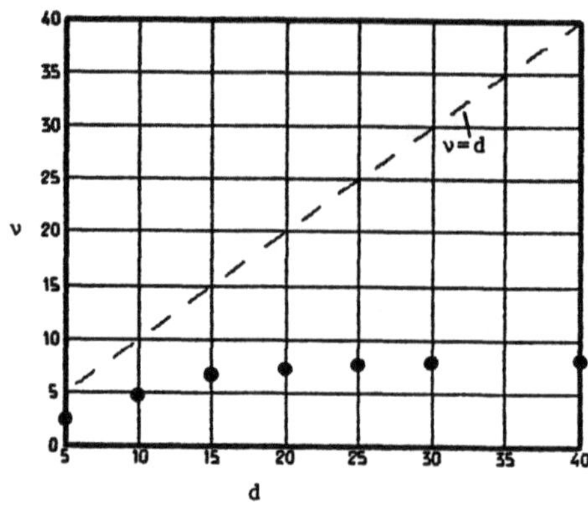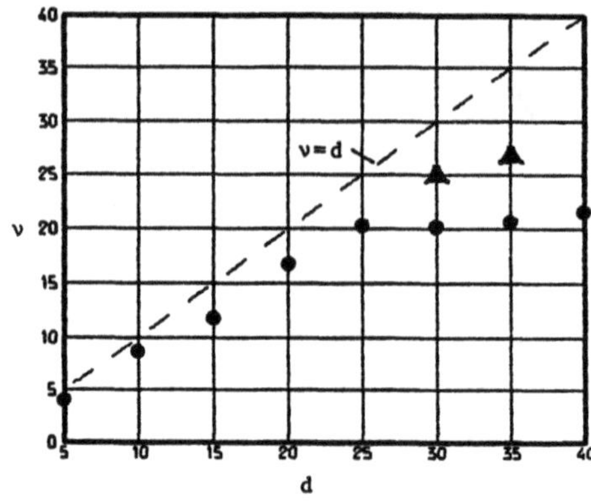

Fig. 4a   The variation of the correlation dimension ν with the embed-
           ding dimension d. Re = 500, approximately 5 diameters down-
           stream of the cylinder. A space-filling attractor is
           expected to have the behavior shown by the dashed line.

Fig. 4b   The variation of the correlation dimension ν with the embed-
           ding dimension d. Re = 2000, approximately 5 diameters
           behind the cylinder. ν = d line holds for a space-filling
           attractor. The ▲'s indicate the values of ν computed for
           the random noise from a commercial random noise generator.
           Notice that the asymptotic value of ν is definitely below
           the noise data, although only by a small margin. The near-
           ness of the noise data to the flow data shows why we cannot
           place too much emphasis on high dimension computations.

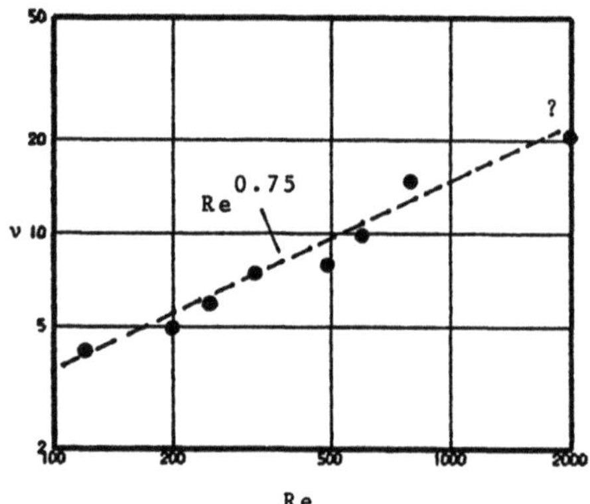

Fig. 5   The variation of the dimension with Reynolds number. Data
          are for the wake of a circular cylinder.

      Figures 4a and b illustrate this last point; Fig. 4b is the upper
limit on the Reynolds number at which some credibility (already rather
low!) can be ascribed to the dimension calculations. If we believe
the numbers obtained from such calculations, we may deduce that a
power law relation like $Re^{3/4}$ is not unlikely (Fig. 5).

      It is worth mentioning that Constantin et al. [8] have placed the
upper bound on the dimension of Navier-Stokes attractors to be
of order $R^{9/4}$ (and higher if self-similarity in the Kolmogorov range
does not obtain!), where the Reynolds number $R = u'L/\nu$, u' being a

226

root-mean-square velocity fluctuation,and L is an integral state of
turbulence. The precise relation between R and the Reynolds numbers
Re used in Table 1 depends on the flow, but it is clear that if the
present finding of a 3/4 - power law is true, it is of undoubted sig-
nificance in spite of our earlier reservations on the meaning of the
dimension obtained in this way.

Fully turbulent flows are characterized by temporal <u>and</u> spatial
chaos. Temporal dynamics is thus merely a part of the whole story;
this in itself is hard to come to grips with, even if the dimension
were to increase 'only' according to a 3/4 power of the Reynolds numbers. Is there then any connection between real turbulent flows and
finite- (and low-) dimensional dynamical models which one hopes one can
construct? (That, presumably, is the practical motivation for studies
of this type.) The answer would have been an unequivocal 'no' were
it not for the fact that some (perhaps strong?) spatial coherence appears to exist at least in some classes of fully turbulent flows. One
might, in some way that remains unclear, be able to decompose the
motion into two components, one of which consists of this coherent
element and the other, its complement. One can then think of a low-
dimensional attractor characterizing the coherent motion, the attractor being made fuzzy by the small scale motion whose effect is to reduce the correlation. Unfortunately, it is not clear whether this
loosely worded picture is consistent with facts.

Elementary tests of this hypothesis can be made if one is able
to separate the incoherent motion from the coherent part. This might
be possible, for example, by some kind of ensemble averaging methods
such as used in [9]. The simplest (by no means the most correct) way
is to filter out linearly in the frequency domain the coherent motion
from the rest. To avoid many conceptual difficulties associated with
filtering as the technique for separating the coherent and incoherent
motions we choose a (relatively) high Reynolds number flow where the
coherent part is clearly contained within a narrow band of frequencies; we then enquire whether the motion associated with this narrow
band is low dimensional.

Figure 6a shows the streamwise velocity fluctuation in the wake
of a circular cylinder, measured about 2 diameters behind the
cylinder and a diameter off-axis; the flow Reynolds number of 10,000
is considered moderately high. Computing the dimension of the attrac-

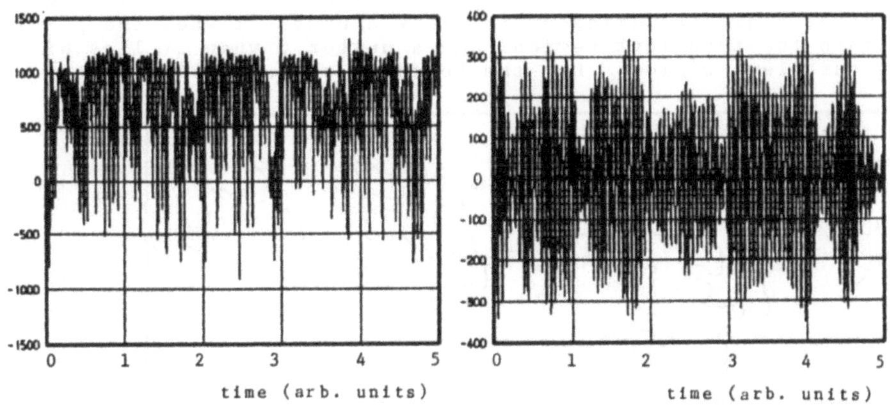

time (arb. units)   time (arb. units)

Figs 6a,b: The total (unfiltered) and the coherent part respectively
            of the streamwise velocity fluctuation in the wake of a
            cylinder; Re = 10,000. Both the ordinate and abscissa
            are arbitrary but the same in the two figures.

tor constructed from this signal is doomed to be meaningless in view
of the remarks made earlier. (If the $Re^{3/4}$ dependence is valid, the
extrapolated estimate for ν is of the order of 30!) We do know from
power spectral measurements that this signal has a peak at a frequency
f of about 550 Hz; this peak, corresponding to a Strouhal number fd/U_o
= 0.21, characterizes the coherent part of the motion. If we bandpass filter this signal between, say, 500 and 600 Hz, the resulting
signature is given in Fig. 6b. Calculations show that the corresponding attractor has a dimension of about 5.5!

It is appropriate to end this discussion with the statement that
the coherent part, as we defined it here, contains a significant
fraction of energy.

4. The fractal geometry of turbulence: a brief note

We have indicated that measurements of attractor dimensions are
beset with increasing uncertainties at increasingly high Reynolds numbers. But there are other fractal dimensions whose measurement becomes increasingly definitive as Reynolds number increases. It is to
a mention of two of these aspects that this section is devoted; more
details should be forthcoming in [10]. The results of this section
are essentially spurred by Mandelbrot's remarks on several occasions
that many facets of turbulence are fractal.

4a. The fractal dimension of the turbulent/non-turbulent interface

Observations suggest that in high Reynolds number free shear
flows (i.e., open flow systems with no constraining boundary) a sharp
front or interface demarcates the turbulent and non-turbulent regions.
Although a completely accepted view of the detailed nature of this
interface does not seem to exist, a visual or spectral study suggests
that contortions over a wide range of scales occur. This leads one
to the natural expectation that the interface is a fractal surface.

By illuminating a thin section of a flow, and by digitizing the
resulting picture, one can evaluate the fractal dimension of the
curve that separates the turbulent from the non-turbulent regions; a
threshold set on the intensity of illumination separates the two regions. The fractal dimension of the surface bounding turbulent regions is then one more than that of the curve.

Several methods can be adopted to measure the fractal dimension
[11]. We shall describe only one rather briefly. Assign to each
point in the digitized image of the flow a number 1 when the point
lies within the turbulent region, and a number 0 when it lies within
the non-turbulent region. Let the boundary shown in Fig. 7 represent

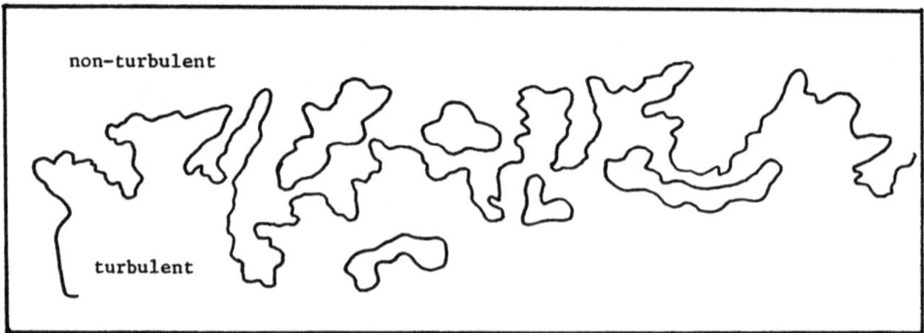

Fig. 7   The boundary between the turbulent and non-turbulent regions.
         If a circle of radius ε drawn around a given point in the
         digitized image crosses the boundary, the point is consider-
         ed to be within a distance ε from the boundary.

the boundary between the 1's and the 0's. Count the number $N_b(\epsilon)$ of
the digitized points which are within a distance ϵ from the boundary.
If this boundary is a fractal of dimension D, then it is easily shown
from the basic definition of D that

$$N_b(\epsilon) \sim \epsilon^{2-D} \ . \tag{1}$$

Measurements to be described in [10] show that (1) holds for
scales ranging from the Kolmogorov scale to a fraction of the integral length scale (but excluding scales of the order of the integral
scale and higher). The measured value of the fractal dimension for
the interface varies between 2.3 and 2.4; there is no identifiable
variation from one type of flow to another.

4b. The fractal dimension of the velocity and scalar dissipation
    fields

Another aspect of turbulence that is a candidate for fractal behavior is its dissipative (or internal or small) structure. It has been
well-known for some time that the small structure of turbulence is
intermittent. The essence of scale-similarity arguments in this context is the following. Within a given field of (fully developed)
turbulence, consider a cube with sides of length L_o, where L_o is an
integral scale of turbulence. If we divide this cube into arbitrarily
large number (n>>1) of smaller cubes of length $L_1 = L_o n^{-1/3}$, the
density of dissipation rate in each of these smaller cubes is distributed according to a probabilistic law. Further subdivision of these
cubes into second-order ones of length $L_2 = L_1^{-1/3}$ leaves the probability distribution unaltered. This similarity extends to all scales
of motion until one reaches sizes directly affected by viscosity.
Clearly, this case cries out for fractal description.

Using methods discussed in [11], we have obtained the results
shown in Table 3.

One concludes from here that the dissipation field is not space-
filling (less space-filling in the high Reynolds number regime) and
that (c) is less space-filling than (b) — a result consistent with
observations in oceanography. Note that the result (b) is only at
slight variance with Mandelbrot's [11] original estimate of 2.6.

Table 3: Summary of the fractal dimensions of the dissipation fields

	Field	Fractal dimension
(a)	Kinetic energy dissipation (low Reynolds number)*	2.9
(b)	Kinetic energy dissipation (high Reynolds number)	2.7
(c)	Passive scalar (e.g., temperature) dissipation (high Reynolds number)	2.6

* The boundary between the low and high Reynolds number regimes
is not well-defined. A convenient boundary occurs at a microscale
Reynolds number of about 150.

Theoretical explanations of these fractal dimensions, as well as of the connections that might exist among them, would be of fundamental interest.

Acknowledgements: I am indebted to Mr. R. Ramshankar and Mr. P.J. Strykowski for their help with programming, and to Dr. J. McMichael and AFOSR for the financial support.

References:

1. K.R. Sreenivasan, In 'Frontiers in Fluid Mechanics' (eds. S.H. Davis and J.L. Lumley, Springer-Verlag 1985), p.41.

2. D. Coles, J. Fluid Mech. 21, 385 (1965).

3. K.R. Sreenivasan & P.J. Strykowski, In 'Turbulence and Chaotic Phenomena in Fluids (ed. T. Tatsumi, North-Holland 1984), p.191.

4. F. Takens, In 'Lecture Notes in Mathematics 898 (eds. D.A. Rand and L.S. Young, Springer-Verlag 1981), p.366.

5. R.J. Deissler & K. Kaneko, Submitted to Phys. Rev. Lett. (Sept. 1985).

6. P. Grassberger & I. Procaccia, Phys. Rev. Lett. 50, 346 (1983).

7. A.Wolf, J.B. Swift, H.L. Swinney, & J.A. Vastano, Physica 16D, 285 (1985).

8. C. Constantin, C. Foias, O.P. Manley & R. Temam, J. Fluid Mech. 150, 427 (1985).

9. K.R. Sreenivasan, R.A. Antonia, & D. Britz, J. Fluid Mech. 94, 745 (1979).

10. K.R. Sreenivasan, The fractal facets of turbulence, in preparation, 1985.

11. B. Mandelbrot, The Fractal Geometry of Nature, Freeman and Co., New York, 1982.

Lasers and Brains: Complex Systems
with Low-Dimensional Attractors

A.M. Albano[1], N.B. Abraham[1], G.C. de Guzman[1]*, M.F.H. Tarroja[1],
D.K. Bandy[†1], R.S. Gioggia[2], P.E. Rapp[3], I.D. Zimmerman[3],
N.N. Greenbaun[4], and T.R. Bashore[5]

[1] Department of Physics, Bryn Mawr College, Bryn Mawr, PA 19090, USA
[2] Department of Physics, Widener University, Chester, PA 19013, USA
[3] Department of Physiology and Biochemistry, The Medical College of
   Pennsylvania, 3300 Henry Ave., Philadelphia, PA 19129, USA
[4] Department of Mathematical Sciences, Trenton State College,
   Trenton, NJ 08625,USA
[5] Department of Psychiatry, The Medical College of Pennsylvania,
   The Eastern Pennsylvania Psychiatric Institute, 3200 Henry Ave.,
Philadelphia, PA 19129, USA

1. Introduction

The quantification of complex dynamical phenomena associated with motions
on strange attractors has made available a tool of considerable power
for the analysis of systems which display aperiodic or apparently random temporal behavior. Until recently, aperiodic phenomena were described primarily in terms of snapshots of time sequences, power spectra,
or correlation functions. These made possible some qualitative or pictorial analyses but did not provide simple numerical criteria suitable for
more quantitative studies. During the past decade, spectral studies have
made possible the identification of a few characteristic routes to apparently chaotic behavior in hydrodynamic[1-3], chemical[4], optical [5-7,
16, 29] , and electronic[8,9] systems. There were very strong indications
that the complex motions, characterized by broadband spectra, to which
these routes led,were in fact motions on strange attractors.

Spectral studies, however, are not suitable for the classification or
comparison of different kinds of complex behavior or even for the discri-
mination between aperiodic deterministic motion, now commonly called
deterministic chaos, and purely stochastic noise. The inability to do
even this kind of taxonomy thus made it quite difficult to do meaningful
quantitative analysis of phenomena that were neither periodic nor purely
random. It was difficult, for instance, to assess the effects of changes
in parameter values on the behavior of experimental systems when such
parameter changes resulted in transitions from one complicated motion to
another. It was equally difficult, if not impossible, to compare predictions of mathematical models with experimental results.

In the past few years, it has become possible to calculate quantities
such as dimensions, entropies and Lyapunov exponents [10-15] which describe the geometry of strange attractors, or which characterize the time
evolution of trajectories on them. A great virtue of these quantities is

---

*Present Address: Department of Physiology and Biochemistry, The Medical College of
   Pennsylvania, 3300 Henry Ave., Philadelphia, PA 19129.
+Present Address: Department of Physics and Atmospheric Science, Drexel University,
   Philadelphia, PA 19104.

that the information which they provide is simple, global, and in some
sense, invariant. A dimension, for instance, contains in a single number
information on the geometry of the entire attractor. This is in stark
contrast with the information in the spectrum of a particular time sequence which is contained in the values of the spectral function over a
wide range of frequencies and which may change from one time sequence to
another, even for motions on the same attractor. Ultimately, spectra do
not give much more than confirmations of aperiodicity. From the point of
view of those analyzing experimental results, a very significant additional virtue of these quantities is that some of them are relatively easily
calculable from sequences of time measurements of single dynamical variables.

In this paper, we discuss the use of the order-2 information dimension, D_2, and the order-2 Kolmogoroff entropy, K_2, to study (a) the output intensity of a single-mode, inhomogeneously-broadened, He-Xe ring
laser, (b) the time intervals between successive action potentials spontaneously generated by single neurons in the motor cortex of a squirrel
monkey, and (c) human electroencephalograms. We have found that, under
certain conditions, each of these extremely complicated systems can be characterized
by attractors of relatively low dimensionalities.

2. Calculational Techniques

For each of the systems considered, the experimental data consisted of sequences of digitized measurements on single experimental variables. In the case of the laser this
variable was the output intensity, measured in intervals of 4 ns for several known values of the discharge current and laser cavity length. Each data set consisted of 512
measurements. For the single neuron measurements on the squirrel monkey, each data set
consisted of a sequence of measurements, from several hundred to a few thousand, of
successive interspike intervals, the time between action potentials, of spontaneously
active neurons in the precentral and postcentral gyri (the areas immediately anterior
and posterior to the central fissure). The electroencephalographic data sets each consisted of a sequence of a few thousand digitized measurements, in intervals of 2 ms, of
the eyes-closed alpha rhythm, the subject either resting or counting backward. The details of each of these experiments are discussed elsewhere [16-19, 32] and will not be
repeated here.

For each system, a time series for a single observable is used to reconstruct phase
portraits of the attractor by means of the embedding technique [20-22]. That is, for
each set of N numbers, $(x_i \; i=1,2,...,N)$, n-dimensional time-delay vectors, $Y(n)_k =
(x_k, x_{k+1}, ..., x_{k+n-1})$, are constructed. For the laser and the electroencephalograms,
$x_j = x(jt)$, with j an integer and t the time interval between measurements. For the
monkey, x_j is merely the jth interspike interval measured.

The structure of the system's attractor is inferred from the structure of the set
of vectors, $Y(n)_k$ by evaluating the correlation sum,

$$C_n(\varepsilon) = (1/N_n) \sum_{j \neq k} \theta(\varepsilon - |Y(n)_k - Y(n)_j|),$$

where $\theta(...)$ is the Heaviside function, and N_n is the number of pairs of n-dimensional
vectors used in the sum. For sufficiently small ε's and large embedding dimensions, n,
the correlation sum scales as [10], $C_n(\varepsilon) \sim \varepsilon^{D_2} e^{-ntK_2}$. Treating $C_n(\varepsilon)$ as a function of
ε and the embedding dimension, n, D_2 and K_2 are obtained from log-log plots of
$C_n(\varepsilon)$ vs ε.

In practice, ε is limited from below by noise, and n is eventually limited from above
by the size of the data set, although a more stringent limit on n is usually set by
constraints on computing time. These limitations require that the limiting procedure be
carried out with some circumspection.

It has been found convenient to numerically differentiate the $\log C_n(\varepsilon)$ vs. $\log \varepsilon$ curves for several values of n and then to plot the slopes so obtained against $\log C_n$ (see, for instance, Figs. 3 and 5). Noise-dominated regions, typically corresponding to small distance scales or small $\log C_n$'s, are characterized by slopes that scale with the embedding dimension. If, beyond this noisy region, there is a range of $\log C_n$'s or of ε's over which the slope remains reasonably constant (a plateau), this is taken as an indication of the existence of a scaling region, in which the attractor shows self-similarity. It is presumed that, in the absence of noise, the scaling region would extend to smaller values of ε, and that the value of the slope at the plateau may be taken as the low-ε limit. If, in addition, the plateaus of plots for several values of n coincide, this is taken to mean that the high-n limit has been reached. If this last condition is not met, one cannot distinguish between random noise and motion on an attractor with a dimension greater than n.

We have restricted our calculations to the order-2 dimension and entropy, because these calculations require relatively modest data sets and computing time, and because of their relative insensitivity to noise -- provided that the noise is not overwhelmingly large [10,23]. The box-counting algorithm for the calculation of the fractal dimension, for instance, has been shown to require excessively large and noise-free data sets and is, in addition, notoriously slow to converge [24].

3. Lasers

Instabilities and transitions to irregular behavior have been observed in lasers from the times of the pioneering experiments in ruby [25,26] and continue to be a problem in current applications of semiconductor lasers. These problems, which once were a nuisance for those engaged in the design of stable lasers have, of late, become the subjects of very intensive studies by laser physicists to the extent that a recent special issue of the Journal of the Optical Society of America was devoted solely to laser instabilities [27]. Spectral studies have led to the identification of some of the scenarios for the approach to chaotic behavior which have also been found in other systems [5-7]. Studies of mathematical models for laser systems have also led to the identification of regions in the models' parameter spaces corresponding to various kinds of periodic and aperiodic behavior [28-31]. In the past couple of years, calculations of dimensions and entropies have been employed to distinguish between deterministic chaos and noise, as well as to compare experimenal results with predictions of mathematical models [17, 32-33].

We report here some results obtained for a single-mode, inhomogeneously broadened Helium-Xenon ring laser that uses the 3.51 micron transition in Xenon. Earlier spectral studies of this system [16] have shown that depending on values of the gas pressures in the laser, the discharge current through it, and the laser cavity length, the laser's behavior could be characterized by a single frequency, f (period 1), by a frequency f' and a subharmonic, f'/2 (period 2), or be apparently chaotic. Fig. 1a shows the principal pulsing frequencies of the laser for several cavity lengths. The cavity length is measured by the voltage across a piezoelectric crystal (V_{PZT}).

Figure 2a shows part of the time series of an apparently chaotic signal, the power spectrum of which is shown in Fig. 2b. The slope of the $\log C_n$ vs $\log \varepsilon$, for n=10 to 20, are plotted against $\log C_n$ in Fig. 3. The "plateau" for values of $\log C_n$ between 8 and 10, corresponds to a slope of 2.7 ± 0.2. This is taken to be the value of D_2 for this data set. K_2's were calculated using ε's in the plateau region. These procedures were repeated for several runs corresponding to the same parameter values, yielding averages which are reported below.

We have compared these results with the predictions of a model for a single-mode, inhomogeneously-broadened laser [28,29]. The model consists of a collection of active two-level atoms in a ring laser cavity. The inhomogeneously-broadened atomic profile is described by a Gaussian distribution of atomic frequencies and the interactions between the atoms and the electric field are described by the Maxwell-Bloch equations. The Gaussian distribution of frequencies describes the Doppler-shifted centers of the atomic emission lines and essentially introduces an infinity of degrees of freedom. In

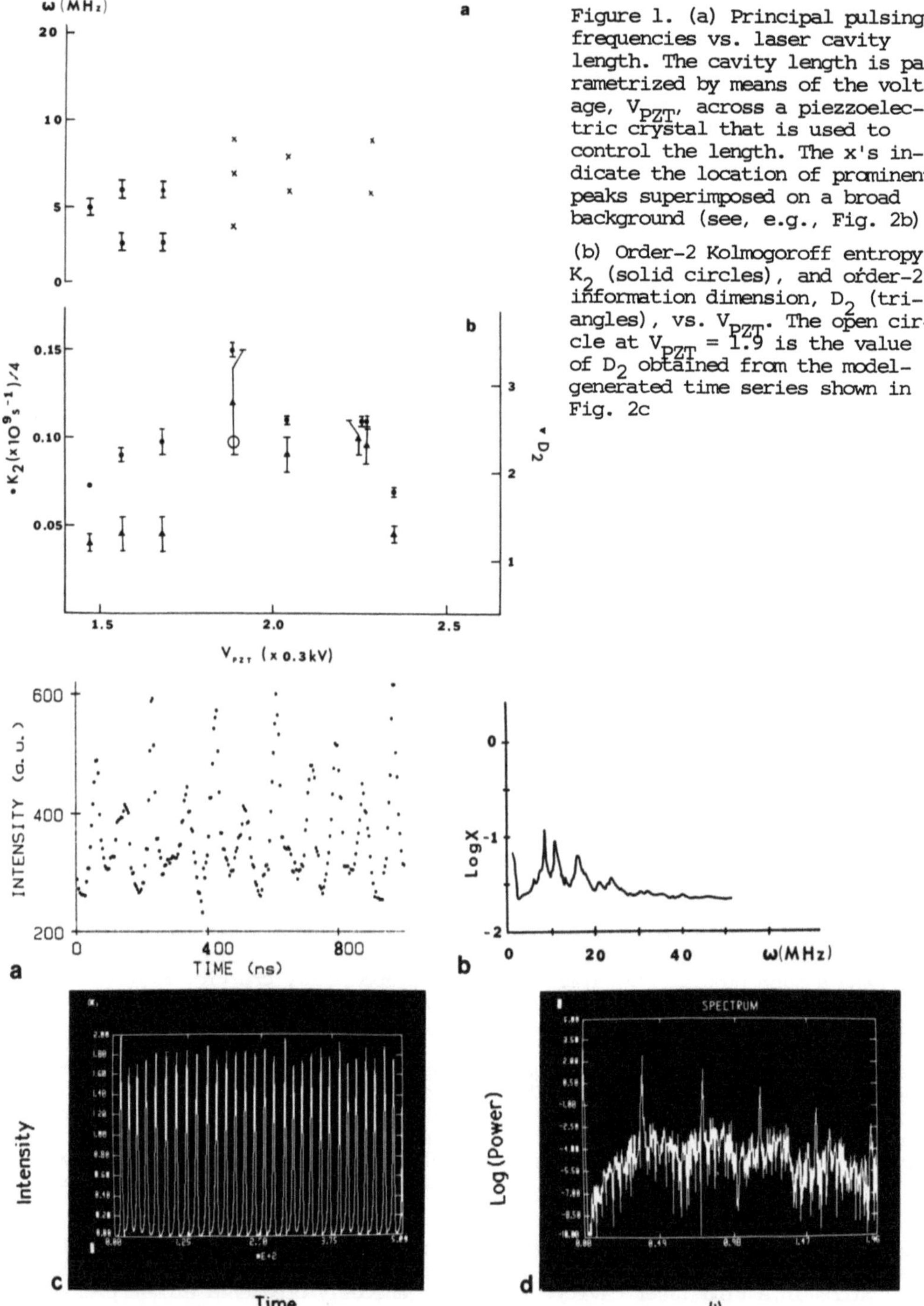

Figure 1. (a) Principal pulsing frequencies vs. laser cavity length. The cavity length is parametrized by means of the voltage, V_{PZT}, across a piezoelectric crystal that is used to control the length. The x's indicate the location of prominent peaks superimposed on a broad background (see, e.g., Fig. 2b).

(b) Order-2 Kolmogoroff entropy, K_2 (solid circles), and order-2 information dimension, D_2 (triangles), vs. V_{PZT}. The open circle at $V_{PZT} = 1.9$ is the value of D_2 obtained from the model-generated time series shown in Fig. 2c

Figure 2. (a) Laser output intensity vs. time and (b) Power spectrum for an experimentally obtained time series. (c) Intensity vs. time and (d) Power spectrum for a time series generated from a model

234

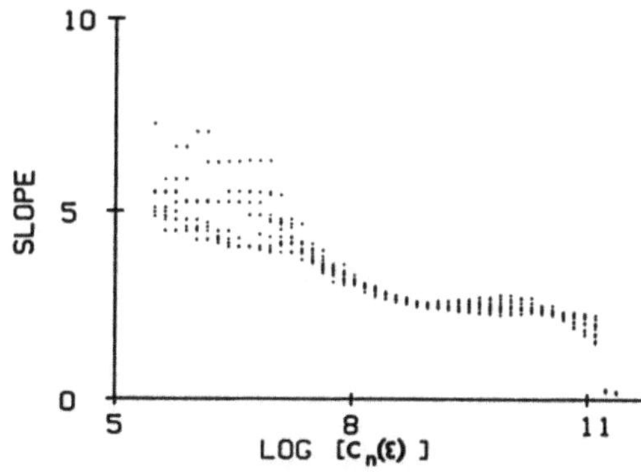

Figure 3. The slopes of the log C_n vs. log ε curves plotted against log C_n for the time series of Fig. 2a, for embedding dimensions, n=10 to 20. Note that the logC_n's here are not normalized

actual calculations, these were approximated by 100 frequency classes of atoms, each characterized by a complex-valued polarization and a (real-valued) population difference. These, together with the complex-valued electric field, lead to 302 coupled equations. The time evolution of the system is followed by numerically solving the Maxwell-Bloch equations and recording the corresponding output intensity (or electric field amplitude) time series at parameter values that approximate one of the experimental situations.

A numerically generated time series and its corresponding power spectrum for parameter values corresponding to those for Figs. 2a and 2b are shown in Figs. 2c and 2d, respectively. The model-generated time series was then subjected to the same analysis for D_2 as that used for the experimental values, resulting in the open circle in Fig. 1b. The agreement between theory and experiment is reassuring, but what is more important in the present context is that such an unambiguous comparison is made possible only because a quantity such as D_2 could be evaluated.

4. Brains

The mammalian central nervous system is perhaps the most complex system that is currently subject to scientific inquiry. In addition to its intrinsic fascination, there are several topics of clinical interest which have motivated studies seeking connections between transitions from fixed point to periodic to chaotic behavior and failures in physiological regulation. These failures have been termed "dynamical diseases" by GLASS AND MACKEY [35] and include respiratory instabilities, cardiac arrythmias, and seizure disorders, among others [34] . If indeed these connections exist, and if the transitions causing physiological failures are caused by changes in external control parameters as they are in some physical and chemical systems, then knowing these connections may lead to the invention of novel forms of clinical intervention for the control or cure of these diseases. For now, these are matters for speculation.

There is a growing body of theoretical work (see ref. 34 and references quoted therein) indicating that single neurons, which are complex nonlinear systems, are capable of demonstrating chaotic behavior at the cellular level, and that neural networks can also display chaotic behavior as a collective phenomenon. The criteria used for the identification of chaotic phenomena in these theoretical works have mainly been pictorial and qualitative, making use of time series snapshots and spectra. Similar criteria have been used to identify chaotic behavior experimentally [34] . The need for simpler quantitative criteria is as obvious here as in the case of the lasers.

We report here on some studies aimed at determining the feasibility of using dimensions to characterize neurological signals. These studies consist of D_2 calculations on (a) the time intervals between action potentials of spontaneously active neurons in the precentral and postcentral gyri of the squirrel monkey [18] , and (b) the alpha rhythm of human electroencephalograms, in one case when the subject has his eyes closed and is resting, in another with eyes closed and counting backwards [34] .

(a) Single Neurons.

A neuron generates an action potential when certain threshold conditions are satisfied.
The time intervals between successive action potentials may be considered as iterates
of a dynamical variable, much as the time intervals between successive drops in SHAW's
faucet [36] , so the phenomenon is describable by a map rather than by a flow.

Interspike data from ten single unit recordings were analyzed. In all these cases,
care was taken to make sure that the data came from neurons which showed no signs of
injury discharges and which showed no variations in their action potential amplitudes.We
found a group of three neurons with relatively low dimensions (D_2 between 2.2 and 3.5),
and two which gave ambiguous indications of dimensions between 5 and 7 but which could
not be resolved using our techniques. The rest behaved indistinguishably from random
noise up to embedding dimension 20, and in one case up to embedding dimension 40. Analysis of the histrograms of the distributions of interspike intervals showed that the
low-dimensional neurons typically were slower than the high-dimensional ones. Histograms of the low-dimensional neurons peaked between 17 and 32 ms, the ambiguous neurons peaked at approximately 5 ms, while the noisy neurons peaked at between 1 and 2 ms.

Figure 4a shows plots of log C_n vs ε for embedding dimensions 10 to 20 for one of
the low-dimensional neurons. Fig. 4b shows the slopes of six of the curves in Fig. 4a
(n=15 to 20) plotted against log C_n. The "plateau" at -4.5<log C_n < -2.5 gives a dimension of 3.5^{+}_{-} 0.1. Figs. 4c and 4d are the corresponding graphs for a noisy neuron
except that Fig. 4c is for embedding dimensions 20 to 40, in steps of two and Fig. 4d
is for embedding dimensions 30 to 40, also in steps of two. In Fig. 4d, no plateau
region is evident,and for log C_n greater than -10, the slopes increase with embedding
dimension,n, even up to n=40.

The ambiguities we found when confronted with data which seemed to indicate structures with dimensions between 5 and 7 agrees with an observation mady by SWINNEY and
GOLLUB [39] that the effectivity of these techniques may currently be limited to calculations of dimensions not too much greater than 4.

(b) Electroencephalograms
    Attempts to quantify the interpretation of EEG records have a long history [37] .
Present clinical techniques seem to be well described by what we have earlier termed

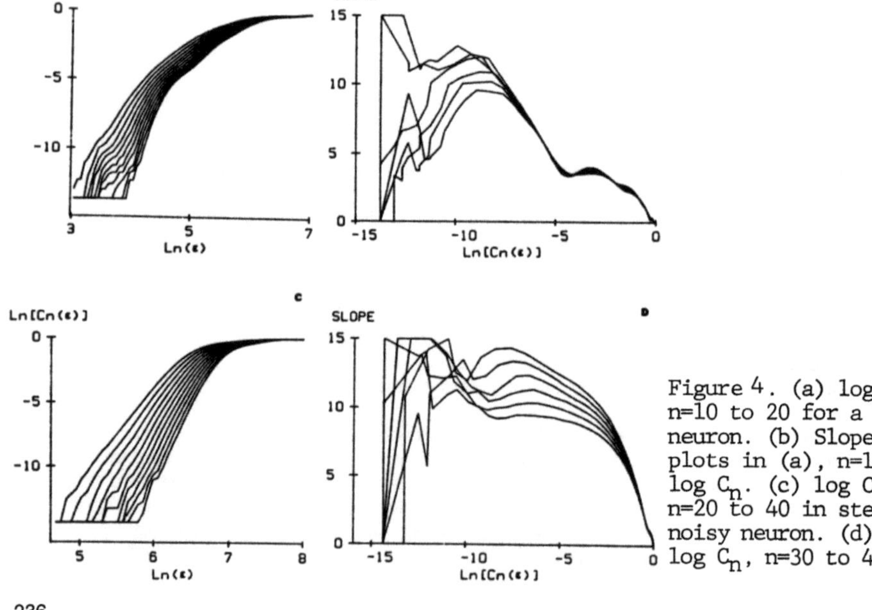

Figure 4. (a) log C_n vs. log ε,
n=10 to 20 for a "low-dimensional"
neuron. (b) Slopes of six of the
plots in (a), n=15 to 20 vs
log C_n. (c) log C_n vs. log ε,
n=20 to 40 in steps of 2, for a
noisy neuron. (d) Slopes vs
log C_n, n=30 to 40 in steps of 2

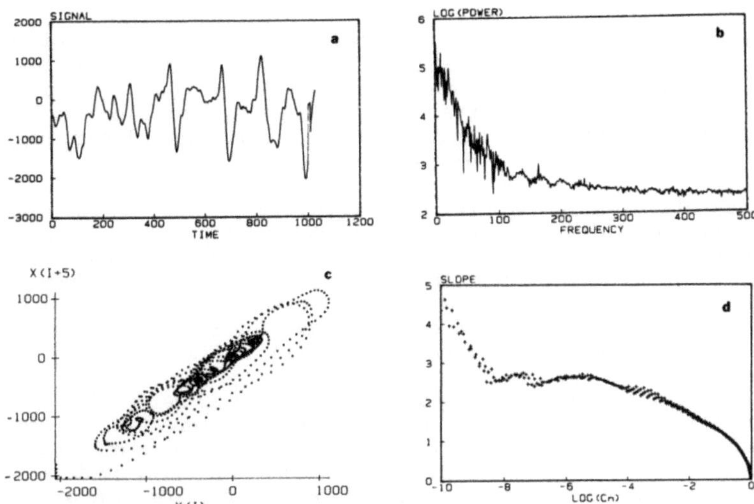

Figure 5. Alpha rhythm of a human EEG. Subject is resting with eyes closed. (a) Time series. Time is in units of 2 ms. (b) Power spectrum. (c) Two-dimensional phase portrait. X(J) is the value of the signal at time, t = 2J ms. (d) Slope vs. log C_n, n=15 to 19. The value of the slope at the plateau is 2.6 ± 0.2

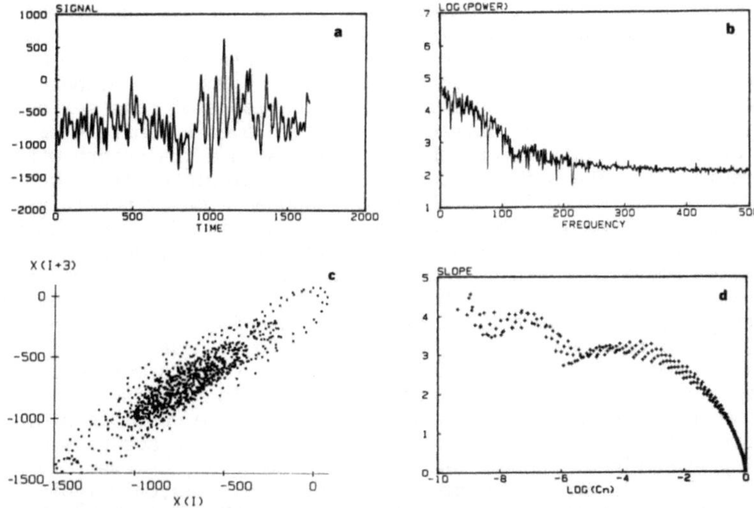

Figure 6. Alpha rhythm of a human EEG. Subject has eyes closed and is counting backward. (a) Time series. Time is in units of 2 ms. (b) Power spectrum. (c) Two-dimensional phase portrait. X(J) is the value of the signal at time, t = 2J ms. (d) Slope vs. log C_n, n = 15 to 19. The graphs do not show an unambiguous plateau.

"pictorial" analyses of time series, while spectral analysis seems to have had little impact on clinical practice [34] . A suggestion to use autocorrelation measurements, made by WIENER a few decades ago [38], seems to have had equally little impact. The obvious similarity between the time series and spectra of EEG signals (Figs. 5 a-b, 6 a-b) and those of signals from chaotic physical systems (Fig. 2) raises the question of whether dimensions, entropies or Lyapunov exponents are appropriate and useful criteria for the quantification of electroencephalographic data. We report here an attempt to test the feasibility of dimension calculations using EEG data.

Figures 5a and 5b show the signal and spectrum, respectively, of 1024 data points
from the alpha rhythm. digitized in intervals of 2 ms, for a subject who is resting
with eyes closed. Fig. 5c is a two-dimensinal phase portrait, $X(I + 5)$ vs $X(I)$, where
I and 5 are in units of the digitization interval (2 ms). Fig. 5d shows slope vs log
C_n for embedding dimensions 15 to 20, calculated using 2900 points. The plateau region,
which is reassuringly broad, gives a D_2 of 2.6 ± 0.2.

A preliminary study was also made on the effect of cognitive activity on the alpha
wave. The subject was instructed to count backwards from 300 in steps of 7. The signal
and its spectrum are shown in Figs. 6a and 6b. Both show qualitative differences from
Figs. 5a and 5b, the signal seems more chaotic, the spectrum broader. A more dramatic
difference is evident in the phase portrait shown in Fig. 6c where we plot $X(I+3)$ vs
$X(I)$ to display a portrait with the same aspect ratio as Fig. 5c. The initiation of
mental activity seems to have washed out the intricate details of Fig. 5c.

Figure 6d shows slope vs log C_n for embedding dimensions 15 to 20 using 1500 data
points. The differences between this and Fig 5d are apparent. There are two regions
which may be interpreted as plateaus, and there is more dispersion. These differences
are not due to the difference in the number of data points used. Analysis of data sets
consisting of 1024 points for the eyes closed, resting case gave results similar to
those shown in Fig. 5d, although with a narrower plateau. We believe that the results
shown in Fig 6d are too ambiguous for a determination of D_2.

5. Conclusions, Problems and Prospects

The examples discussed above provide merely "anecdotal evidence" of the usefulness of
dimensions and entropies as quantitative criteria for the analysis of chaotic behavior. Nevertheless, it is obvious that these and related quantities make possible the
considerations of questions concerning aperiodic phenomena which would be either un-
answerable or meaningless were we to limit ourselves to comparisons of time series and
spectra. No single trajectory on a strange attractor, observed over a finite time interval, can adequately describe the attractor, and comparing such a trajectory with
another is virtually meaningless. The same is true of spectra calculated from such trajectories. The use of dimensions, entropies and Lyapunov exponents makes possible a
"taxonomy" of chaotic behavior. Even if such a taxonomy does not immediately lead to
a better understanding of the behavior in terms of, say, mathematical models that describe it, at least it makes possible a search for empirical correlations between patterns of behavior and external influences. In the case of the nervous system, one could
speculate that such correlations may eventually lead to diagnostic tools or clinical
practices, even if they do not necessarily lead to a more profound understanding of the
brain. In any event, dimensions and similar quantities provide criteria for comparing
experiments with mathematical models when such mathematical models do exist. In the
case of laser systems, we have seen how a dimension calculation made possible a con-
frontation between theory and experiment in a situation where it could not have been
done as ambiguously using traditional techniques. A single dimension, however, does
not specify an attractor. There is a need for more definitive criteria,calculable from
time series measurements that do. Perhaps, a suggestion made by PROCACCIA [40] elsewhere in this volume provides one such criterion.

The work on the nervous system described here is of a most preliminary nature. It
remains to be seen whether results similar to those discussed above are reproducible
even for a single person, let alone for population classes. Much work still needs to
be done, and is underway.

Ultimately, the role of these quantities in understanding chaotic phenomena will
depend on the ease and the reliability with which they can be calculated from data
sets of reasonable size. It is known that to explore a fixed range of scales on an
attractor, the data requirements varies exponentially with the dimension [39] . Analysis of high-dimensional attractors may well require such large data sets and such
exhorbitant computing times as to be impractical. It is clear that more efficient
calculational techniques need to be developed.

6. Acknowledgments

We gratefully acknowledge financial support from the following institutions: Alfred P. Sloan Foundation (N.B.A); National Science Foundation ECS82-10263 (N.B.A., A.M.A. and M.F.H.T.); National Institutes of Health, Epilepsy Branch of the National Institute of Neurological and Communicative Disorders and Stroke NS19716 (P.E.R.); National Science Foundation (I.D.Z.); The Faculty and Institutional Development Program of Trenton State College (N.N.G.); National Institute of Mental Health MH40627-01 and the Neuroscience Development Program of the Medical College of Pennsylvania/Eastern Pennsylvania Psychiatric Institute (T.R.B.).

7. References

1. J.P.Gollub and S.V.Benson: J. Fluid Mech. $\underline{100}$, 449 (1980)
2. M. Gorman, L.A.Reith and H.L.Swinney: Ann. N.Y. Acad. Sci. $\underline{357}$, 10 (1980)
3. A. Liebschaber, S. Fauve and C. Laroche: Physica $\underline{7D}$, 73 (1983)
4. J.C. Roux: Physica $\underline{7D}$, 89 (1983)
5. R.S. Gioggia and N.B. Abraham: Phys. Rev. A $\underline{29}$, 1304 (1984); Opt. Commun. $\underline{47}$, 278 (1983)
6. F.T. Arecchi, R. Meucci, G. Puccioni and J. Tredice: Phys.Rev. Lett. $\underline{49}$, 1217 (1982)
7. C.O. Weiss and H. King: Opt. Commun. $\underline{44}$, 59 (1982)
8. G.A. Held, C. Jeffries and E.E. Haller: Phys. Rev. Lett. $\underline{52}$, 1037 (1984)
9. G. Gibson and C. Jeffries: Phys. Rev. A $\underline{29}$, (1984)
10. P. Grassberger and I. Procaccia: Physica $\underline{9D}$, 189 (1983); Phys. Rev. Lett. $\underline{50}$, 349 (1983); Phys. Rev. A $\underline{28}$, 2591 (1983); Physica $\underline{13D}$, 34 (1984)
11. A. Ben Mizrachi, I. Procaccia and P. Grassberger: Phys. Rev. A $\underline{29}$, 975 (1984)
12. A. Cohen and I. Procaccia: Phys. Rev. A $\underline{31}$, 1872 (1985)
13. A. Wolf and J. Swift: "Progress in computing Lyapunov exponents from experimental data", in Statistical Physics and Chaos in Fusion Plasmas, W. Horton and L. Reichl, eds. (Wiley, New York, 1984)
14. A. Wolf, J. Swift, H.L. Swinney and J. Vastano: Physica $\underline{15D}$, (in press).
15. A. Brandstatter, J. Swift, H.L. Swinney and A. Wolf, "A strange attractor in a Couette-Taylor experiment", in: Turbulence and Chaotic Phenomena in Fluids, Proc. IUTAM Symposium (Kyoto), T. Tatsumi, ed. (North-Holland, Amsterdam, 1983)
16. L. M. Hoffer, T.H. Chyba, and N. B. Abraham: J. Opt. Soc. Am. B $\underline{2}$, 102 (1985)
17. A. M. Albano, J. Abounadi, T.C. Chyba, C.E. Searle, S. Yong, R.S. Gioggia and N. B. Abraham: J. Opt. Soc. Am. B $\underline{2}$, 47 (1985)
18. P. E. Rapp, I.D. Zimmerman, A.M. Albano, G.C. de Guzman and N.N. Greenbaum: Phys. Lett. $\underline{110A}$, 335 (1985)
19. I.D. Zimmerman and N.R. Kreisman: Nature, Lond. $\underline{227}$, 1361 (1970)
20. N.H. Packard, J.P. Crutchfield, J.D. Farmer and R.S. Shaw: Phys. Rev. Lett. $\underline{45}$, 712 (1980)
21. J.C. Roux, R.H. Simoyi and H.L. Swinney: Physica $\underline{8D}$, 257 (1983)
22. J.D. Farmer: Physica $\underline{4D}$, 336 (1982)
23. N.B. Abraham, A.M. Albano, G.C. de Guzman, M.F.H. Tarroja, S. Yong, S.P. Adams and R. S. Gioggia: "Low-dimensional attractor for a single-mode Xe-He ring laser", in Proceedings of the Georgia Tech Conference on Chaotic Dyamics (1985) M.F. Barnsley, ed. (to be published); N.B. Abraham, A.M. Albano, B. Das, G. de Guzman, S. Yong, R.S. Gioggia, G.P. Puccioni and J.R. Tredice: Phys. Lett. A (submitted)
24. H.S. Greenside, A. Wolf, J. Swift and T. Pignataro: Phys. Rev. A $\underline{25}$, 3453 (1982)
25. N.B. Abraham, L.A. Lugiato and L.M. Narducci: J. Opt. Soc. Am. B $\underline{2}$, 7 (1985)
26. N.B. Abraham: "A new focus on laser instabilities and chaos", in Laser Focus (May, 1983)
27. J. Opt. Soc. Am. B $\underline{2}$
28. D. K. Bandy, L.M. Narducci, L.A. Lugiato and N.B. Abraham: J. Opt. Soc. Am. B $\underline{2}$, 56 (1985)
29. M.F.H. Tarroja, D.K. Bandy, T. Isaacs, N.B. Abraham, R.S. Gioggia, S.P. Adams, L.M. Narducci, L.A. Lugiato: "Experimental and theoretical studies of bifurcations by measuring the optical spectrum of a ring laser", in Proceedings of the International Meeting on Instabilities and Dynamics of Lasers and Nonlinear Optical Systems (Rochester, 1985) R.W. Boyd, M. Raymer and L.M. Narducci, eds. (Cambridge University Press, to be published)

30. P. Mandel: J. Opt. Soc. Am. B **2**, 112 (1985)
31. M.L. Minden and L.W. Casperson: J. Opt. Soc. Am. B **2**, 120 (1985)
32. N.B. Abraham, A.M. Albano, D.K. Bandy, B. Das, G.C. de Guzman, T. Isaacs, M.F.H. Tarroja, S. Yong, S.P. Adams and R.S. Gioggia:"Low-dimensional attractors for a single-mode Xe-He Ring laser", in Proceedings of the International Meeting on Instabilities and Dynamics of Lasers and Nonlinear Optical Systems (Rochester, 1985) R.W. Boyd, M. Raymer and L.M. Narducci, eds. (Cambridge University Press, to be published)
33. F.T. Arecchi, J.R. Tredicce, G.P. Puccioni, A. Poggi and W. Gadomski: Phys. Rev. A (submitted)
34. P.E. Rapp, I.D. Zimmerman, A.M. Albano, G.C. de Guzman, N.N. Greenbaun and T. R. Bashore: "Experimental studies of chaotic neural behavior: cellular activity and electroencephalographic signals", in Nonlinear Oscillations in Chemistry and Biology, H.G. Othmer, ed. (Springer-Verlag, New York, to be published)
35. L. Glass and M.C. Mackey: Ann. N.Y. Acad. Sci. **316**, 214 (1979)
36. R.S. Shaw: The Dripping Faucet (Ariel Press, Santa Cruz, 1984)
37. P.Y. Ktonas: CRC Critical Reviews of Biomedical Engineering **9**, 39 (1983)
38. N. Wiener: Cybernetics (Wiley, New York, 1948)
39. H.L. Swinney and J.P. Gollub: Physica D (to be published)
40. see I. Procaccia's article elsewhere in this volume.

Evidence of Chaotic Dynamics of
Brain Activity During the Sleep Cycle

A. Babloyantz

Faculte des sciences, Universite Libre de Bruxelles, Campus Plaine CP 231,
Boulevard du Triomphe, B-1050 Bruxelles, Belgium

1. Introduction

The study of complex systems may be performed by analysing experimental data recorded as a series of measurements in time of a pertinent and easily accessible variable of the system. In most cases, such variables describe a global or averaged property of the system.

Several papers in this volume are devoted to the analysis of such times series and determination of dimensionality of strange attractors. These methods have been fruitful in such fields as hydrodynamics (1) chemistry (2) and climatic variability (3,4).

A time series may be obtained by recording at regular time intervals the mean electrical activity of a portion of the mammalian cortex. The analysis of the time series recorded from electroencephalogram or EEG may answer the following questions.   .

(i)   Is it possible to identify an attractor for a given EEG ? In other words, can the salient features of neuronal activities be viewed as the manifestation of a deterministic dynamics (possibly very complex one) or rather, do they contain an irreducible stochastic element?

(ii) Provided an attractor exists, what is its dimensionality d? How does d evolve as brain activity changes?

(iii) What is the minimum embedding dimension of the phase space which contains the attractor. This number defines the minimum number of variables that must be considered in the description of the underlying dynamics.

2. EEG Attractors

Mammalian brain is certainly one of the most complex systems encountered in nature. It is made of billions of cells endowed with individual electrical activity and interconnected in a highly intricate network.

The average electrical activity of a portion of this network may be recorded in course of time and is called electroencephalogram (EEG). The EEG reflects the sum of elemental self-sustained neuronal activities of a relatively long period (of the order of 0.5 to 40 Hz). Recordings from the human brain show that to various stages of brain activity there correspond characteristic electrical wave forms. For example a resting and alert brain shows activity of an average frequency of about 10 cycles per second and amplitudes of the order of 10 microvolts (α waves). During a normal night's sleep waves give way to other repeated cycles of activity each marked by several stages.

In stage one, the individual drifts in and out of sleep. In
stage two, the sleeper is disturbed by the slightest noise. In stage
three a loud noise would be needed to rouse the sleeper. Finally,
the deep sleep of stage four sets in. Afterwards the cycle is rever-
sed back through stages three and two. After this stage the sleeper
enters the phase of rapid eye movement sleep (REM) in which he
dreams. This episode is followed by stage two and a new cycle
begins. The sleep cycles continue through the night; however the
periods of REM get longer and those of deep sleep shorter.

As sleep sets in, the fast waves gradually give way to slower
and higher amplitude waves. At deep sleep stage four the average
frequency is of the order of 3-5 cycles per second and amplitudes are
of several hundred microvolts (δ waves). During REM sleep intense
bursts of high-frequency activity appear.

In a recent paper the EEG of deep sleep stage four, α waves,
sleep stage two and the REM sleep, considered as time series, were
analysed (5).

The first step is to define, using the time series the appropriate variables spanning the phase space. These variables are obtained
by shifting the original time series by a fixed τ (τ = m Δt, where
Δt is the interval between successive samplings (6)). Phase trajectories may be constructed with the help of these variables.

The phase portrait of the awake subject is densely filled and
occupies a small portion of the phase space (Fig 1a). The representative point undergoes deviations from some mean position in practi-
cally all directions. At the sleep stage two, already a tendency
towards a privileged direction is seen, and a larger portion of the
phase space is visited (Fig 1b). This tendency is amplified in the
sleep stage four and one sees preferential pathways, suggesting the
existence of reproducible relationships between instantaneous values
of the pertinent variables (Fig 1c). This phase portrait is the largest and exhibits a maximum "coherence" which diminishes again when
REM sleep sets in (Fig 1d).

A universal attractor for different REM episodes of a single
night and a given individual seemes unlikely, as the REM episodes are
associated with intense brain activity and generation of dreams.
Fig. (1.e) shows a second REM episode in the sleep cycle of the same
individual who's EEG recording was used in Fig. 1.d.

3. <u>Dimensionality Analysis</u>.

The dimensionality of the phase portraits is analysed by computing
the integral correlation function according to the algorithm developed by Grassberger and Procaccia (7,8).

If $\vec{X}_i$ stands for a point of phase space and if θ is the Heaviside function :

$$C(r) = \frac{1}{N^2} \sum_{\substack{i,j=1 \\ i \neq j}}^{N} \theta(r - |\vec{X}_i - \vec{X}_j|)$$

and for small r

$$C(r) \simeq r^d$$

The dimensionality d is computed by considering successively
higher values of the embedding dimension n of the phase space. If

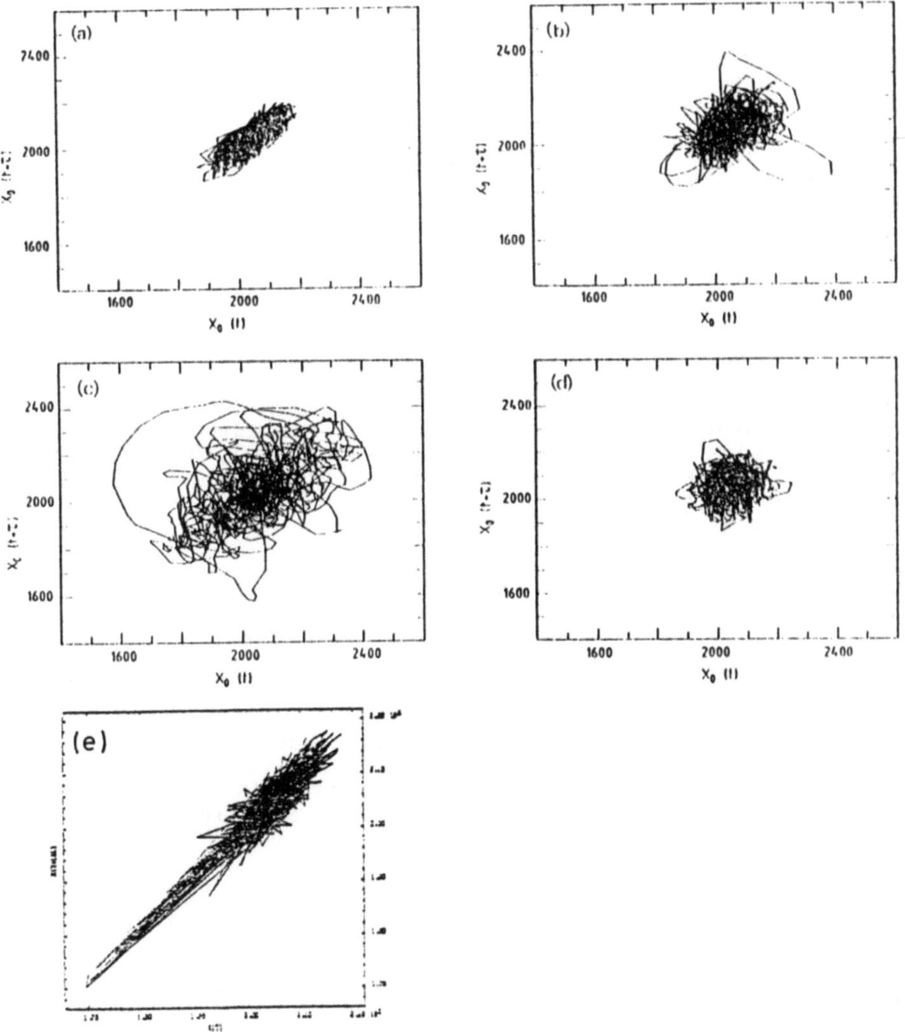

Fig.1.   Two-dimensional phase portraits derived from the EEG of (a)
         an awake subject, (b) sleep stage two, (c) sleep stage four,
         (d,e) REM sleep.   The time series $X_0(t)$ is made of N=4000
         equidistant   points.   The   central   EEG   derivation   C4-A1
         according to the Jasper system was recorded with PDP 11-44,
         100 Hz for 40 s.   The value of the shift from 1a to 1e is
         $\tau = 10\,\Delta t$.

the d versus n dependence is saturated beyond some relatively small
n, the system represented by the time series should possess an
attractor.   The saturation value d is regarded as the dimensionality
of the attractor represented by the time series.   The values of n
beyond which saturation is observed provides the minimum number of
variables necessary to model the behavior represented by the attractor (see Fig.2).

     The analysis of time series shows that the phase portraits of
sleep stage two, recorded from two individuals, and sleep stage four,
recorded from three individuals, show the existence of chaotic

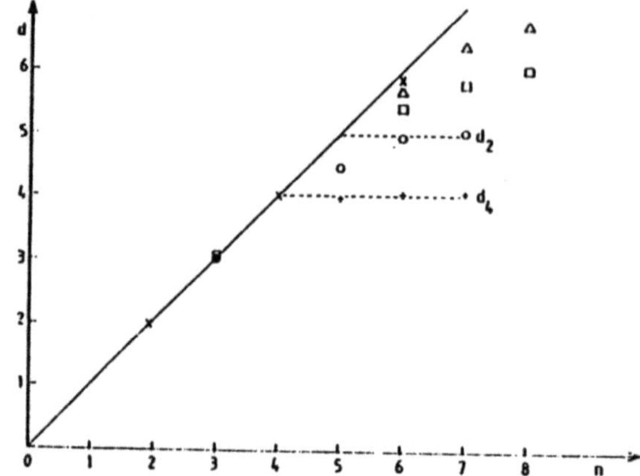

Fig.2. Dependance of dimensionality d on the number of phase space
        variables n for a white noise signal (x), the EEG attractor
        of an awake subject ($\triangle$), sleep-stage two (o) sleep stage
        four (+) and REM sleep ($\square$); for the same number of data
        points as in Fig.1.

attractors. No satisfactory saturation was observed for awake state
and REM sleep in a space of 10 variables. These results are shown in
table (1).

        Using the algorithm developed by A. Wolf (9) (see this volume)
the chaotic nature of the attractors may also be assessed from the
computation of the largest positive Lyapounov exponents. We have
evaluated the largest positive Lyapounov exponents for stage two and
stage four of deep sleep. For stage two we find a positive value of
λ_2 between 0.4 and 0.8. The inverse of this quantity gives the
limits of predictability of the long-term behavior of the system.
For stage four we find also a positive number $0.3 < \lambda_4 < 0.6$.

Table 1. Embedding dimension of the phase space and dimensionality
          of the attractors for various stages of brain dynamics.

Stages of Brain Dynamics	Embedding Dimension p	Attractor Dimension d.
Awake	$n>9$	No saturation
Sleep 2	6	5.03 ± 0.07
		5.0 ± 0.1
Sleep 4	5	4.05 ± 0.05
		4.08 ± 0.05
		4.4 ± 0.1
REM	$n>9$	No saturation

244

The presence of positive Lyapounov exponents and noninteger
values of dimensionality d establish the presence of chaotic
attractors during sleep stage two and sleep stage four. This in turn
implies the existence of deterministic dynamics, which may be
described by a limited number of variables. The fact that the
dimensionality decreases as the deep sleep sets in, implies that the
dynamics becomes more coherent during the deep sleep.

4. Conclusions

We have shown that from a routine EEG recording, the dynamics of
brain activity could be reconstructed. The fact that chaotic attractors could be identified for several stages of normal and pathologi-
cal brain activity indicates the presence of deterministic dynamics
of a complex nature. This property should be related to the ability
of the brain to generate and process information.

Unlike periodic phenomena,which are characterized by a limited
number of frequencies, chaotic dynamics show a broad band spectrum.
Thus, chaotic dynamics increases the resonance capacity of the
brain. In other words, although globally a chaotic attractor shows
asymptotic stability, there is an internal instability reflected by
the presence of positive Lyapounov exponents. This results in a great
sensitivity to the initial conditions, thus an extremely rich response to external input.

The topological properties of the attractors and their quantification by means of dimensionality analysis may be an appropriate tool
in the classification of brain activity, thus a possible diagnostic
tool. For example, various forms of epileptic seizures could be
classified according to their degree of coherence.

References
----

1. Brandstater A., Swift J., Swinney H.L.& Wolf A.(1983) Phys.
    Rev. Lett. 51,1442-1445.

2. Roux J.C., Simoyi R.M.& Swinney H.L.(1983) Physica 8D,257-266.

3. Nicolis C.& Nicolis G.(1984) Nature 311,529-532.

4. Nicolis C.& Nicolis G.(1985) Proc. Natl. Acad. Sci. USA in
    press.
5. Babloyantz A., Nicolis C.& Salazar M.(1985) Phys. Lett. IIIA,-
    152-156.

6. Takens F.(1981) in Lectures notes in mathematics 898 Ed. Rand
    D.A.& Young L.S. Springer Berlin.

7. Grassberger P.& Procaccia I.(1983) Phys. Rev. Lett. 50,346-349.

8. Grassberger P.& Procaccia I.(1983) Physica 9D,189-208.

9. Wolf A., Swift J.B., Swinney H.L.& Vastano J.A.(1985) Physica
    16D,285-317.

Problems Associated with Dimensional Analysis of Electroencephalogram Data

S.P. Layne, G. Mayer-Kress, and J. Holzfuss

Center for Nonlinear Studies, Los Alamos National Laboratory,
Los Alamos, NM 87545, USA

We begin with a basic introduction to the electroencephalogram and discuss some of the EEG's clinical uses. Next we introduce a practical application of dimensional analysis to the EEG by asking the question: *How does the "dimension" of the EEG change with general anesthesia ?* Finally we discuss major problems associated with dimensional analysis of the EEG.

1. Introduction to the EEG

Electrical activity originating from the brain was first discovered in 1929. In spite of this long history, slow progress has been made in understanding and interpretating the EEG. Currently, there is no ordered atlas of EEG findings versus brain activity -- there are only tendencies for EEG findings versus brain activity. In addition, very few EEG centers are equipped with computers for detailed analysis and so today, EEG interpretation is primarily by skilled visual inspection.

The microscopic genesis of the EEG is not understood. Macroscopically, electrical signals appear to originate from collections of dendritic endings on neurons [1]. These collections, in turn, generate signals that are coordinated in space and time near the surface of the brain. Beyond this, there is some attenuation of the brain's electrical activity by the intervening skull and scalp. At the scalp, the EEG is measured as a μ-Volt potential and to record this weak potential, it must be amplified by several orders of magnitude. Generally, EEG signals are recorded by contacting a standard pattern of electrodes, with conductive paste, to rubbed scalp. Pronounced artifacts often originate from slight movements of the electrodes and from contraction of muscles below the electrodes. Since the EEG is a weak signal in a sea of noise, the importance of skilled electrode placement and inspection for artifacts cannot be overestimated [2,3].

The EEG's frequency spectrum ranges from 0.5 to 100 Hz, with most of the power concentrated in the range from 1 to 30 Hz. By visual inspection, the EEG looks like a complex wave that is aperiodic. The appearance of this signal fluctuates throughout the day and changes cyclically with sleep. There are four general classifications of EEG frequencies. They were established primarily by discovery, as early instrumentation improved, and are of limited intrinsic value: 1 - 3 Hz is termed *delta,* 4 - 7 Hz *theta,* 8 - 14 Hz *alpha,* and >14 Hz *beta.* In addition, the overall appearance of the EEG is modified by drugs. For example:

Cocaine	Increases alpha rhythm
Amphetamines	Decreases amplitude
Thorazine	Decreases alpha rhythm
Barbiturates	Increases beta rhythm

Often, one or more drugs so completely alters the normal pattern of the EEG that it becomes difficult or impossible to interpret. In these instances, diagnostic EEG's are performed only after a period of drug privation.

2. General anesthesia and the EEG

Realizing that drugs alter the EEG, attempts have been made to assess the depth of general anesthesia by analyzing brain waves in real-time. However, two difficult and related questions are involved: 1. *What is the choice for the measure?* and 2. *What is the sensitivity of the measure?* This attempt to assess anesthesic depth by *quantitative* means is important because physicians currently rely on *qualitative* physical signs. Although signs involving flinching, tearing, blood pressure, respiration rate, pupil diameter and heart rate are useful, they do not provide specific information on the well-being of the brain.

To our knowledge, Hanley and Walts [4] have done the most successful work to assess anesthetic depth by computer analysis of the EEG. In this study, a variety of auto-spectral and cross-spectral relationships were correlated: 1. Power density, mean frequency, and band width of selected frequencies for a given lead, and 2. Phase relationships and shared electrical activities between different leads. Anesthesia was induced by a single agent and depth was evaluated by measuring blood concentrations of drug with a mass spectrometer. In order to establish self-controls and dose-related effects, the EEG was analyzed before anesthesia, during light anesthesia, and during medium anesthesia.

The above quantitative analysis involving the EEG is promising because: 1. The EEG is an overall measure of brain actvity and general anesthetics globally affect the brain, 2. Certain drugs induce dose-related and frequency-related changes in the EEG, and 3. Overwhelming doses of some general anesthetics diminish the EEG to a flat line. During routine surgery, however, general anesthetics are not given in sufficient amounts to reduce the EEG to a flat line. Only in special cases involving prolonged anoxia, such as drowning, are drug comas induced to minimize brain damage. In these instances, sufficient anesthetic is administered so that the EEG is not active for several days.

Attempts to assess brain function and depth have been most successful when one drug is used to induce general anesthesia. In most surgical cases, however, several drugs are used to induce anesthesia which results in complex and unpredictable EEG changes. In these instances, numerical analysis of the EEG has been less accurate and rather difficult to interpret. These limitations have motivated us to consider a different approach.

3. Rationale for dimensional analysis of the EEG

In this section, we use "dimension" as a qualitative term. In this regard, "dimension" corresponds to the number of independent variables needed to specify the activity of a system at a given instant. It also corresponds to a measure for the number of active modes modulating a physical process and therefore, it is a measure of complexity. In the next section, we will describe dimension by outlining our numerical procedures.

Dimensional analysis of the EEG during general anesthesia is an interesting approach for several reasons: 1. Large doses of some anesthetics reduce the EEG to a flat line, therefore the dimension of the EEG may decline with anesthesia, 2. It provides a single number rather than a series of correlations for assessing depth, 3. The dimension provides information about the complexity of the EEG signal which cannot be extracted from the power spectrum, 4. Dimension may be calculated before anesthesia and during anesthesia for the same person, which provides a method of standardization, and 5. The EEG may be more than "noise" in cerebral tissue; it may play a causal role in information processing and storage [5]. In this regard, there has been conjecture that the brain's electrical activity reflects a complex dynamical system which may be evaluated by the tools of nonlinear dynamics [6,7].

Our original goal was to evaluate anesthetic depth for a series of 5 to 10 patients by dimensional analysis. This goal turned out to be more ambitious than we thought. It has been very difficult to obtain clean EEG records from the operating room. Noise is prominent due to electrocautery and to movement of the patient's head by operating room personnel. In addition, specialized EEG equipment must be used to reduce noise and to accommodate limited space in the room.

In the following section, we discuss the problems associated with dimensional analysis of the EEG. We choose one EEG record from a single patient, in order to study the method but not to draw general conclusions. For simplicity, we consider only two states: awake but quiet, and medium anesthesia. The EEG data we use comes from Hanley and Walts [4]. It was selected because anesthesia was induced by a single agent, and because of its uninterupted length and lack of artifacts [8].

4. Dimensional analysis of the EEG

We analyze data that was derived from two standard EEG leads, as shown in Fig. 1. Lead T3-C3 records activity primarily from the motor cortex, and lead P3-O1 records activity primarily from the optical cortex of the brain.

248

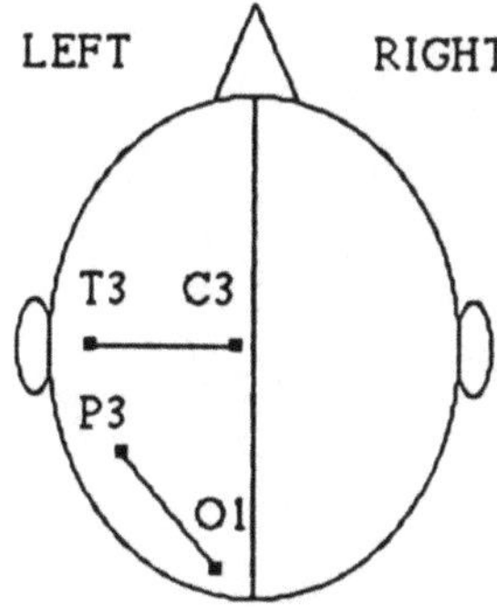

Fig. 1. Two standard EEG leads from the International 10-20 system

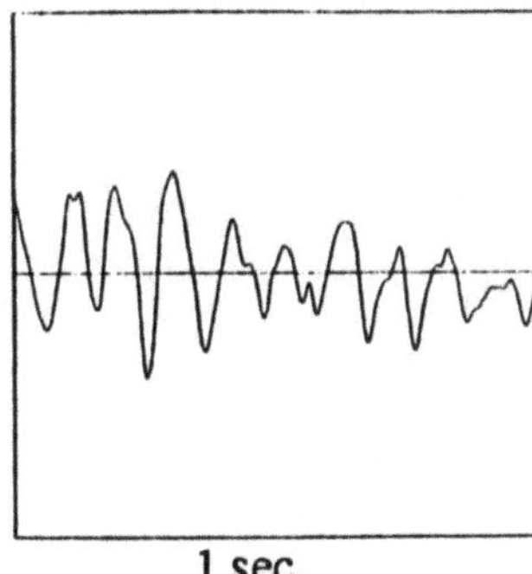

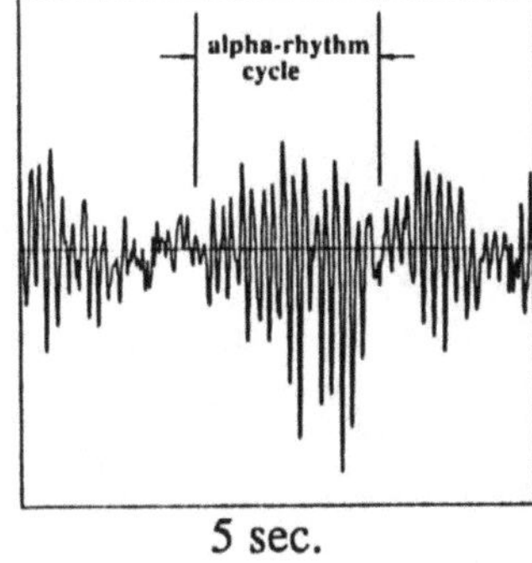

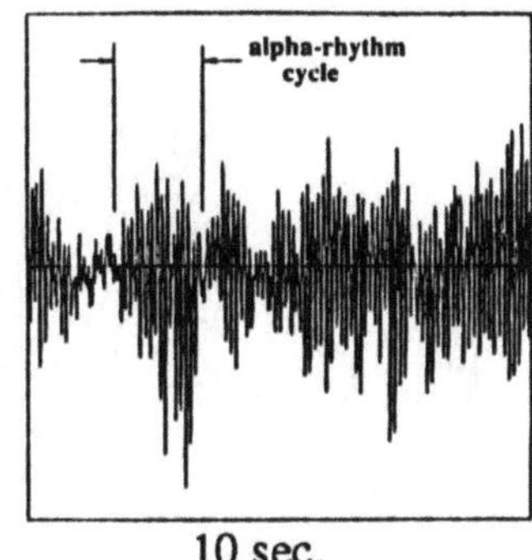

<table>
<tr><td>1 sec.</td><td>5 sec.</td><td>10 sec.</td></tr>
</table>

Fig. 2. Awake but qiuet EEG data from P3-O1. The length of each time-series is indicated above

In order to convey a feeling for the appearance of the EEG, Fig. 2 shows three time-series of different lengths. All three time-series start from the same location in time and were recorded from lead P3-O1, while the patient was awake but quiet. In most people, the optical cortex generates a strong alpha-rhythm in states of quiet consciousness. This rhythm appears as cycles of large amplitude waves that come and go with time. Longer time-series (not shown) demonstrate that this overall "cyclic" activity is aperiodic. At the same time, lead T3-C3 does not demonstrate a prominent alpha-rhythm.

Fourier transform reveals several additional characteristics, see Fig. 3. The EEG is composed of multiple frequencies that vary with time, and a time-average over several minutes shows no dominant bands. Even the prominent alpha-rhythm of lead P3-O1, Fig. 3c, is a broad band that varies in width and intensity. Each lead has a different and somewhat characteristic power spectrum that depends on the state of consciousness (note the lack of alpha-rhythm in C3-T3). With medium (fluroxene) anesthesia, the power spectrum reveals more energy at higher (15 to 25 Hz) frequencies, and with deeper anesthesia (not shown) the power spectrum shifts to lower (1 to 8 Hz) frequencies.

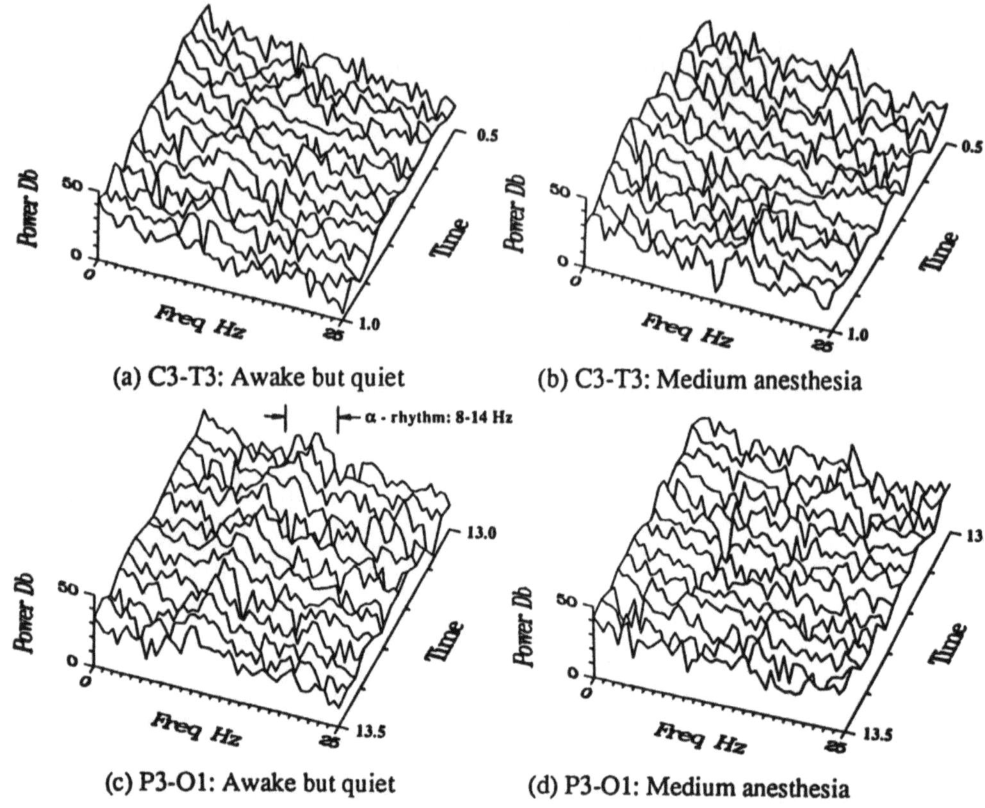

(a) C3-T3: Awake but quiet   (b) C3-T3: Medium anesthesia

(c) P3-O1: Awake but quiet   (d) P3-O1: Medium anesthesia

Fig. 3. Time-resolved Fourier spectra of the EEG. Each track (or spectrum) transforms 2.05 seconds of EEG data at a resolution of 0.5 Hz. Time (in minutes) indicated the position in the EEG record.

In order to calculate the "dimension" of the EEG, it is first necessary to reconstruct an "attractor" from the time-series data. The usual procedure for this is to construct vectors $V(t_k)$ in an m-dimensional phase-space by taking time-delayed samples of scalar data $x(t_k)$. In our case, phase portraits of dimension m are constructed such that: $V(t_k)=\{x(t_k), x(t_k+\Delta T), x(t_k+2\Delta T), ..., x(t_k+(m-1)\Delta T\}$, where t_k is discrete time with k running from 1 to the number of data points and ΔT is a constant time-delay [9]. For an ideal data set, the size of ΔT is essentially arbitrary. For a realistic data set, however, the quality of the phase portrait depends strongly on ΔT. We have chosen values for ΔT that correspond to the first minimum in mutual information. This procedure selects time-delay coordinates that are generally independent and therefore aids in the construction of an "attractor" from limited time-series data [10]. From a number of mutual information calculations, we find different values for ΔT at each lead: 40 ms for C3-T3, and 20 ms for P3-O1. These values are used for subsequent phase portrait construction and dimensional analysis.

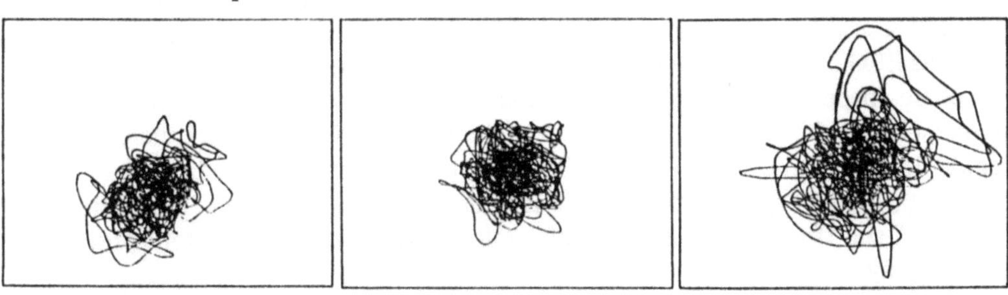

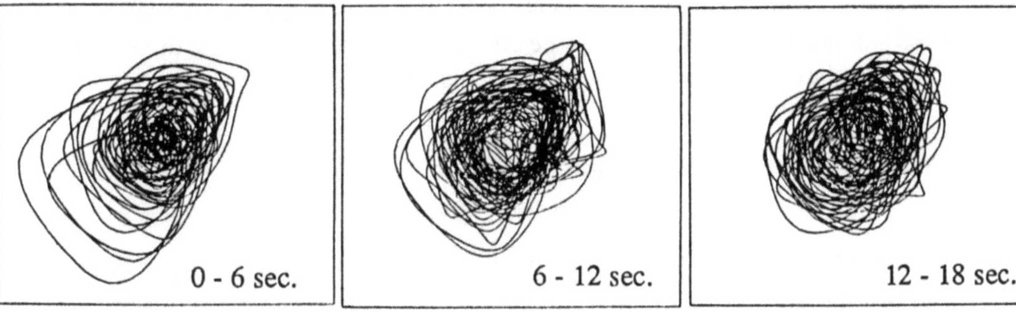

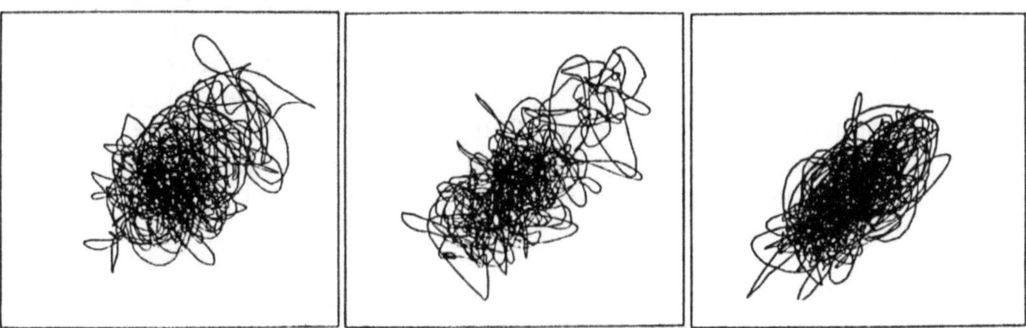

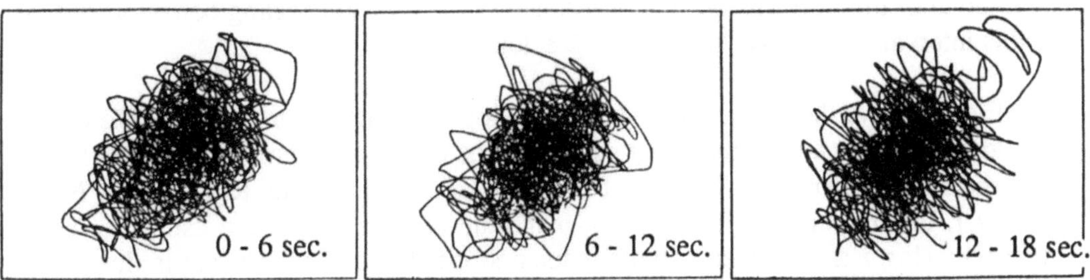

Fig. 4. Two-dimensional phase portraits, $V(t_k+\Delta T)$ vs $V(t_k)$, of EEG data. Each portrait is generated by 3000 data points and consecutive portraits (in time) appear from left to right. The top two portraits are form "awake but quiet" EEG data; the bottom two portraits are from "medium anesthesia" EEG data

Conventional two-dimensional phase portraits ($m=2$) of the EEG are shown in Fig. 4. These phase portraits reveal chaotic orbits with diverging trajectories. With time, the EEG changes its average position and on occasion makes large excursions in the phase-space. Such complex trajectories are related to the multiple and variable frequencies of the EEG. The dominant alpha-rhythm of lead P3-O1 yields a more regular-looking structure but again with diverging trajectories. Such behavior implies a <u>non-stationary</u> system of high dimension. In such instances, the notions of "attractor" and "fractal dimension" (which refer to stationary or asymptotic properties) do not apply.

To compute the "dimension" of the EEG, we use the method of Grassberger and Procaccia [11]. This method is now in standard use, since it requires modest computer time and storage. For a sufficiently large embedding dimension m, the dimension d of an attractor behaves as a power of r for small distances r such that

$$\log C(r) \propto d \log r \quad \text{and} \quad C(r) \equiv \lim_{k \to \infty} 1/n_{ref} \sum_{j=1}^{n_{ref}} 1/k \sum_{i=1}^{k} \Theta(r - |V_i - V_j|),$$

where Θ equals 1 for positive and 0 for negative arguments, and n_{ref} equals the number of reference points in the calculation. For all our calculations, we follow two procedures: 1. Reconstruct sets of phase-space vectors such that $\{V\} \equiv \{V(t_k)\}$ for each embedding dimension m ranging from 1 to 20, and 2. Calculate the dimension d for each embedding dimension m by averaging over 200 equally spaced reference points in the "attractor". We average over that many reference points because $C(r)$ is sensitive to the local structure of the "attractor". Certain regions of the attractor are more "important" in the sense that they are visited more often than others. By using 200 reference points, an average for the various regions is calculated. The details of this procedure and our codes are described in reference [12].

Fig. 5 shows the result of such procedures. It plots the logarithm of the average number of vectors n (which are separated by a distance $< r$) versus the logarithm of the distance r. The value of the "dimension" d is obtained by fitting a straight line to an interval of this "dimension" curve that shows the relevant scaling behavior. In Fig. 5, this interval occurs over the region $-2 \leq \log r \leq -1$.

Fig. 6 shows the "dimension" (or local slope) of Fig. 5 versus $\log r$. We obtain this local slope by a weighted least squares fit of length $L = 7$ unit distances in $\log r$. The maximum "dimension" occurs for $\log r \approx -1.8$ and is equal to 7.7 ± 7.4. Note, however, that the error is of the same magnitude as the dimension itself. For short distances ($\log r \leq -3$) we see that the local slope $d \approx 1$. This artifact is due to an excessive digitalization rate of 500 Hz for the EEG. Such a rapid rate generates large numbers of vectors along a short segment of the trajectory. For large distances ($\log r \geq 0.5$) we see that the local slope $d \approx 0$, which reflects the finite size of the "attractor".

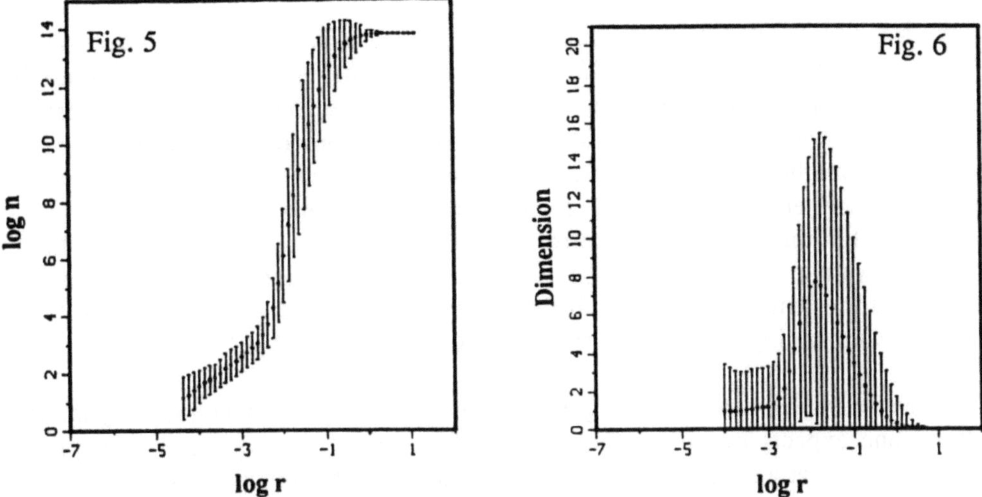

Fig. 5. Dimension curve, with embedding dimension m=20, for typical EEG data. This calculation is based on 15000 data points k (30 seconds of data) and 200 reference points. Error bars indicate standard deviations in the distribution of the average number of vectors n at distance r. Figs. 5 through 8 are from the same awake but quite EEG data (lead T3-C3)

Fig. 6. Local slope d of the "dimension" curve vs log r. Note the asymptotic values of $d \approx 1$ for $r \to 0$ and $d = 0$ for $r \to \infty$

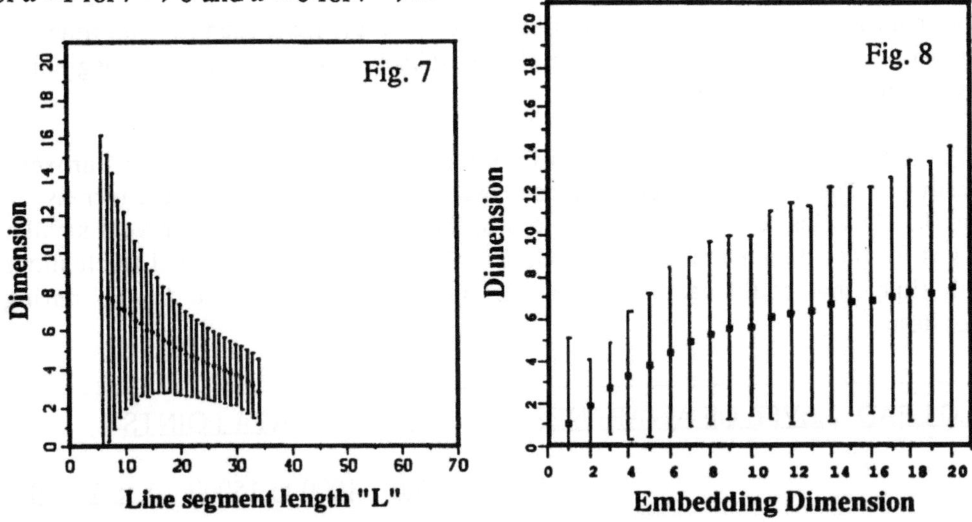

Fig. 7. Dimension and error-estimate as a function of length L. The result is generated by a weighted least squares fit to Fig. 5. A scaling factor of 2 corresponds to 8 unit divisions in r on the horizontal axis

Fig. 8. Observed "dimension" as a function of embedding dimension m. Dimension values at each m were calculated by a weighted least squares fit. The length of the line segments L for each fit ranged from 7 to 9 unit divisions in log r on the "dimension curve".

We also investigated the question: *How does "dimension" depend on the length of a line segment L that is fit to the "dimension" curve?* In Fig. 7, we see a decrease in the observed dimension d as L increases. At the same time, the size of the uncertainty in d decreases, which indicates a systematic error in the procedure. This error occurs because the weighted least square fit is constrained to shorter ranges of slope by line segments of longer length. For EEG calculations, dimension and error estimates were obtained by limiting the maximum error in the weighted fit, see reference [11]. In all cases, a maximum error of 10% appeared to be reasonable constraint. This constraint determined the length L over which the dimension curves scale and generally, we found relevant scaling over a range that is close to a factor of 2 [13].

When a complex time-series is generated by a low-dimensional chaotic dynamical system, one expects that dimension d (as a function of embedding dimension m) should level off at a finite value. In Fig. 8, we see that this behavior is not observed for the EEG data. This lack of saturation at $m = 20$ indicates that low-dimensional models cannot account for the dynamics of the EEG.

5. Summary of calculations for the EEG

The table below summarizes the results of our "dimension" calculations for the EEG. In addition to calculating "dimension" for longer data sets of 15000 points, we also calculated "dimension" for shorter subsets of 4000 data points. We considered these shorter subsets in order to see how the "dimension" of the EEG changes with time and number of data points. The data for the shorter subsets were taken from the beginning and the end of the longer sets.

To our surprise, the "dimension" of the EEG appears to increase with fluroxene anesthesia. The most prominent increase is seen in lead P3-O1, while the increase in lead T3-C3 is less convincing. In all cases, the shorter data subsets reveal smaller values for the "dimension" of the EEG. The shorter subsets also reveal fluctuations with time, that could easily account for increases seen in lead C3-T3 during anesthesia.

EEG LEAD	LEVEL OF AWARENES	SET OF DATA POINTS		
		1 to 4000	11000 to 15000	1 to 15000
$T_3 - C_3$	AWAKE BUT QUIET	6.5 ± 4.5	6.9 ± 6.7	7.7 ± 7.4
	MEDIUM ANESTHESIA	6.8 ± 5.0	7.8 ± 5.0	8.4 ± 6.4
$P_3 - O_1$	AWAKE BUT QUIET	5.5 ± 5.1	6.3 ± 3.6	6.6 ± 5.0
	MEDIUM ANESTHESIA	7.4 ± 4.5	7.4 ± 4.4	8.6 ± 5.6

From these results, it is clear that neither 4000 nor 15000 data points sample the full complexity of the EEG. Therefore, the "dimension" of the EEG has potential relevance only in a comparative sense. Moreover, this comparison must be made for each lead and not between leads. We feel that reports which try to explain EEG dynamics by low-dimensional chaotic attractors are too optimistic [14].

6. Conclusions

1. The EEG is both non-stationary and high-dimensional. Therefore, we cannot speak of dimensions in the strict sense. Our calculations of "dimension" have meaning only in a comparative sense.

2. Our use of the word "dimension" is not strictly accurate. More appropriate terms might be "apparent dimension" or "dimensional complexity".

3. Our findings do not conclusively show whether the "dimension" of the EEG is a sensitive tool for measuring anesthetic depth. For fluroxene anesthesia, the "dimension" appears to increase,but there is significant uncertainty in our calculations. Further work on representative EEG records is necessary.

4. In general, new approaches to the EEG hold promise,since limited progress has been made in finding new methods for analysis and diagnosis.

7. Acknowledgement

We thank Alwyn C. Scott for first suggesting the idea of anesthetic monitoring. We also thank John Hanley for providing the EEG data and for valuable discussion. One of us (G.M.-K.) acknowledges expanding conversations with Cindy Ehlers and Arnold Mandell. Finally, we thank Andrew Fraser for the mutual information code, and Erica Jen for early versions of the dimension codes.

8. References and Notes

1. R. Elul: Int. Rev. Neurobiol. 15, 227 (1972)
2. J. Hanley: "Electroencephalography in Psychiatric Disorders: Part I", in Directions in Psychiatry vol. 4, lesson 7, pp. 1-8 (1984)
3. J. Hanley: "Electroencephalography in Psychiatric Disorders: Part II", in Directions in Psychiatry vol. 4, lesson 8, pp. 1-8 (1984)
4. J. Hanley & L. Walts: Unpublished results form a series of clinical trials with gaseous anesthetic agents at UCLA from 1974 - 1976 (Private communication: April 12, 1985)
5. W. R. Adey: "The Influences of Impressed Electrical Fields at EEG Frequencies on Brain Behavior", in Behavior and Brain Electrical Activity, N. Burch & H. Altshuler eds., Plenum, pp. 363-390 (1974)

6.   J. D. Farmer: "Dimension, Fractal Measures, and Chaotic Dynamics", in Evolution of Order and Chaos in Physics, Chemistry and Biology, H. Haken ed., Springer-Verlag, pp. 228-246 (1982)

7.   A. Mandell: private communication

8.   The patient was a 45 year old female. No premedication was prior to surgery. Anesthesia was induced by inhalation of fluroxene. The EEG record lasts for 15 minutes: 5 min. awake but quiet, 5 min. light anesthesia, and 5 min. medium anesthesia. Analog data from the EEG was digitalized at 500 Hz.

9.   N. H. Packard, J. P. Crutchfield, J. D. Farmer & R. S. Shaw: Phys. Rev. Lett. 45, 712 (1980)

10. A. M. Fraser & H. L. Swinney: Phys. Rev. A 33 (1986); A.M. Fraser: this volume

11. P. Grassberger & I. Procaccia: Phys. Rev. Lett. 50, 346 (1983)

12. J. Holzfuss & G. Mayer-Kress: this volume

13. One might argue that "attractors" which look similar over a scaling factor of ≈ 2 are not really self-similar. Since we are operating in 20-dimensional space, a scaling factor of ≈ 2 is rather large and denotes similarity.

14. A. Babloyantz , C. Nicolis & M. Salazar: Phys. Lett. 111A, 152 (1985); A. Babloyantz: this volume

Index of Contributors

Springer